Basic Mathematics for Chemists

Second Edition

Basic Mathematics for Chemists

Second Edition

Peter Tebbutt

JOHN WILEY & SONS
Chichester · New York · Weinheim · Brisbane · Singapore · Toronto

Baffins Lane, Chichester,
West Sussex PO19 1UD, England

National 01243 779777
International (+44) 1243 779777

Reprinted October 1998

e-mail (for orders and customer service enquiries): cs-books@wiley.co.uk.
Visit our Home Page on http://www.wiley.co.uk
or http://www.wiley.com

Other Wiley Editorial Offices

John Wiley & Sons, Inc., 605 Third Avenue,
New York, NY 10158-0012, USA

WILEY-VCH Verlag GmbH, Pappelallee 3,
D-69469 Weinheim, Germany

Jacaranda Wiley Ltd, 33 Park Road, Milton,
Queensland 4064, Australia

John Wiley & Sons (Asia) Pte Ltd, 2 Clementi Loop #02-01,
Jin Xing Distripark, Singapore 129809

John Wiley & Sons (Canada) Ltd, 22 Worcester Road,
Rexdale, Ontario M9W 1L1, Canada

British Library Cataloguing in Publication Data

A catalogue record for this book is available from the British Library

ISBN 0 471 97283 5; 0 471 97284 3 (pbk)

Typeset in 10/12pt Times by Vision Typesetting, Manchester
Printed and bound in Great Britain by Bookcraft (Bath) Ltd
This book is printed on acid-free paper responsibly manufactured from sustainable forestation, for which at least two trees are planted for each one used for paper production.

For Susan, Lily, Grace and Chloe

Contents

Preface to the First Edition

Many students embarking on a degree or other higher level courses in chemistry do so without any training in advanced high school mathematics. It is almost certain, however, that they will need to use some of this mathematics in their chemistry course. Although there are many books of mathematics for scientists and science students, few of these are specifically for chemists and even fewer are aimed at students not already familiar with advanced mathematical concepts. This book has been written to address this problem.

The 'basic' in the title is relative. Some of the material is advanced compared to what might be expected in a high school course but basic compared with that expected in a university course. The book is, however, meant to bridge a gap between school mathematics and that used in advanced chemistry courses. It therefore begins with elementary concepts, equations and graphs, and moves on to more advanced topics such as differentiation, integration and the solution of differential equations.

In attempting to address advanced topics in as simple a fashion as possible there are obvious omissions in what might be expected of a book of mathematics for chemists. The reader will find nothing of vectors, matrix algebra and only a mention of quantum mechanics. Non-purists and innocents may find this an advantage, others may of course disagree.

Topics that are covered are as follows. Chapter 1 begins with an introduction to the rudimentary concepts of equations and the all important skill of equation manipulation. Also included are derivation and manipulation of units. Chapter 2 introduces some functions not included in Chapter 1, logarithms, exponential functions and trigonometric functions. Use of these functions is illustrated as much as possible with examples commonly found in chemistry courses.

Chapters 3, 4 and 5 form the core of the book and should really be read in conjunction with each other. Chapter 3 introduces differentiation and Chapter 4 integration. Chapter 5 (Differential Equations) consists almost entirely of examples, utilising the material of the previous two chapters.

No book on chemistry would be complete without some mention of statistics and this is supplied in Chapter 6. The first part of the chapter provides the briefest of descriptions of the use of statistics in experimental chemistry. The second part introduces some of the statistical concepts used in theoretical chemistry and concludes with a derivation of the ubiquitous Boltzmann equation.

It is true that the only way to learn mathematics properly is to do it. To this end a number of problems are included in each chapter. These are, however, meant to be illustrative and are not extensive. The keen student will find more than enough examples to keep him or her busy in currently available chemistry textbooks. Those who are not so keen will doubtless have problems thrust their way regardless.

Finally, I wish to thank my colleague Dr Don Jenkins for his proof reading parts of the manuscript, Mr Chris Carline for his helpful comments, Mr Melvin Tebbutt for his word processing skills and my family for their support.

PT
October 1993

Preface to the Second Edition

Since the publication of the first edition of this book there has been a wider recognition that more students than ever start advanced chemistry courses without experience of the mathematics necessary for a full understanding of high level chemistry. As a result the student now has a much wider choice of textbooks which range from the basic level to the advanced, and there are even computer programs available for those who can afford the hardware, the software and the accompanying text. Amidst all of this the second edition has been produced with the same aim as the first—to present some basic concepts, perhaps even the minimum necessary—in the context of the chemistry in which they are used. The degree to which this works varies greatly—some simple mathematical concepts are used in advanced chemistry and there is no point in baffling the student with chemistry for the sole purpose of learning a simple piece of mathematics.

The major differences between this and the first edition are as follows. There has been a rearrangement of the text and I have split the former Chapter 6 into two. Experimental statistics now appears after the introductory chapters because the mathematics used is no more advanced than that in Chapters 1 and 2. The core of calculus is now Chapters 4, 5 and 6. Statistics for theoretical chemistry is now Chapter 7. These changes are mostly superficial though all the chapters now have additional problems.

The biggest change is in the addition of a new chapter introducing four different concepts—complex numbers, vectors, determinants and matrices. Each of these could reasonably have a chapter to itself but I have tried to maintain the spirit of the first edition in covering these only as far as the rest of the material in the book allows. I have, however, attempted to show that each of these has its own kind of algebra and notation although the mechanics of use is based on simple mathematical principles. There is still no more than a mention of quantum mechanics although you will perhaps find it is mentioned a few more times than in the first edition.

Critics of the first edition usually maintained that it 'did not go far enough'. It now goes a little further though these same critics will still find that it still does not go far enough—but then it is not intended to. Many other books do this and my critics should read the preface to the first edition to find the reason why this one does not.

It remains for me to thank those who have helped in the production of this new edition.

In addition to those mentioned in the first edition I wish to thank Professor Jeremy M. Hutson of the University of Durham, Dr Roger M. Nix of Queen Mary and Westfield College, University of London and Dr Gareth J. Price of the University of Bath for their helpful comments on the new chapter. Most thanks should go to my family—Susan, Lily, Grace and Chloe—for their continued support and encouragement. Finally I should also thank Andy Slade of John Wiley & Sons for his seemingly infinite patience.

PT
September 1997

1 Equations, Functions and Graphs

1.1 Introduction

In this chapter we introduce some basic terminology and skills used in dealing with equations; using them, rearranging them and solving them. Much of the material will already be familiar but is included as an introduction to methods used in later chapters. One aim of this book is to put this mathematics in a setting based on its usage in chemistry courses. To this end we also include the use and manipulation of units and the derivation of equations from chemical equations.

In reading any physical chemistry textbook you will find that much involves the more complex ideas described in later chapters. There is, however, a great deal which can be achieved using elementary mathematics. Unfortunately, this is often omitted as assumed knowledge. In this chapter we hope to begin to fill some of the gaps created by these omissions.

1.2 Equations and Functions

We begin with a typical example of an equation,

$$y = 3x + 1 \tag{1.1}$$

This already embodies much mathematical convention which we will now define.

Variables. The equation uses both letters and numbers. The letters represent numbers but we are allowed to give them different values. Because the values may vary they are called *variables*. Mathematical equations in science use letters from both the Roman (a to z, lower and upper case) and the Greek alphabet (listed in Appendix 1). In this book letters representing variables will be written in italic script, i.e., *italic*.

In computing the equation (1.1) we use the value of x to calculate the value of y. Therefore the value of y is dependent on the value of x. In this example y is called the *dependent* variable and x is called the *independent* variable.

Terms. Each part of the equation, the y, the $3x$ and the 1 may be referred to as a *term*. The term $3x$ is a shorthand form of three multiplied by x or $3 \times x$. This not only makes the equation more compact it also acts to avoid any ambiguity about the order in which it should be computed (see Section 1.5, Priorities). Sometimes a full stop (period) or dot may be placed between the variables, and this also acts as a shorthand form of multiplication.

Constants. The numbers involved in equations do not change in value and may be referred to as *constants*. Equations may contain constants that are, nevertheless, represented by letters, e.g.,

$$y = mx + c \tag{1.2}$$

In such cases it should always be stated that m and c are constants.

Sometimes constants are predefined and letters used as shorthand. A number of 'universal constants' are used routinely. Examples are the letters R for the gas constant, k for the Boltzmann constant, N for the Avogadro constant and h for the Planck constant. When these letters are used we always know (with experience) what they mean. The numerical values alone would not be particularly informative.

In some circumstances constants arise in the derivation of equations. When this occurs the numerical values may only be found by solution of the equation, by experiment or both.

The Form of an Equation. We may write an equation containing only letters to define the general form of the equation. This is a name which describes the terms in it. Equations (1.1) and (1.2) are both in the form of *first degree* or *linear* equations. Other types of equation will be defined in due course.

The Solution to an Equation. Finding the solution to an equation usually involves finding numerical values to the unknown quantities. Alternatively the solution may be another equation or a different form of the same equation which tells us more explicitly the information we require. There are many methods for solving equations and we will describe some of them in this chapter.

Functions. In general terms we would say that in equations (1.1) and (1.2) y is a function of x. In shorthand this is written,

$$y = \mathrm{f}(x) \tag{1.3}$$

This tells us that y is related to x by some mathematical equation. We could define the function of x explicitly,

$$\mathrm{f}(x) = 3x + 1 \tag{1.4}$$

This method of defining a mathematical relationship is useful when relating experimental variables to each other. If we are to do this we must define which is the dependent and which is the independent variable. This will probably be decided when we do the experiment. For example, in a kinetic experiment we will follow the change in concentration of a reactant as it changes with time. Time is the independent variable and concentration, which we have to measure at predetermined time intervals, depends on the time intervals we choose. In an experiment to investigate the validity of the Beer–Lambert law we measure the amount of light absorbed by different concentrations of a salt in solution. We prepare solutions of known concentration in advance and then measure the amount of light absorbed by each one. We then try and find the relationship describing the amount of light absorbed as a function of concentration. If the function we find is the same as that described by the Beer–Lambert Law then we might deem that the law is valid.

Therefore the dependent variable is the unknown quantity which we have to measure, the independent variable is the quantity which we control.* Once we have decided what the experiment is to be the mathematics is used to determine the relationship outlined in the experiment.

Table 1.1 Calculated values of y from values of x related by equation (1.1)

Value of x	Calculated value of y
0	1
1	4
2	7
3	10
5	16
10	31
100	301

Using equation (1.1), given any value of x we can calculate a corresponding value of y. For example, if we know that x is equal to one then y is given by three multiplied by one, which is equal to three, plus one, which is equal to four. If we had more than one unknown quantity in the equation we would not be able to find a solution. Table 1.1 lists values of x and corresponding values of y.

1.3 Graphs, Plots and Coordinates

1.3.1 Definitions

Using equation (1.1) we have successfully generated pairs of numbers, values of x and y which are listed in Table 1.1. An important way of representing these number pairs is in the form of a graph.

A graph is a two-dimensional diagram in the form of a square grid (Figure 1.1). The edges of the grid, called the *axes* are labelled according to the variables they represent. Most commonly the horizontal *axis* represents the value of x (the independent variable) and the vertical axis represents y (the dependent variable). They may therefore be referred to as the x and the y axes. A more general nomenclature for the horizontal axis is the *abscissa* and the y axis is the *ordinate*. This allows us to use any letter variables we choose yet let the reader know which axis we mean.

*Note that time, in the experimental sense, is very much under our control. We determine when the experiment starts, when it ends and the intervals between measurements.

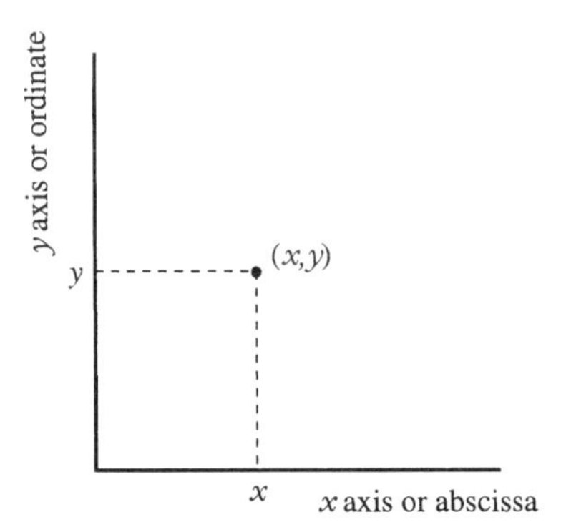

Figure 1.1 Plotting the general point (x, y) by moving along the abscissa a distance x and up the ordinate a distance y

A function such as equation (1.1) is represented by marking the value of x along the abscissa and then moving upwards, parallel to the ordinate, a distance corresponding to the related value of y. This position is then marked with a dot or possibly a cross, as shown in Figure 1.1.

The corresponding values of x and y are then referred to as *coordinates* and written in the form (x,y), e.g., (1,4) from Table 1.1. These coordinates are then plotted as shown in Figure 1.2. This shows a plot of several coordinates which fit equation (1.1). Note that they all appear to fall on a straight line. Points which we plot in this way follow the normal rules by which we deal with numbers. That is they may be whole (*integer*) numbers such as 1, 2, 3 etc., or they may be fractional. For fractional numbers we may deal in common fractions ($\frac{1}{2}$, $\frac{1}{4}$, $\frac{3}{4}$) or decimal representations (0.5, 0.25, 0.75). Graphically the latter is the more common form. Some numbers cannot be represented by a complete fraction or limited set of decimal digits. These *irrational* numbers must be approximated, even when plotted on a graph. The number referred to as π (Greek letter 'pi') is such a number and occurs frequently in all sciences. It is defined as the ratio of the area of a circle to the square of its radius but is sometimes approximated by the fraction 22/7. Pressing the 'π' button on your calculator reveals the number 3.1415927. A computer allowed to reveal more digits will show that this number goes on and on. To approximate it on a graph depends on the scale of the graph but 3.14 is commonly used.

Thus it is possible to calculate all of the intermediate, non-integer coordinates using equation (1.1) and plot the graph in the usual manner. As shown on the inset to Figure 1.3 they still lie in the same straight line and we show this by actually drawing the line through the points. Alternatively we may omit the points and just draw a line which represents the function.

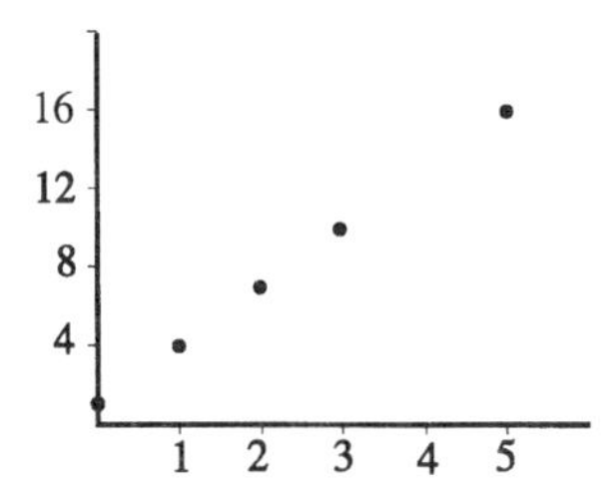

Figure 1.2 Several coordinates calculated from equation (1.1), listed in Table 1.1

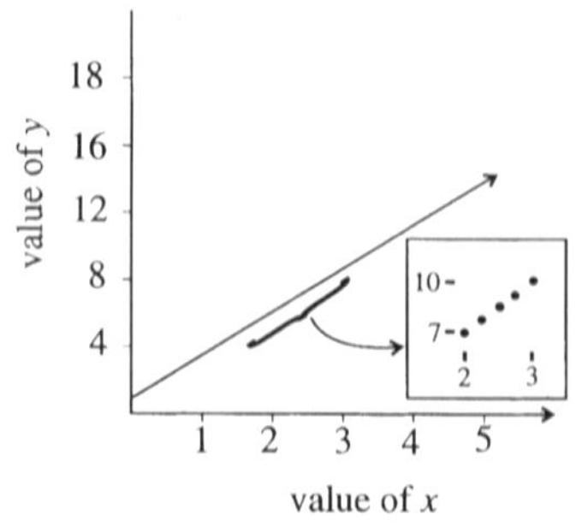

Figure 1.3 Fractional coordinates my also be plotted. That the function may be computed for all points may be represented by drawing a straight line

As with single numbers co-ordinates may also have negative values. As shown in Figure 1.4 a graph may be drawn to include positive and negative numbers of both x and y. The four areas of this type of plot, defined by the now crossed axes are called *quadrants*. The point defined by the coordinate (0,0) is called the origin.

Functions which can be represented by a solid line are referred to as *continuous*. Occasionally a function generates a value of the dependent variable which is infinite in magnitude. Such a function is

$$y = \frac{1}{x} \tag{1.5}$$

When x is equal to zero y is infinite in value. If we continue past zero and to a negative value of x (Figure 1.5) the plot returns to real values from a negative value of infinity. Functions such as this which show gaps are called *discontinuous*. Note also that this line is curved. Functions which give rise to straight line plots are *linear* and functions such as equation (1.5) for which the plots are curved are *non-linear*.

1.3.2 Properties of Straight Line Functions

One of the most important properties of the line which represents equation (1.1) is that it slopes. Like a road which goes up or down it has a *gradient*. Indeed, the gradient is often

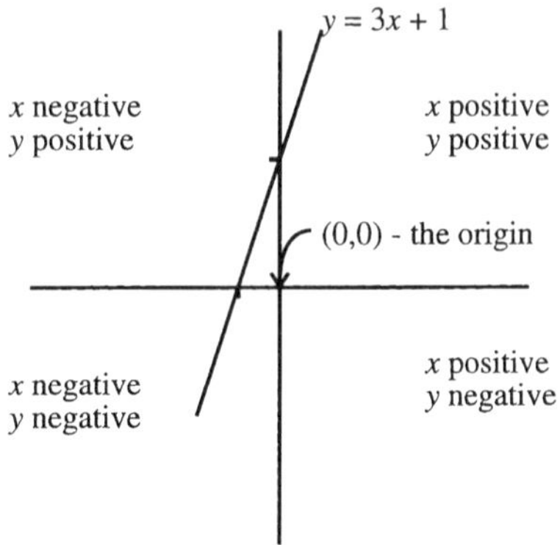

Figure 1.4 Negative numbers may also be represented by drawing crossed axes. The four sections are called quadrants

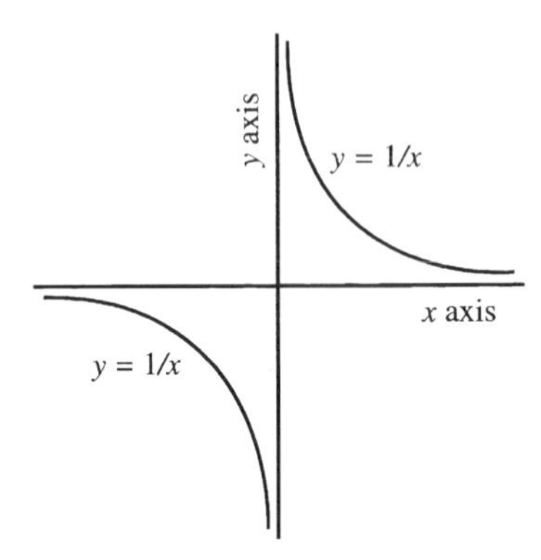

Figure 1.5 Some functions such as $1/x$ cannot be computed for certain values. At $x = 0$ y is infinite in magnitude and at this point the function is discontinuous

referred to as the *slope*. The value of the slope is defined as the increase in y per unit increase in value of x.

In Table 1.1 the value of x increases by one between two and one. The corresponding increase in y is equal to

$$\begin{aligned} \text{change in } y &= 7 - 4 \\ &= 3 \end{aligned} \tag{1.6}$$

The gradient is therefore equal to three. We arrive at the same number by dividing any change in y by the corresponding change in x. For example y changes from seven to 301 as x changes from two to 100. We calculate the slope from the equation,

$$\text{Slope} = \frac{\text{change in } y}{\text{change in } x} \tag{1.7}$$

Therefore

$$\begin{aligned} \text{Slope} &= \frac{301 - 7}{100 - 2} \\ &= \frac{294}{98} \\ &= 3 \end{aligned} \tag{1.8}$$

Note that the value of the slope is equal to the number which multiplies x in the original equation. This or any number which is used to multiply a variable is called a *coefficient*. In this case the coefficient may be referred to as the slope or the gradient. This is the letter m in the general form, equation (1.2). So we can write an equation defining the value of m. Taking equation (1.7) we define the 'change in y' by taking the difference between two arbitrarily chosen values of the variable y. We label these y_2 and y_1. The corresponding values of x are x_2 and x_1. The definition of the slope m is

$$\text{Slope, } m = \frac{y_2 - y_1}{x_2 - x_1} \tag{1.9}$$

The difference between two quantities may also be represented by the Δ (Greek capital delta) notation. This will be discussed in more detail in Chapter 4 (Differential Calculus) but for now we assume that whenever this symbol precedes a variable it means that we find the difference between two values of the variable. In the following equation we use Δy and Δx, and we assume that the differences we calculate are the corresponding differences as in equation (1.9).

$$m = \frac{\Delta y}{\Delta x} \tag{1.10}$$

Another important property of a straight line graph is the point where the line crosses the y axis. This is called the y axis *intercept* and is where x has zero value.

$$y = m \cdot 0 + c \tag{1.11}$$

The product $m \cdot 0$, as with anything multiplied by zero, is zero and thus

$$y = c \tag{1.12}$$

The intercept is the constant c in equation (1.2). The physical significance of the intercept depends on the system under investigation. For example, the equation describing the dependence of concentration on time for a zero order reaction may be written as

$$c = c_0 - kt \tag{1.13}$$

The graph representing this equation (Figure 1.6) has a negative slope (going down from left to right) reflecting the fact that there is a minus sign in front of the term containing the independent variable, t. The y axis intercept has a value c_0, the value of c when t is equal to zero. This is the initial concentration of the reactant. Later in this chapter we will meet Kohlrauch's Law (p. 26). In this case the intercept is an important fundamental parameter, typical of the system under investigation. As it cannot be measured directly it must be calculated by plotting a graph and extending the plotted line beyond the measurable data to estimate the intercept.

In some cases the intercept may be equal to zero. *The Beer–Lambert Law* relates the absorbance, A, of light by a species in solution to its concentration, c, by the equation

$$A = \varepsilon \cdot c \cdot l \tag{1.14}$$

where ε is a constant called the *molar absorption coefficient* and l is the length of solution (called the *path length*) through which the light passes. Both ε and l are constants and the product εl is equivalent to the constant m in equation (1.2). The equation has no

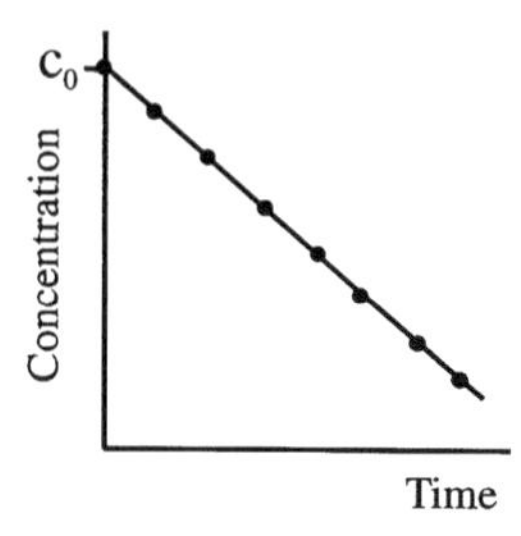

Figure 1.6 Plots of concentration versus time following the loss of a reactant have a negative slope

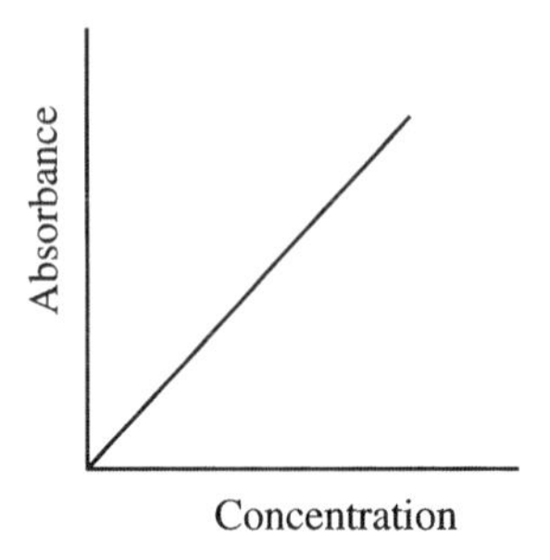

Figure 1.7 The Beer–Lambert law predicts a plot with zero intercept

additional constant, i.e., c in (1.2), and so the intercept when concentration is zero should be y equals zero (Figure 1.7).

1.3.3 The Use of Graphs to Represent Experimental Data

Graphs are particularly useful for representing experimental data. If we are making pairs of measurements then it is a simple matter to plot one with respect to the other. We must decide which measurement is to be the independent and which the dependent variable and then plot the coordinates accordingly. To do this we do not have to know the mathematical relationship between the two. Rather, the graph is a means of determining the nature of the relationship.

If, for example, all of the points lie on a straight line then we know that the relationship is of the form of equation (1.2). We can draw a straight line through the experimental points and extend it to the axis. By calculating the slope of the line we can calculate the experimental value of m and estimate the value of c.

It is likely that experimental data will not fall exactly on a straight line but that there will be some scatter, as shown in Figure 1.8. In such cases we must use some judgement in deciding how the line should be drawn. Although there are mathematical methods of calculating the slope and intercept from raw data (Chapter 3) it is often possible to make a good estimate of these values by drawing the line (called the *line of best fit*) using a rule and your own discretion. It may even be possible to estimate the accuracy of your estimate by

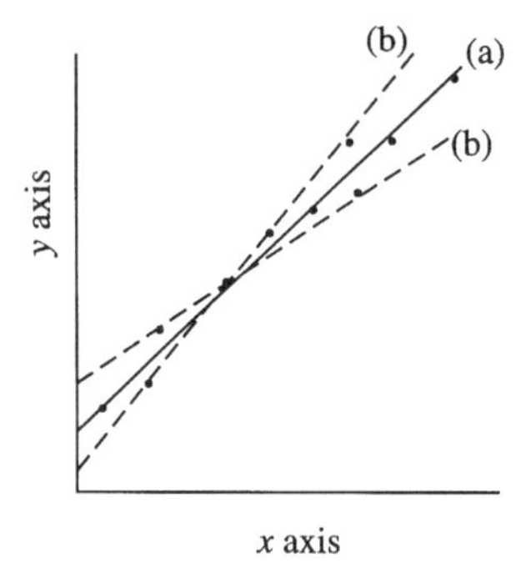

Figure 1.8 Experimental data rarely falls on a straight line, even when theory suggests it should. There are many reasons for this (Chapter 3) but assuming the theory is correct we can estimate the best straight line, (a). We can also estimate the accuracy of this by drawing the two worst fits to the data, (b), one with slope greater and one with slope less than the best fit

drawing not only the line of best fit but two lines of *worst fit*, one with very high slope and one with very low slope (Figure 1.8).

When the relationship is non-linear the points fall on a curve and the nature of their relationship is harder to determine. Much of the work in the remaining chapters of this book describes the methods used to find mathematical relationships between experimental variables. It is possible to use a 'trial and error' method whereby one guesses the equation which relates two variables. A plot is then made according to the equation and the 'goodness of fit' is estimated. For example if the guess is that the relationship is linear but the data fall on a curve then the fit would be poor. This method is fairly haphazard and the quality of fit may be difficult to judge as the difference between curved functions may be quite subtle.

The preferred method is that one suggests a way in which the chemistry operates and then derives a mathematical equation from this. For example, in attempting to find the mechanism of a chemical reaction one suggests a route by which the reaction takes place (the mechanism) and then derives a 'rate' equation. This is usually done by deriving and solving differential equations (Chapter 6). This obviously takes some experience and it is usually sufficient for students to understand the method through a small number of relatively simple cases. Chapters 4, 5 and 6 are dedicated to these methods and describe a number of typical examples.

If the fit of the experimental data to the theoretical plot (drawn by simply putting a range of numbers in the equation suggested) is poor then one must reassess the proposed mechanism and reassess the mathematics appropriately. Although this might seem like taking two steps backwards the advantages outweigh the time factor involved. One major advantage is that the meanings of constants arise as part of the derivation. In the trial-and-error method one must assign the meaning of constants afterwards.

1.3.4 Calibration Plots

If we know that the relationship between our experimental variables is reproducible then we may use a graph for estimating unknown quantities. For example if we wish to find the concentration of a species in solution then we begin by making several solutions of known concentration. The absorbance of each of these is measured and a plot made of absorbance versus concentration. This is the calibration plot. If we then take a solution of unknown concentration and measure the absorbance we can read off the concentration as shown in Figure 1.9. If we use such a plot to find values within the range of the calibration it is called *interpolation*. Extending outside the range of data (by drawing a straight line for example) is called *extrapolation*.

1.3.5 Using Computers—Plotting Programs and Spreadsheets

There are a number of programs available for plotting data on a computer. These range from dedicated scientific software to business-oriented *spreadsheets*. They are all used in similar ways and usually have useful on-line tutorials to help new users. Spreadsheets, although designed with needs of business in mind have all the necessary capability for data manipulation and presentation necessary for the chemist.

Most applications consist of an array of rectangles, called *cells*, and one value is typed into each. Typically x values are in one column and y values in an adjacent column. At the

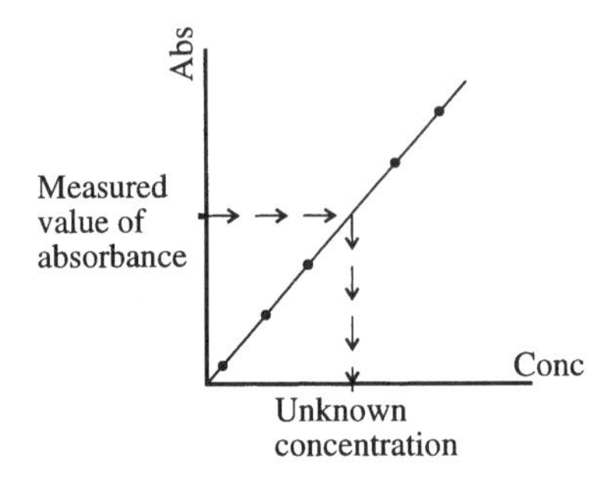

Figure 1.9 Calibration plots may be used to find unknown quantities. For example, a plot of the Beer–Lambert law using known concentrations is made and the absorbance of the unknown solution measured. The value of concentration is then read off the plot as shown in this figure

click of a mouse button the data will then be plotted with more elegance than could ever be achieved by hand. It is also possible to change data, i.e., derive a function of measured data and plot that instead. It will then be possible to calculate and draw the best straight line through the data and calculate the slope and intercept.

However useful computers may be one should not overlook the importance of manual graph plotting skills. One reason is that at the time this book is published it is unlikely that you will have a computer available in an examination where you may have to plot a graph. The other is that in the laboratory, plotting data as an experiment proceeds can be used to follow its progress. Although there are uncertainties (what are scales of the axes?) you will be able to see directly whether you have enough data. If it is supposed to fall on a curve then insufficient data may actually appear linear. A linear plot with some scatter may appear curved. Additional points highlight aberrant data by pulling the best fit back into line. You do not want to waste time collecting more results than are actually necessary and you should be able to assess this from your rough plot. Producing this as part of your report will show that you have thought about the experiment in advance.

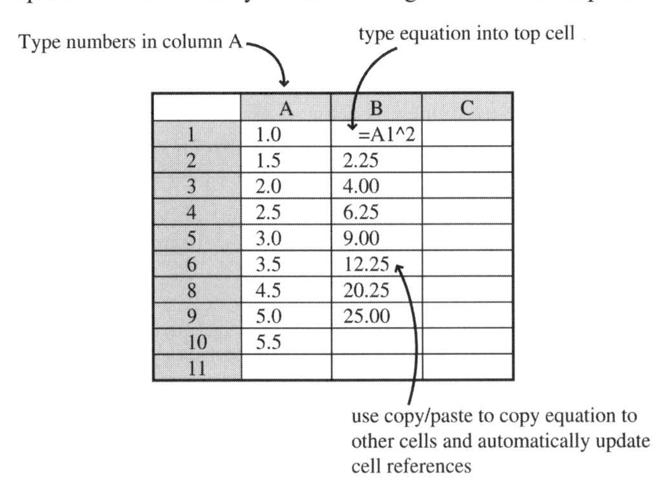

	A	B	C
1	1.0	=A1^2	
2	1.5	2.25	
3	2.0	4.00	
4	2.5	6.25	
5	3.0	9.00	
6	3.5	12.25	
8	4.5	20.25	
9	5.0	25.00	
10	5.5		
11			

Figure 1.10 Schematic diagram of a spreadsheet. Numbers may be entered in columns or rows of cells. Derived values may be found using equations typed in other cells. The function may be copied and pasted into other cells with cell addresses automatically updated. Most spreadsheets have automatic plotting facilities to make plotting easy

Although you may wish to present your final version neatly typed with graphs from a plotting program it is nevertheless good practice to make and keep all measurements and rough plots.

1.4 Powers, Indices and Exponents

1.4.1 Definitions

If we multiply a number (constant or variable) by itself a number of times then we can substitute the notation of multiplication by a superscript called the *index* (plural *indices*) or *exponent*. For example if we define a function of x, $f(x)$ as

$$f(x) = x \times x \tag{1.15}$$

then this can be written as

$$f(x) = x^2 \tag{1.16}$$

which reads as 'function of x is equal to x squared'. If we multiply x by itself three times ($x \times x \times x$) then this would be written as x^3 which reads 'x cubed'. When the self multiplication is repeated four times then the notation x^4 is read as '*x raised to the power of four*' or '*x raised to the fourth power*'. Fifth and higher powers continue in this vein. This terminology is general and lends itself to the use of fractional and negative powers and powers denoted by constants or variables such as x^n. Thus x^2 may also be read as 'x raised to the power two' and x^n as 'x raised to the nth power'.

An equation containing terms up to the power n are called *polynomial* equations *of the degree n*. Therefore an equation in which the highest power of x is two is called a second degree equation. One in which the highest power of x is four is called a fourth degree equation. Equations in x^2 and x^3 are also referred to as *quadratic* and *cubic* equations respectively.

1.4.2 Properties

Indices have the following important properties.

Multiplication. Multiplication of numbers raised to given powers follows the rule defined by the equation (1.17).

$$x^a \times x^b = x^{(a+b)} \tag{1.17}$$

We can illustrate this by multiplying x^2 and x^3.

$$\begin{aligned} x^2 \times x^3 &= (x \times x) \times (x \times x \times x) \\ &= x \times x \times x \times x \times x \\ &= x^5 \end{aligned} \tag{1.18}$$

where x^5 is of course the same as $x^{(2+3)}$.

Division. Division of numbers raised to given powers follows the rule defined by equation (1.19).

$$\frac{x^a}{x^b} = x^{(a-b)} \tag{1.19}$$

This can be illustrated by performing the division x^6/x^2.

$$\begin{aligned} \frac{x^6}{x^2} &= \frac{x \times x \times x \times x \times x \times x}{x \times x} \\ &= x \times x \times x \times x \\ &= x^4 \end{aligned} \tag{1.20}$$

where x^4 is equal to $x^{(6-2)}$.

Note that these rules only work for the variable or constant defined by x. We could not perform the same sums on y^n being multiplied by x^m. As is normally the case we can only add together quantities of the same thing. $x + x$ may be the same as $2x$ but $x + y$ is always $x + y$. For example we cannot multiply 2^2 and 3^2 using the rule on multiplication. We must compute both squares and then add the results.

Powers of Powers. If a number x is raised to some power and the result is raised again to another power then the final result is defined by equation (1.21)

$$(x^n)^m = x^{nm} \tag{1.21}$$

We can show that this is true by the following example.

$$\begin{aligned} (x^2)^3 &= (x \times x)^3 \\ &= (x \times x) \times (x \times x) \times (x \times x) \\ &= x^6 \end{aligned} \tag{1.22}$$

where the index six is equal to the product of two and three.

The Power of One. Anything raised to the power of one is equal to itself. Thus x may be written as x^1.

The Power of Zero. Anything raised to the power of zero is equal to one. This follows from the rule on division. If we perform x^a/x^a we would expect the result to be equal to one, since any number divided by itself is equal to one.

$$\begin{aligned} \frac{x^a}{x^a} &= x^{(a-a)} \\ 1 &= x^0 \end{aligned} \tag{1.23}$$

hence

$$x^0 = 1 \tag{1.24}$$

Negative Indices. These are defined by equation (1.25).

$$x^{-n} = \frac{1}{x^n} \tag{1.25}$$

This can be shown to be true because '1' is the same as x^0 and so we can rewrite the right hand side of equation (1.25),

$$\frac{x^0}{x^n} = x^{(0-n)} = x^{-n} \tag{1.26}$$

Fractional Indices. These may be thought of as representing the inverse function of raising a number to some power. The *inverse* of a function does the opposite of that function, as addition is the opposite of subtraction. That is, the function $x^{1/2}$ is the inverse of x^2. The latter is called finding the square of the variable x and the former is called finding the *square root*. This is finding the number which when multiplied by itself results in the number of interest. We know that the square of two (2^2) is equal to four and therefore the square root of four ($4^{1/2}$) is equal to two. The inverse of x^3 is similarly $x^{1/3}$ and this is the number which when multiplied by itself three times results in the number represented by x. Finding these numbers is not as straightforward as finding squares or higher functions but fortunately all the hard work has already been done. In the past they have all been listed in books of *Mathematical Tables*. These listed not only squares, cubes and the corresponding roots but many other mathematical functions such as logarithms and geometric functions (Chapter 2). The possession of a pocket calculator makes the finding of a square or a root of a number easy. The user merely presses the correct button after input of the number of interest. Scientific calculators usually have a button labelled x^y for finding any power and an inverse button labelled $x^{1/y}$ for finding a root.

Roots are named in a similar manner to their associated powers and so the number which multiplied by itself four times gives x is the fourth root of x.

Another form of notation for finding roots is the *radical* notation,

$$x^{1/2} = \sqrt{x} \tag{1.27}$$

Cubic and higher roots are labelled with the degree of the root being sought,

$$x^{1/3} = \sqrt[3]{x} \tag{1.28}$$

Although useful for indicating the root of a whole function (rather than a single variable) this notation does not lend itself to mathematical manipulation and so the fractional power is preferred. We can show the logic of this notation by multiplying $x^{1/2}$ by itself.

$$x^{1/2} \times x^{1/2} = x^{(1/2+1/2)} = x^1 \tag{1.29}$$

Therefore $x^{1/2}$ is by definition the square root of x.

In Figure 1.11 we plot the graph of $y = x^2$. It can be seen that for every value of y there are two values of x, one positive and one negative. This reflects the normal rules of multiplication which state that the product of two negative numbers is a positive number. If we are trying to find the numbers which when multiplied together produce a positive number then we must account for this rule. Therefore when calculating the square root of a positive number ($4^{1/2}$ for example) there are two possible results, $+2$ and -2. This may be written as ± 2 which reads as 'plus or minus two'.

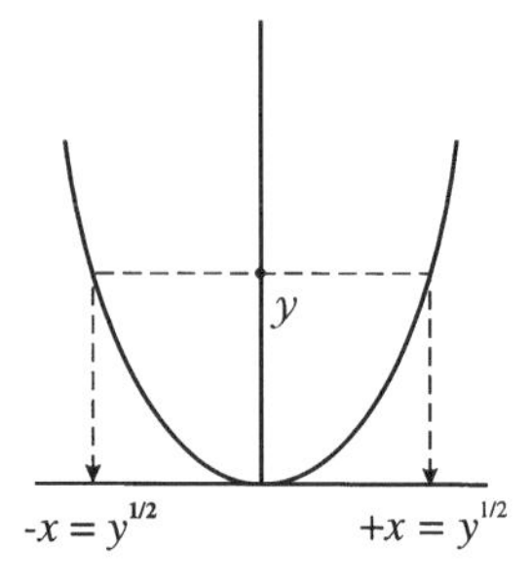

Figure 1.11 Plot of $y = x^2$ showing that for every value of y there are two values of x

The practicalities of the real world often mean that only the positive square root is valid. For example the solubility product (K_{sp}) of silver chloride is defined as

$$K_{sp} = [Ag^+][Cl^-] \tag{1.30}$$

Since the number of chloride ions must be the same as the number of silver ions (for electroneutrality) this may be written as

$$K_{sp} = [Ag^+]^2 \tag{1.31}$$

If we were told that K_{sp} had a value of 0.000 000 000 1 we could calculate the numerical value of concentration of silver ions by finding the square root of this number, i.e.,

$$0.000\,000\,000\,1 = [Ag^+]^2 \tag{1.32}$$

Therefore

$$(0.000\,000\,000\,1)^{1/2} = [Ag^+] \tag{1.33}$$

which computes to

$$[Ag^+] = \pm\,0.000\,01 \tag{1.34}$$

Clearly the concentration cannot have a negative value and so the result must be the positive root of the solubility product.

1.4.3 The Floating Point

In chemistry we frequently deal with very large and very small numbers. In the example above we use the figure 0.000 000 000 1 for the value of the solubility product of silver chloride. The value of the Avagadro constant which represents the number of atoms per mole may be written in the form 602 213 670 000 000 000 000 000. It is undoubtedly inconvenient either to write this number down every time we use it, or to put it into a calculator in this form. The *floating point* or *exponential* notation is a way of solving this problem.

This notation reduces every number to one digit in front of the decimal point and the remaining digits follow after. This number is then multiplied by ten raised to some power in order to produce the equivalent of the original figure.

Example—the number 23

The number 23 is reduced to one digit in front of the decimal point, 2.3. In order to make this equivalent to 23 it must be multiplied by 10 or 10^1. It is written as 2.3×10^1.

Example—the number 2300

This is also reduced to 2.3 and then multiplied by one thousand, or 10^3. It is written 2.3×10^3.

Example—the number 0.023

Once again this is reduced to 2.3 and then divided by 100. This is the same as dividing by ten squared or multiplying by 10^{-2}. It is therefore written as 2.3×10^{-2}.

Example—solubility product of AgCl

This was given above as 0.000 000 000 1. This is the same as 1/100 000 000 00 or 1×10^{-10}.

Example—the Avogadro constant

This is normally written in the form $6.022\,136\,7 \times 10^{23}$.

This form of notation is used routinely in chemistry and students should be familiar with its usage. Some points to remember are listed below.

Addition and Subtraction. To add or subtract two numbers given in floating point notation we must make sure that they are both raised to the same power of ten. Should we want to add 2.3×10^2 and 1.6×10^3 we convert the numbers so that they are in exactly the same form. 1.6×10^3 is exactly the same as 16×10^2 and we can now add 16 and 2.3. The result is given as 18.3×10^2 which is normally written as 1.83×10^3.

Recalling how we are taught at an early age to perform summations this is no more than making sure that all of our hundreds, tens and units are all in the correct columns and aligned when doing the sum.

$$\begin{array}{r} 1600 \\ 230 \\ \hline 1830 \\ \hline \end{array} \tag{1.35}$$

Of course using a calculator the sum is performed directly.

Multiplication and Division. We perform the multiplication or division of the numbers (the coefficient of 10^n) directly and then calculate the resulting value of 10^n using the rules on multiplication and division of indices.

Example—product of 2.30×10^2 by 5.30×10^3

We first multiply 2.3 and 5.3, the result of which is 12.19. The product of 10^2 and 10^3 is

$10^{(2+3)}$ or 10^5. The complete result is therefore 12.19×10^5. This would then be reduced to 1.219×10^6.

Example—division of 2.30×10^2 by 1.30×10^3

Divide 2.30 by 1.30, result 1.77. 10^2 divided by 10^3 is $10^{(2-3)}$ or 10^{-1}. The overall result is therefore 1.77×10^{-1}.

1.5 Priorities

When using a mathematical equation to calculate the value of a variable it is not simply a case of moving from left to right as the equation is written. For example, if we were to use the following equation to calculate a value for the variable y,

$$y = 2 \times 3 + 1 \tag{1.36}$$

we may choose to first multiply two by three, result six, and then add one, result seven. We may, however, also write this equation as

$$y = 1 + 2 \times 3 \tag{1.37}$$

Moving from left to right we add one and two, result three and multiply by three, result nine. But what is there to stop us performing the calculation as above and getting seven as the result? To avoid ambiguities and to ensure that equations (1.36) and (1.37) give the same result we follow a set of *priorities*. That is, we perform the functions with highest priority before those with lower priority however the equation may be written. The priorities in decreasing order are:

Brackets
Orders or powers
Division and
Multiplication
Addition and
Subtraction

Brackets are used, as we have already seen in this chapter, to enclose collections of functions. Any sum or computation within a pair of brackets should be performed before any other. Sometimes several pairs of brackets are nested within each other and when this occurs we start at the inside and work outwards. For example, we have the equation

$$y = ((2 + 3) \times (4 + 3) + 1) \times 6 \tag{1.38}$$

To calculate y we take the following steps. Begin by adding two to three and four to three,

$$\begin{aligned} y &= ((5 \times 7) + 1) \times 6 \\ &= (35 + 1) \times 6 \\ &= 36 \times 6 \\ &= 216 \end{aligned} \tag{1.39}$$

Next in priority is raising a number to some power. Therefore the addition of 2^2 to 3^3 follows the steps,

$$\begin{aligned} y &= 2^2 + 3^3 \\ &= 4 + 27 \\ &= 31 \end{aligned} \tag{1.40}$$

Following this, multiplication and division have equal priority. Any ambiguity in performing these calculations may be avoided by replacing the '×' and '÷' or 'slash' (/) signs by the conventional shorthand. We have already met the use of a full stop or the omission of a space or sign to indicate a product. If we write an equation in the form

$$y = 1 + 2x \tag{1.41}$$

it is clear that we multiply two by x before adding one. If we wished to override this priority and add one to two before multiplying by x we would use brackets,

$$y = (1 + 2)x \tag{1.42}$$

Note that we can now omit the multiplication sign between the right bracket and the variable. This illustrates the trend to enclose functions involving additions and subtractions inside brackets rather than multiplications.

In the case of division the equation may be better written on two lines rather than one, in the form of a fraction. For example, the equation

$$y = 1/x + 3 \tag{1.43}$$

would normally be read as one divided by x, the result of which is added to three. If we wished to divide one by the sum of three and x we would either have to use brackets, $1/(x + 3)$, or use the form of a fraction

$$y = \frac{1}{x + 3} \tag{1.44}$$

Finally addition and subtraction are equal last in the list of priorities. They are only performed first when enclosed in brackets.

1.6 Factors

Factors are those numbers which may be divided into another number. For example, two and three are both factors of six. Twelve has more factors: two, three, four and six. Sometimes factors of functions may be found and used in the manipulation and solution of equations. Finding the factors is called *factorising*.

Example—factors of 2x + 2y

The factors of the product $2x$ are two and x, the factors of $2y$ are two and y. They have a common factor, two. If we divide each term in the sum by this common factor this then leaves another factor of the sum which we enclose in brackets: for example, writing the function of the title in the form of a function $f(z)$ of the two variables x and y.

$$z = 2x + 2y \tag{1.45}$$

We remove the most obvious common factor and rewrite the equation as the product of this and the remaining factor. Equation (1.45) becomes

$$z = 2(x + y) \tag{1.46}$$

This is relatively trivial and finding the factors of more complicated functions may be much more difficult.

Example—solving an equation ($0 = x^2 - 1$)

To indicate how this process may be used to solve an equation Table 1.2 lists the factors of a number of quadratic functions. The first three are equations of two variables and the lower three are the same equations where y is equal to one.

Figure 1.12 shows a plot of the equation $y = x^2 - 1$. By inspection of this plot we see that the curve intersects the x axis in two places, at $x = -1$ and $x = +1$. Since the x axis is also the line where y is equal to zero these two values must also be the solutions to the equation $0 = x^2 - 1$. We can confirm this result by solving the equation directly. From Table 1.2 we see that the function $x^2 - 1$ is the same as $(x + 1)(x - 1)$. If either of these two bracketed terms becomes equal to zero then the whole function becomes zero and we have our solution. The first becomes zero when x is equal to -1. The second becomes zero when x is equal to $+1$. Therefore the solutions to the function in the title are that x is equal to ± 1. More methods of solving equations are discussed in Section 1.9 (Solving Equations).

Table 1.2 Some functions and their factors

Function	Factorised function
$x^2 + 2xy + y^2$	$(x + y)(x + y)$
$x^2 - y^2$	$(x - y)(x + y)$
$x^2 - 2xy + y^2$	$(x - y)(x - y)$
$x^2 + 2x + 1$	$(x + 1)(x + 1)$
$x^2 - 1$	$(x + 1)(x - 1)$
$x^2 - 2x + 1$	$(x - 1)(x - 1)$

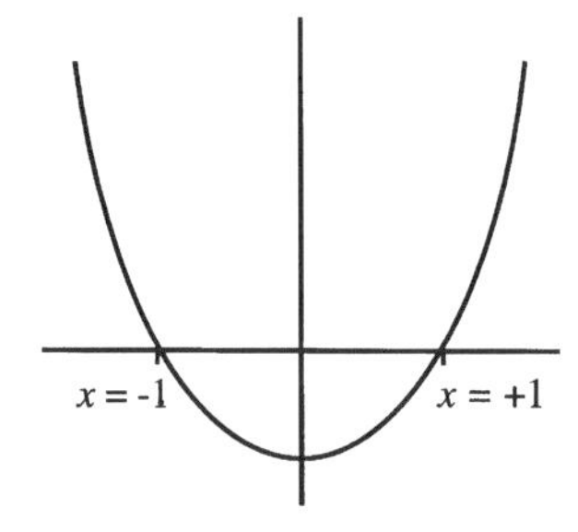

Figure 1.12 Plot of $y = x^2 - 1$ showing that there are two roots or solutions when y is zero

1.7 The Binomial Theorem and Pascal's Triangle

It is possible to show that the factors in the right hand column of Table 1.2 are correct by working backwards, starting with the products in bracketed terms. Multiplying out these products is called *expanding* the function. To do this we first note that the function connecting the two sets of brackets is multiplication. Thus we multiply every term inside the first pair of brackets by every term inside the second pair, taking into account the signs associated with each term.

Example—expanding $(x + 1)(x + 1)$

The first term on the left is x and this is multiplied by the first term, x, and then the second term, 1, on the right.

$$x(x + 1) = x^2 + x \tag{1.47}$$

We then multiply each term on the right by one.

$$1(x + 1) = x + 1 \tag{1.48}$$

The results are then added together.

$$x^2 + x + x + 1 = x^2 + 2x + 1 \tag{1.49}$$

This is the result given in Table 1.2.

Sometimes the 'expanded' version is more compact than the factorised version. This happens when we have positive and negative numbers in the brackets and these cancel, as shown in the next example.

Example—expanding $(x + 1)(x - 1)$

Again we start by multiplying each term on the right by the first term on the left (x).

$$x(x - 1) = x^2 - x \tag{1.50}$$

Next we multiply each term on the right by the second term on the left (1).

$$1(x - 1) = x - 1 \tag{1.51}$$

Adding the results we obtain

$$x^2 - x + x - 1 = x^2 - 1 \tag{1.52}$$

Of particular interest is the expansion of the functions $(x + y)(x + y)$ and $(1 + x)(1 + x)$. These may also be written as $(x + y)^2$ and $(1 + x)^2$ respectively. What happens if we try to calculate higher powers of these functions?

Example—expansion of $(1 + x)^3$

The expansion of $(1 + x)^3$ may be found by multiplying out the square $(1 + x)^2$ and $(1 + x)$. The former is given in Table 1.2. Following the method above the result is

$$1(1 + 2x + x^2) = 1 + 2x + x^2 \tag{1.53}$$

plus

$$x(1 + 2x + x^2) = x + 2x^2 + x^3 \tag{1.54}$$

which is

$$1 + 2x + x^2 + x + 2x^2 + x^3 = 1 + 3x + 3x^2 + x^3 \tag{1.55}$$

This method could be repeated to find further results and Table 1.3 lists these up to the fifth degree. Repeated calculations would become very tedious but in Table 1.3 we can already see a pattern emerging.

First note that in the expanded version the exponent of x increases in steps of one from term to term across the function. The index of the far right term is equal to that of the original bracketed function. That is, we have a polynomial equation of degree n where n is the value of the exponent of the bracketed term.

Secondly the coefficient of x goes up and down across the function. Not only that, the coefficients of different expansions are related to each other. As they are written in this table each coefficient is the sum of the two nearest coefficients in the equation above. This becomes more apparent if we remove the x terms and use only the coefficients as shown in Table 1.4. Each number is the sum of the two nearest numbers in the row above. This table shows the first eleven rows of what is known as *Pascal's triangle.*

This expansion of functions is known as *binomial expansion* and the coefficients in Pascal's triangle are the *binomial coefficients.* Using Pascal's triangle and the observation that the degree of the x terms increases it should be possible to predict the expanded version of any $(1 + x)^n$ function where n is a positive integer.

Table 1.3 Expansion of $(1 + x)^n$

$(1 + x)^n$	Expanded version
$(1 + x)^0$	1
$(1 + x)^1$	$1 + x$
$(1 + x)^2$	$1 + 2x + x^2$
$(1 + x)^3$	$1 + 3x + 3x^2 + x^3$
$(1 + x)^4$	$1 + 4x + 6x^2 + 4x^3 + x^4$
$(1 + x)^5$	$1 + 5x + 10x^2 + 10x^3 + 5x^4 + x^5$

Example—expansion of $(1 + x)^6$

We should have terms in x^0, x^1, x^2, x^3, x^4, x^5 and x^6. The coefficients of these should be given by Pascal's triangle, i.e., 1, 6, 15, 20, 15, 6 and 1. The expansion is therefore

$$(1 + x)^6 = 1 + 6x + 15x^2 + 20x^3 + 15x^4 + 6x^5 + x^6 \tag{1.56}$$

An alternative way of producing this expansion is to use the *binomial theorem* or *binomial series.* This is an equation which allows us to calculate the coefficients of the binomial expansion. The equation is

$$(1 + x)^n = 1 + nx + \frac{n(n-1)}{2.1}x^2 + \frac{n(n-1)(n-2)}{3.2.1}x^3 \cdots x^n \tag{1.57}$$

The products 2.1 and 3.2.1 are usually written as 2! and 3! respectively. This is known as *factorial* notation. 2! reads as 'two factorial', 3! as 'three factorial' and so on. Note that when we reach the x^n term the top part of the fraction or *numerator* is $n(n-1)(n-2)(n-3)\ldots 1$ and this is equal to $n!$. The bottom part of the fraction or *denominator* is also $n!$ and so the two cancel producing a coefficient of unity.

Table 1.4 Pascal's triangle

1
1 1
1 2 1
1 3 3 1
1 4 6 4 1
1 5 10 10 5 1
1 6 15 20 15 6 1
1 7 21 35 35 21 7 1
1 8 28 56 70 56 28 8 1
1 9 36 84 126 126 84 36 9 1
1 10 45 120 210 252 210 120 45 10 1

It is possible to use equation (1.57) to calculate real values of squares. For example, 3^2 may be written as $(1+2)^2$ and so we may put both x and n equal to two in equation (1.57). The result, as expected, is nine. In the age of the pocket calculator this method seems archaic but the binomial expansion may be used to solve theoretical problems and in manipulating equations. For example, if x is very small compared to one, say 0.01, then the squared and higher powers of x become very small.

Example—binomial expansion of $(1.01)^3$

This is calculated using x equal to 0.01 in the binomial series.

$$\begin{aligned}(1+0.01)^3 &= 1 + 3(0.01) + \frac{3.2}{2!}(0.01)^2 + \frac{3.2.1}{3!}(0.01)^3 \\ &= 1 + 0.03 + 0.0003 + 0.000\,001 \qquad (1.58) \\ &= 1.030\,301\end{aligned}$$

Now the last two terms in the second row of this equation are very small compared to the first two and the result could reasonably be approximated by ignoring these altogether. The result 1.03 is a good estimate of the true result. It is therefore possible, if we know that the value of x is very small, to ignore squared and higher terms in the binomial expansion and write an approximation,

$$(1+x)^n \approx 1 + nx \qquad (1.59)$$

where the curly equals sign means *approximately equal to.*

Example—approximation of $(0.99)^2$

This is the same as writing $(1-0.01)^2$. Using the approximation (equation (1.59)) with x equal to -0.01 we obtain

$$\begin{aligned}(1 - 0.01)^2 &\approx 1 - 2(0.01)\\ &\approx 1 - 0.02\\ &\approx 0.98\end{aligned} \tag{1.60}$$

Check this with your calculator and you will find that it is very close to the true square of 0.99 which is 0.9801.

The best use of this is not in approximating results which can be found exactly using a calculator but in making approximations when deriving equations. If an equation arises which contains a function $(1 + x)^n$ and we know that x is small then we can make the approximation of equation (1.59). In doing this we have successfully replaced a fairly complicated polynomial with a simple linear function.

Equations such as (1.57) are called *series* and are widely used in mathematics. Some more will be presented in Chapter 2.

The binomial theorem itself also occurs widely in science and mathematics, for example in probability theory and in explaining the distribution of experimental data (Chapter 3). Pascal's triangle is manifested quite clearly in nuclear magnetic resonance spectroscopy. Radiofrequency radiation absorbed by a single proton in a magnetic field gives rise to a single line absorption, Figure 1.13(a). If the proton is adjacent to another, chemically different proton the presence of this other nucleus shows itself by a splitting of the peak into two equal peaks, centred where the original peak would be, Figure 1.13(b). If there are two chemically equivalent protons adjacent to the proton of interest then three peaks are observed. The heights of these are proportional to the binomial coefficients for the expansion of $(1 + x)^2$, Figure 1.13(c). Three adjacent protons give rise to four peaks with peak heights in proportion to the binomial coefficients for $(1 + x)^3$, that is 1: 3: 3: 1.

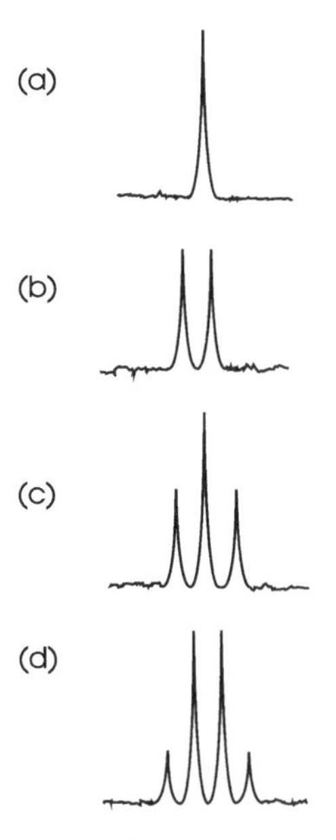

Figure 1.13 The binomial coefficients are important in many areas of chemistry. In NMR spectroscopy the number and relative heights of absorption peaks are found to follow the pattern of the coefficients. See text for explanation

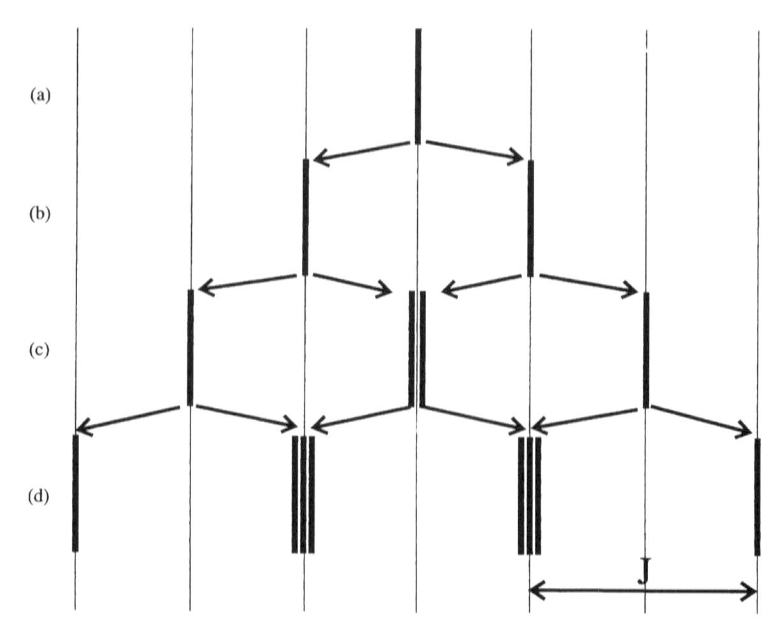

Figure 1.14 The binomial coefficients in NMR spectroscopy. Imagine a single peak which is split into two by a distance J. If these two peaks are again split by J then the two central peaks overlap giving a single peak with twice the peak height. The observed spectrum has three peaks with height ratio 1:2:1. If these are split again the 'overlaps' lead to observed peak heights of 1:3:3:1

How this arises can be demonstrated easily using a piece of graph paper. Start at the top and in the centre draw a vertical line. This represents the absorption line observed for a single proton. If this proton were adjacent to another, chemically different proton then we would observe two lines centred on the original line. We see two lines because the new proton gives the proton under observation two new environments in which it is interacting with the radiation. Draw two lines below the first, say two centimetres apart, Figure 1.14. It is now important to note that the splitting due to the adjacent protons will always be the same. Therefore if we now assume that there are two adjacent protons the first will give the 1:1 splitting just seen and the second will effectively act on these new lines. Below each of the two lines draw two new ones, centred on these and split by the same two centimetres. You observe that two lines now coincide in the middle. We therefore see the middle line with height double the two outer lines. We have produced the third line of binomial coefficients. Continuing further we also produce further lines with heights in the same height ratio as given by Pascal's triangle.

1.8 Manipulating Equations

It is often the case that an equation is not in the form most readily used. For example, in using equation (1.1) we might find that we could measure y quite easily but not x. We could possibly draw a graph of the function and read off values of x for given values of y, but this may be impractical as well as inaccurate. Instead we rearrange the equation. We start with an equation giving y as a function of x and end with an equation giving x as a function of y. Manipulation of equations is one of the most important skills for any mathematician. It is used not only to reverse the roles of dependent and independent variables but routinely in solving equations and in developing mathematical theory. In this section we look at some of the ways that the manipulation of equations is achieved.

1.8.1 The Rule of the Game

In manipulating equations there is only one rule which we have to follow. The remainder is down to experience and using the imagination. This rule is

Whatever we do to one side of the equation we must also do to the other.

As long as we do this then the equation remains true. To demonstrate this rule, and the general approach to rearranging equations, we will look at rearranging equation (1.1).

Example—equation (1.1) as f (y)

Equation (1.1) is

$$y = 3x + 1 \tag{1.1}$$

To put this in the form giving x as a function of y we must isolate x on one side of the equals sign with all the other terms on the other. Initially we want to remove the 'plus one' on the right hand side. We therefore subtract one from this side. To take account of the all-important rule we must also subtract one from the left hand side. This ensures that the equation remains the same and is still true.

$$y - 1 = 3x + 1 - 1 \tag{1.61}$$

The ones on the right cancel to become zero and we have

$$y - 1 = 3x \tag{1.62}$$

x is now isolated but is still in the form of a function, $3x$. To eliminate the three we must divide by three. Again we must do this, not only to the right hand side, but to every term on the left hand side.

$$\frac{y}{3} - \frac{1}{3} = \frac{3x}{3} \tag{1.63}$$

3/3 on the right cancels to leave x, as we desire, all alone.

$$\frac{y}{3} - \frac{1}{3} = x \tag{1.64}$$

This is still in the form of a linear equation as defined by equation (1.2) but we may choose at this point to combine the two fractions on the left into a single fraction. We could also write the equation the other way around so that the newly dependent variable is on the left.

$$x = \frac{y - 1}{3} \tag{1.65}$$

This example is relatively trivial and it is no doubt the more complicated functions which cause problems. The strategy is, however, the same for all rearrangements. We seek to isolate the variable, constant or number of interest and banish all the others to the far side of the equals sign. We must take a systematic approach, removing one term at a time. Although experience allows us to take short cuts or do several steps at one go it is better, if you are unsure, to do the rearrangement step by step and make certain that no terms are lost along the way. Having said this there is one short cut which can be used straight away—using inverse functions.

1.8.2 The Short Cut: Inverse Functions

All functions have their *inverse* or opposite functions. Subtraction is the opposite of addition and division is the opposite of multiplication. In this chapter we have also met the square root function which is the inverse of taking the square or raising something to the power of two.

In rearranging equation (1.1) above we see that we have moved the terms we no longer want associated with x to the left hand side of the equals sign. Rather than carefully subtracting bits here and there and dividing through it is as if we have just lifted these numbers off of the page and set them down on the other side. In passing through the equals sign it is as if they pass through a looking glass and become their own opposite. So in moving the 'one' from the right to the left it takes with it the associated function, addition, which then becomes subtraction on the other side. In moving the three its function, multiplication, becomes division.

This inversion of a function applies to all functions in rearrangements where they move across the equals sign. Always remember that a variable, number or constant takes its associated function with it wherever it goes. We illustrate this with some examples.

Example—the number of moles of solute in a solution

The definition of molar concentration is the number of moles of solute per unit volume (litre) of solution. We use the symbols N for the number of moles and V for the volume of solution in moles per litre (symbol, dm^3, cubic decimetre, the SI unit which is the same as the litre). Written as an equation this means that the concentration, c, is given by

$$c = \frac{N}{V} \tag{1.66}$$

In relating a stoichiometric chemical equation to the amounts of substances we might actually measure out in the laboratory it is useful to know the number of moles rather than the concentration. We therefore need to isolate N as the dependent variable. We therefore move V, whose function is division. In crossing the equals sign this becomes multiplication. Note that initially N is *divided by* V and at the end c is *multiplied by* V. This gives

$$cV = N \tag{1.67}$$

which is exactly the same as

$$N = cV \tag{1.68}$$

Example—rearrangement of Kohlrauch's law

Kohlrauch's law is an equation which relates the molar conductivity of a salt, Λ_m, to its concentration, c. The equation is

$$\Lambda_m = \Lambda_m^0 - K\sqrt{c} \tag{1.69}$$

where K is a constant and Λ_m^0 is the molar conductivity at infinite dilution. If we plot Λ_m against the square root of c rather than c itself this in the form of a linear equation where K is the slope and Λ_m^0 is the intercept when concentration is zero (a more down-to-earth definition of infinite dilution).

Suppose, then, we were given values of molar conductivity, molar conductivity at infinite dilution and the constant K, how do we calculate c?

We rearrange equation (1.69) so that c is the dependent variable, all alone on one side of the equals sign. The easiest thing to get rid of is the term which does not involve c at all and so we whisk Λ_m^0 away to the other side of the equals sign. In doing this the function, addition, becomes subtraction. Note that unless there is a minus sign associated with a variable we can always assume that it is positive and therefore the function is addition. The equation becomes

$$\Lambda_m - \Lambda_m^0 = -K\sqrt{c} \tag{1.70}$$

We now have a multiplying factor, $-K$, and a square root on the right hand side. We must remove these one at a time. The first to go is the lower priority multiplication. In taking $-K$ (which is really two multiplications, -1 and K) to the other side it becomes a division by $-K$.

$$-\frac{\Lambda_m}{K} + \frac{\Lambda_m^0}{K} = \sqrt{c} \tag{1.71}$$

Note that we have divided each term by $-K$ and that the $-\Lambda_m^0$ term becomes positive on division by a negative number. We can rewrite the sum on the left noting that a sum such as $-x + y$ is the same as $y - x$.

$$\frac{\Lambda_m^0}{K} - \frac{\Lambda_m}{K} = \sqrt{c} \tag{1.72}$$

Finally we have to remove the square root function. In doing this it not only becomes a square but it is the whole of the left hand side which is raised to the power of two. As implied in Section 1.7 the square of a sum is not the same as the sum of individual squares.

$$\left(\frac{\Lambda_m^0}{K} - \frac{\Lambda_m}{K}\right)^2 = c \tag{1.73}$$

If we were intending to perform a calculation we might also choose to combine the fractions on the left. This would also serve to remind us that in transferring the square root function it acts to give the square of the whole of the opposite side.

$$\left(\frac{\Lambda_m^0 - \Lambda_m}{K}\right)^2 = c \tag{1.74}$$

1.8.3 Dealing with Fractions

Not all rearrangements involve movement of functions through the equals sign. Some simplifications may be made on one side only and this is particularly the case when fractions with combined functions are involved. When dealing with these it is essential to remember elementary rules of addition, subtraction, multiplication and division of fractions. For example

$$z = \frac{y}{1 - x} \tag{1.75}$$

is *not* the same as

$$z = \frac{y}{1} - \frac{y}{x} \tag{1.76}$$

but

$$z = \frac{1 - x}{y} \tag{1.77}$$

is the same as

$$z = \frac{1}{y} - \frac{x}{y} \tag{1.78}$$

There are a number of steps we may take in simplifying or manipulating fractions. Some of these are: turn them upside down, combine them or build them up part by part. What is done will probably be dictated by the result required. It may be that we have to take several steps to get the required result.

In turning something upside down or *inverting* it we must again remember elementary algebra. By inverting a fraction we mean turning

$$y = \frac{1}{x} \tag{1.79}$$

into

$$\frac{1}{y} = x \tag{1.80}$$

This may be thought of as dividing through by y followed by multiplication by x. We cannot invert individual terms but must invert the whole equation. If we wish to invert

$$y = \frac{1}{x} + 2 \tag{1.81}$$

it is *not* given by

$$\frac{1}{y} = \frac{x}{1} + \frac{1}{2} \tag{1.82}$$

but by

$$\frac{1}{y} = \frac{1}{1/x + 2} \tag{1.83}$$

It is possible to avoid such problems by combining the terms on the right to produce a single fraction. We can do this by multiplying through by one, which of course does not change anything. However, if we make the one which multiplies the whole number on the right equal to x/x we can then combine the two terms. Equation (1.81) becomes

$$y = \frac{1}{x} + \frac{2x}{x} \tag{1.84}$$

The right hand side may now be combined as a single fraction

$$\frac{1}{x} + \frac{2x}{x} = \frac{1 + 2x}{x} \tag{1.85}$$

The inverted form of equation (1.81) is therefore

$$\frac{1}{y} = \frac{x}{1 + 2x} \tag{1.86}$$

We can show that this is the same as equation (1.83) by dividing every term on the top and on the bottom of the right hand side by x (which is effectively the same as multiplying by x/x again).

$$\frac{x/x}{1/x + 2x/x} = \frac{1}{1/x + 2} \tag{1.87}$$

Below are some examples of such manipulations.

Example—degree of dissociation and conductivity

When a salt CA dissociates into its constituent ions C^+ and A^-,

$$CA \rightleftharpoons C^+ + A^- \tag{1.88}$$

we can define the equilibrium constant in terms of the degree of dissociation, α. If the salt CA has an initial concentration c then at equilibrium the concentration will have fallen from c by a fraction defined by α, that is a total of αc. The equilibrium concentration is therefore $c - \alpha c$, which factorises to $c(1 - \alpha)$. The concentration of each ion will have increased from zero to αc (since one cation and one anion are produced from one molecule). The equilibrium constant is defined as the ratio of product of concentrations of products to the product of reactants,

$$K = \frac{\text{product (concentrations of products)}}{\text{product (concentrations of reactants)}} \tag{1.89}$$

which from equation (1.88) is

$$K = \frac{[C^+][A^-]}{[CA]} \tag{1.90}$$

We can replace the concentrations with α terms as described above

$$K = \frac{(\alpha c)(\alpha c)}{(1-\alpha)c} = \frac{\alpha^2 c^2}{(1-\alpha)c} = \frac{\alpha^2 c}{1-\alpha} \qquad (1.91)$$

So far so good, but what we wish to do is use this equation in a form which utilises measurable quantities. Because we are observing the dissociation into ions the conductivity is a suitable quantity and we can define α in terms of the molar conductivity Λ_m and the molar conductivity at infinite dilution, Λ_m^0.

$$\alpha = \frac{\Lambda_m}{\Lambda_m^0} \qquad (1.92)$$

Before we substitute conductivities for α it helps to turn equation (1.91) into a more useful form. As we have already seen, a linear equation is preferred because it is the most easily drawn and easily verified. We begin by attempting to isolate α and we obtain

$$\frac{K}{c} = \frac{\alpha^2}{1-\alpha} \qquad (1.93)$$

As we are going to substitute for α a simplification might be to remove the squared term. First of all we turn the fraction on the right upside down to give,

$$\frac{c}{K} = \frac{1-\alpha}{\alpha^2} \qquad (1.94)$$

and write the right hand side as two fractions

$$\frac{c}{K} = \frac{1}{\alpha^2} - \frac{\alpha}{\alpha^2} = \frac{1}{\alpha^2} - \frac{1}{\alpha} \qquad (1.95)$$

If we multiply through by α we get

$$\frac{\alpha c}{K} = \frac{1}{\alpha} - 1 \qquad (1.96)$$

which is the same as

$$\frac{\alpha c}{K} + 1 = \frac{1}{\alpha} \qquad (1.97)$$

This is now linear if we plot $1/\alpha$ against αc. The slope should be $1/K$ and the intercept equal to one.

We can now substitute for α.

$$\frac{\Lambda_m c}{\Lambda_m^0 K} + 1 = \frac{1}{\Lambda_m/\Lambda_m^0} \tag{1.98}$$
$$= \frac{\Lambda_m^0}{\Lambda_m}$$

Dividing through by Λ_m^0 we obtain

$$\frac{\Lambda_m c}{(\Lambda_m^0)^2 K} + \frac{1}{\Lambda_m^0} = \frac{1}{\Lambda_m} \tag{1.99}$$

We may even choose to write this with the dependent (measurable) variable on the left with the set quantity, c, on the right.

$$\frac{1}{\Lambda_m} = \frac{\Lambda_m c}{(\Lambda_m^0)^2 K} + \frac{1}{\Lambda_m^0} \tag{1.100}$$

A plot of $1/\Lambda_m$ against $\Lambda_m c$ should be linear and from the plot we can find two unknown parameters, the dissociation constant of the salt and the molar conductivity at infinite dilution. The latter comes directly from the intercept at zero concentration. The slope is given by

$$\text{slope} = \frac{1}{(\Lambda_m^0)^2 K} \tag{1.101}$$

and we have to calculate K from the value of Λ_m^0 derived from the intercept. Conveniently we use the square of the intercept.

$$(\text{intercept})^2 = \frac{1}{(\Lambda_m^0)^2} \tag{1.102}$$

Therefore

$$\frac{(\text{intercept})^2}{\text{slope}} = \frac{(\Lambda_m^0)^2 K}{(\Lambda_m^0)^2} = \text{K} \tag{1.103}$$

In this example we simplified a function containing a squared term by inverting a fraction which then allowed us to split this into two fractions. This allowed the removal of a squared term on the bottom part of a fraction not by taking a square root (which would have led to a more complicated function) but by multiplying through by the function which was squared. Although the result, equation (1.96), is still complicated to look at it is linear and we may substitute measurable quantities for a theoretical variable, α.

Example—kinetics of consecutive reactions

In Chapter 6 we present the derivation of the equations describing reactions of the kind

$$\text{A} \xrightarrow{k_a} \text{B} \xrightarrow{k_b} \text{C} \tag{1.104}$$

where k_a and k_b are the rate constants for the first and second reactions. The concentrations of A, B and C are A, B and C respectively and the initial concentration of A is A_0. In Chapter 5 we show how to produce expressions for A and B which are of the form

$$A = A_0 a \quad (1.105)$$

$$B = A_0 k_a \left\{ \frac{a - b}{k_b - k_a} \right\} \quad (1.106)$$

where a and b are functions of time. As only A is present initially then the total concentrations of all reactants and products must be equal to A_0 at all times, that is

$$A_0 = A + B + C \quad (1.107)$$

We can show that the concentration of C is given by

$$C = A_0 \left\{ 1 + \left(\frac{k_a b - k_b a}{k_b - k_a} \right) \right\} \quad (1.108)$$

This is often presented directly but there are several steps involved. The first is to rearrange equation (1.107) for C,

$$C = A_0 - A - B \quad (1.109)$$

and then to substitute the expressions for A and B, equations (1.105) and (1.106).

$$C = A_0 - A_0 a - A_0 k_a \left\{ \frac{a - b}{k_b - k_a} \right\} \quad (1.110)$$

A_0 is a common factor and may be removed outside a large bracket.

$$C = A_0 \left\{ 1 - a - k_a \left(\frac{a - b}{k_b - k_a} \right) \right\} \quad (1.111)$$

This is beginning to look a little like equation (1.108) but we need to combine the two right hand terms inside the large curly brackets. To do this we first note that the last term may be written as

$$k_a \left(\frac{a - b}{k_b - k_a} \right) = \frac{k_a(a - b)}{k_b - k_a} \quad (1.112)$$

We transform the a term by multiplying by one, but in a form which resembles the other term, namely

$$1 = \frac{k_b - k_a}{k_b - k_a} \quad (1.113)$$

Focusing on the whole of the bracketed term we observe

$$1 - \frac{a \cdot (k_b - k_a)}{k_b - k_a} - \frac{k_a(a - b)}{k_b - k_a} = 1 - \frac{a \cdot k_b}{k_b - k_a} + \frac{a \cdot k_a}{k_b - k_a} - \frac{a \cdot k_a}{k_b - k_a} + \frac{b \cdot k_a}{k_b - k_a} \quad (1.114)$$

The terms in $a \cdot k_a$ cancel producing

$$1 - \frac{a \cdot k_b}{k_b - k_a} + \frac{b \cdot k_a}{k_b - k_a} \quad (1.115)$$

which is the same as

$$1 + \frac{b \cdot k_a}{k_b - k_a} - \frac{a \cdot k_b}{k_b - k_a} \tag{1.116}$$

Combining fractions we have

$$1 + \frac{b \cdot k_a - a \cdot k_b}{k_b - k_a} \tag{1.117}$$

This can then be substituted back into equation (1.110) and we have our result, equation (1.107).

Example—building up a fraction from parts

We can determine equilibrium constants using spectroscopic techniques to measure the concentrations of species involved. For a simple equilibrium,

$$\mathrm{X} \rightleftharpoons \mathrm{Y} \tag{1.118}$$

the equilibrium constant is defined as the ratio of concentrations of Y and X. As in the previous example we may use the degree of dissociation to link the concentrations. If we started with concentration c of X at equilibrium it will have fallen to $c(1 - \alpha)$ and the concentration of Y rises to $c\alpha$. The equilibrium constant is therefore given by

$$\begin{aligned} K &= \frac{c\alpha}{c(1-\alpha)} \\ &= \frac{\alpha}{1-\alpha} \end{aligned} \tag{1.119}$$

If there are no complicating factors (such as the presence of other absorbing species) the total absorbance A will be the sum of the absorbances of the individual species. The derivation leads to an equation of the form

$$A = A_x(1 - \alpha) + A_y\alpha \tag{1.120}$$

where A_x and A_y are the absorbances of solutions containing only X and only Y. The problem is then to find an expression of K in terms of measurable quantities A, A_x and A_y.

We begin by rearranging equation (1.120) to give α as a function of absorbances. We will then be able to substitute these back into equation (1.119). Initially, therefore, we expand the terms involving α and then regroup by factorising.

$$A = A_x - A_x\alpha + A_y\alpha \tag{1.121}$$

and

$$A = A_x + \alpha(A_y - A_x) \tag{1.122}$$

We take A_x to the left and divide through by $(A_y - A_x)$ to give an equation for α.

$$\alpha = \frac{A - A_x}{A_y - A_x} \tag{1.123}$$

This gives us the top part of the fraction in equation (1.119). To find the bottom we now need an expression for $(1 - \alpha)$. We need to determine

$$1 - \frac{A - A_x}{A_y - A_x} \tag{1.124}$$

To do this we make the substitution

$$1 = \frac{A_y - A_x}{A_y - A_x} \tag{1.125}$$

This produces an equation involving two fractions with the same denominator,

$$\begin{aligned} 1 - \alpha &= \frac{A_y - A_x}{A_y - A_x} - \frac{A - A_x}{A_y - A_x} \\ &= \frac{A_y - A_x - A + A_x}{A_y - A_x} \\ &= \frac{A_y - A}{A_y - A_x} \end{aligned} \tag{1.126}$$

We now divide the result in equation (1.123) by that in equation (1.126) and we have our equation for K.

$$\begin{aligned} K &= \frac{\alpha}{1 - \alpha} \\ &= \frac{A - A_x}{A_y - A_x} \div \frac{A_y - A}{A_y - A_x} \\ &= \frac{A - A_x}{A_y - A_x} \times \frac{A_y - A_x}{A_y - A} \\ &= \frac{A - A_x}{A_y - A} \end{aligned} \tag{1.127}$$

We now have an expression for K exclusively in measurable quantities, A, A_x, A_y.

1.9 Solving Equations

Finding the solution to an equation generally means finding whatever information you need from it. In some cases this may mean finding numerical values for variables, in others another form of the equation may be suitable. These ends may be achieved by rearranging the equation, perhaps using some of the methods described above. There are, however, certain instances when these methods are insufficient. One of these is where there is more than one unknown to be found. Another is when we have a quadratic equation, which may have more than one solution.

1.9.1 Simultaneous Equations

If we have two equations we know to be true, it is possible to find values of two unknown quantities by solving them simultaneously, i.e., together. This amounts to finding values

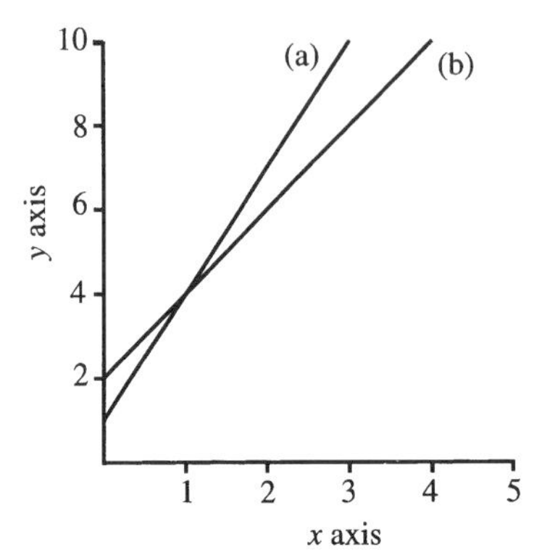

Figure 1.15 Plots of (a) $y = 3x + 1$ and (b) $y = 2x + 2$ intersect at the point (1,4) showing that these two equations have a common solution at this point

of x and y where the two plots representing the equations intersect. This is most easily shown for straight line plots which always intersect if the lines are not parallel. Figure 1.15 shows plots for the equations (1.1) and (1.128)

$$y = 3x + 1 \quad (1.1)$$

$$y = 2x + 2 \quad (1.128)$$

This shows clearly that the lines intersect at the point (1, 4). We can say that the two equations have a common solution when $x = 1$ and $y = 4$. You will see that this is true by putting these values into the equations and noting that they are both correct.

This graphical approach is suitable for simple integer values and simple equations. For equations derived from experiment it might not be accurate enough. Fortunately there are other approaches. These are called *elimination* and *substitution*.

Elimination, as the name implies, involves eliminating one of the variables from the equation. This then allows you to solve the resultant equation for the remaining variable.

Example—solution of equations (1.1) and (1.128)

Examining equations (1.1) and (1.128) we see that they both have the term y on the left hand side. We may eliminate this by subtracting y from the left hand side of, say, equation (1.1). To comply with the rule of equation manipulation we must also subtract y from the right hand side of this equation.

$$y - y = 3x - 1 - y \quad (1.129)$$

We cannot subtract y directly from the right hand side but in equation (1.128) we have an expression in x which is equal to y. We may therefore subtract this function.

$$y - y = 3x + 1 - 2x - 2 \quad (1.130)$$

or

$$0 = x - 1 \quad (1.131)$$

which rearranges to $x = 1$.

Doing this is the same as if we had performed a subtraction involving each side of both equations.

$$\begin{aligned} y &= 3x + 1 \\ -\,(y &= 2x + 2) \\ \hline 0 &= x - 1 \end{aligned} \qquad (1.132)$$

We can also eliminate the terms in x but this is more difficult because they have different coefficients. We can however make them have the same coefficient by multiplying equation (1.1) by two,

$$2y = 6x + 2 \qquad (1.133)$$

and multiplying equation (1.128) by three

$$3y = 6x + 6 \qquad (1.134)$$

and performing the subtraction, equation (1.133) from equation (1.134).

$$\begin{aligned} 3y &= 6x + 6 \\ -\,(2y &= 6x + 2) \\ \hline y &= 0 + 4 \end{aligned} \qquad (1.135)$$

The result, as before, is $y = 4$.

Substitution involves solving one of the equations for one variable and substituting that expression in the other equation. Equation (1.1) already gives y explicitly as a function of x. We can substitute this function into equation (1.128).

$$3x + 1 = 2x + 2 \qquad (1.136)$$

This rearranges to give

$$3x - 2x = 2 - 1 \qquad (1.137)$$

which again gives the result $x = 1$. Now we have a value for x we can use in either of the two equations to calculate y.

$$y = (2 \times 1) + 2 = 4 \qquad (1.138)$$

It is often the case that we will combine these methods to solve the two equations. For the example given it would have been quicker to eliminate y and then substitute the value of x back into one of the equations to calculate y.

Simultaneous equations may come in any number. The general rule is that to find n unknowns we need to solve n equations. Thus to find three unknown quantities we need to solve three equations.

Example—proof of equation (1.9), definition of slope of a graph

For a straight line defined by $y = mx + c$ we can imagine two sets of coordinates, (x_1, y_1) and (x_2, y_2)

$$y_1 = mx_1 + c \tag{1.139}$$

$$y_2 = mx_2 + c \tag{1.140}$$

We can eliminate c by subtracting equation (1.139) from equation (1.140).

$$\begin{array}{r} y_2 = mx_2 + c \\ -\ (y_1 = mx_1 + c) \\ \hline y_2 - y_1 = mx_2 - mx_1 + 0 \end{array} \tag{1.141}$$

This may be written

$$y_2 - y_1 = m(x_2 - x_1) \tag{1.142}$$

which on final rearrangement gives equation (1.9).

$$m = \frac{y_2 - y_1}{x_2 - x_1} \tag{1.9}$$

Example—absorption by a multicomponent solution

If two species, A and B, in a solution both absorb visible light one would expect the absorption of light by each to follow the Beer–Lambert law,

$$A_A = \varepsilon_A c_A l \tag{1.143}$$

$$A_B = \varepsilon_B c_B l \tag{1.144}$$

where the subscripts refer to species A and B. If the two species do not react or interact in any way which might produce additional absorption of light then the total absorbance should be given by

$$A = A_A + A_B \tag{1.145}$$

If we measure the absorbance of the mixture at two wavelengths λ_1 and λ_2 we can write equations for the total absorbance at each wavelength, A_1 and A_2. Substituting equations (1.141) and (1.142) into two equations for A_1 and A_2 we get,

$$A_1 = \varepsilon_{A1} c_A l + \varepsilon_{B1} c_B l \tag{1.146}$$

$$A_2 = \varepsilon_{A2} c_A l + \varepsilon_{B2} c_B l \tag{1.147}$$

where the additional subscripts 1 and 2 label parameters at wavelengths λ_1 and λ_2. The coefficients $\varepsilon_{A1}l$, $\varepsilon_{A2}l$, $\varepsilon_{B1}l$ and $\varepsilon_{B2}l$ are all measured using calibration plots for the pure components and A_1 and A_2 are measured from the mixture. In equations (1.146) and (1.147) we have two unknowns, the concentrations c_A and c_B which may be found by simultaneous solution of these equations.

1.9.2 Quadratic Equations

In Section 1.6 we showed how some quadratic equations may be factorised and the factors used to determine the roots of the equation when y is set equal to zero. The examples given were simple enough that the roots might easily be recognised. In equations such as

$$2x^2 + 3x - 2 = 0 \tag{1.148}$$

the roots are not so obvious. The equation may be factorised to produce

$$(2x - 1)(x + 2) = 0 \tag{1.149}$$

Then the equation is true when $2x$ is equal to one, or x is equal to one half, and when x is equal to -2.

There is an equation which may be applied for the solution of quadratics. For an equation of the form

$$ax^2 + bx + c = 0 \tag{1.150}$$

where a, b and c are constants, solutions for x are given by

$$x = \frac{-b \pm \sqrt{b^2 - 4ac}}{2a} \tag{1.151}$$

The $\pm$ sign reminds us that the square root function can have two results. We therefore take the absolute value of the root* and perform two functions, addition to $(-b)$ and subtraction from $(-b)$.

In equation (1.148) the values of a, b and c are 2, 3 and -2 respectively. If we put these numbers into equation (1.151) we obtain

$$\begin{aligned} x &= \frac{-3 \pm \sqrt{3^2 - (4 \times 2 \times (-2))}}{2 \times 2} \\ &= \frac{-3 \pm \sqrt{9 + 16}}{4} \\ &= \frac{-3 \pm \sqrt{25}}{4} \\ &= \frac{-3 \pm 5}{4} \end{aligned} \tag{1.152}$$

There are two answers, $(-3 - 5)/4$, equal to -2, $(-3 + 5)/4$ and equal to $+\frac{1}{2}$.

The term inside the square root is called the *discriminant*. Its value is important in indicating the nature of the roots. If it is positive, as in the example above, it means we can calculate the roots and they are rational. If it is zero it means that the roots are identical (check this with equation (1.49)). If it is negative, we cannot take the square root of a negative number and there is no root in the conventional sense. 'Unconventional' roots are discussed in Chapter 8. Figure 1.16 shows a plot of equation (1.153).

$$y = x^2 + x + 1 \tag{1.153}$$

The discriminant is equal to -3. It can be seen from the plot that the curve reaches a minimum value of y at x equal to -0.5 and that it does not cross the x axis at all. That is, there is no solution to this equation where y is equal to zero.

*That is, just the number.

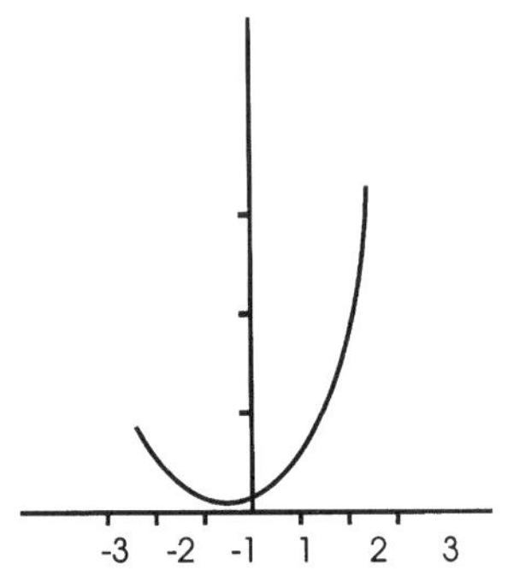

Figure 1.16 Plot of $y = x^2 + x + 1$. The curve does not intersect the x axis (the line $y = 0$) and therefore there are no roots for the equation when $y = 0$

Example—degree of dissociation and equilibrium constant

In equation (1.93) we related the equilibrium constant for the dissociation of a salt in terms of the degree of dissociation and the concentration.

$$\frac{K}{c} = \frac{\alpha^2}{1-\alpha} \tag{1.93}$$

This rearranges to give a quadratic in α,

$$\frac{K}{c}\cdot(1-\alpha) = \alpha^2 \tag{1.154a}$$

$$\frac{K}{c} - \frac{K}{c}\alpha = \alpha^2 \tag{1.154b}$$

$$\alpha^2 + \frac{K}{c}\alpha - \frac{K}{c} = 0 \tag{1.154c}$$

Comparing with equation (1.150) we see that

$$a = 1 \tag{1.155}$$

$$b = \frac{K}{c} \tag{1.156}$$

$$c = \frac{-K}{c} \tag{1.157}$$

Substituting these values in equation (1.151), noting the minus sign in c, we obtain an equation for α.

$$\alpha = \frac{\frac{K}{c} \pm \sqrt{\frac{K^2}{c^2} + 4\frac{K}{c}}}{2} \tag{1.158}$$

We showed on pp. 28–30 that it is possible to estimate, from measurements of conductivity, the equilibrium constant K. By knowing this and the concentration it is also possible to calculate α, the degree of dissociation.

1.10 Manipulating Units

1.10.1 Units? What Units?

In using mathematics to describe the real world we do so using the concept of a *quantity*. For example, if we travel we move a *distance* which is the name for the amount we move. A quantity always has both a number, representing its magnitude, and a unit. Suppose we drive three kilometres. The quantity distance is the product of the number 'three' and the unit, 'kilometre'. To put this in mathematical terms we give the quantity a symbol, e.g., s, and the unit a symbol, e.g., km for the kilometre. We can then write an equation such as

$$s = 3\,\mathrm{km} \tag{1.159}$$

When we do calculations or plot graphs we do so with pure numbers. The only number we have in equation (1.159) is 3 and to represent this in a calculation we must rearrange the equation,

$$3 = \frac{s}{\mathrm{km}} \tag{1.160}$$

This need to keep numbers separate from quantities becomes more apparent if we consider quantities which are derived from other quantities. For example, the velocity of the car being used to transport us is defined as the ratio of the distance travelled, s, to the time, t, taken to travel that distance.

$$v = \frac{s}{t} \tag{1.161}$$

This is, strictly speaking, incorrect as we can only do such mathematics on numbers and so we must divide each of these quantities by the unit to produce a pure number.

$$v/\mathrm{units} = \frac{s/\mathrm{km}}{t/\mathrm{h}} \tag{1.162}$$

To ensure that the equation is balanced we check that the units on each side cancel each other out. Indeed, if we did not know the units of velocity we could rearrange this equation to find them. Rearranging equation (1.162)

$$\begin{aligned} \frac{v}{\mathrm{units}} &= \frac{s}{\mathrm{km}} \div \frac{t}{\mathrm{h}} \\ &= \frac{s}{t} \times \frac{\mathrm{h}}{\mathrm{km}} \end{aligned} \tag{1.163}$$

From our original equation we know that $v = s/t$ and so we can remove these.

$$\frac{1}{\mathrm{units}} = \frac{\mathrm{h}}{\mathrm{km}} \tag{1.164}$$

and

$$\text{unit} = \frac{\text{km}}{\text{h}} = \text{km}\,\text{h}^{-1} \tag{1.165}$$

which reads as 'kilometres *per* hour'. Note the common use of the negative index instead of the fraction or '/' (slash).

Given some figures we can now calculate the velocity. If we travel three kilometres (3 km) in one tenth of an hour (0.1 h), then

$$v/(\text{km}\,\text{h}^{-1}) = \frac{s/\text{km}}{t/\text{h}} = \frac{3}{0.1} = 30 \tag{1.166}$$

Rearrangement of the final result gives

$$\text{velocity, } v = 30\,\text{km}\,\text{h}^{-1} \tag{1.167}$$

Almost everything we do in chemistry considers a quantity. From the relatively simple mass of substance to quantities such as rate constants we must always take account of units. Whenever we present an equation using symbols or numbers we should divide the quantity by the unit otherwise they are, strictly speaking, meaningless.

We now consider two important topics relevant to the use of units and quantities. These are *dimensional analysis* and a method of accounting for the use of different units, *quantity calculus.*

1.10.2 Dimensional Analysis

It is possible to break down all quantities into a relatively small number of fundamental quantities. These are length, symbol L, mass, symbol M, and time, symbol T. In chemistry we also have to consider the amount of substance defined by the *mole* usually abbreviated as *mol*, and the temperature, defined by the *degree* (commonly referred to by its most common unit, the kelvin, K).

These are fundamental because they are the most basic units available and cannot be broken down further. In considering an equation of quantities we can define each part in terms of these fundamental quantities.

Each fundamental quantity has its corresponding SI unit of measurement: length has as its unit the metre, m; mass has the kilogram, kg; and time the second, s. We use these units to define derived units. For example, volume V of a rectangular box is defined by the equation

$$V = \text{length} \times \text{height} \times \text{depth} \tag{1.168}$$

each of which is a measurement of distance or length, L. We therefore define volume by

$$V = L \times L \times L = L^3 \tag{1.169}$$

We can now use this to derive the unit for volume because each distance is measured in metres, m. That is, the unit of volume is m × m × m or m^3.

The density of a substance is the ratio of the mass (the quantity of matter it contains) to the volume it occupies.

$$\text{density}, \rho = \frac{\text{mass}}{\text{volume}} \tag{1.170}$$

To find the units we break down each part into its fundamental quantities.

$$\text{dimensions of density}, \rho = \frac{M}{L^3} = ML^{-3} \tag{1.171}$$

We can now substitute the SI units for the dimensions:

$$\text{unit of density}, \rho = \frac{M}{L^3} \equiv \frac{\text{kg}}{\text{m}^3} = \text{kg m}^{-3} \tag{1.172}$$

As above, we may use the division by a quantity raised to some power or the negative index to represent the relationship between the parts of the derived unit. Another example is the unit of concentration.

$$\text{concentration} = \frac{\text{number of moles}}{\text{volume}} = \frac{\text{mole}}{L^3} \tag{1.173}$$

The common unit of volume for chemists is the cubic decimetre (the SI translation of the litre), written as dm^3. The unit of concentration is therefore

$$\text{unit of concentration}, c = \frac{mol}{dm^3} = mol\, dm^{-3} \tag{1.174}$$

Table 1.5 lists some common quantities and their fundamental dimensions. These may be familiar from school physics but are sometimes useful to chemists.

Table 1.5 Some fundamental quantities and derived units

Measurement	Definition	Dimensions	Units	SI unit
Length, l		L	m	metre
Mass, m		M	kg	kilogram
Time, t		T	s	second
Area, A	$A = l \times l$	L^2	m^2	m^2
Volume, V	$V = l \times l \times l$	L^3	m^3	m^3
Velocity, v	$v = l/t$	LT^{-1}	m s^{-1}	m s^{-1}
Acceleration, a	$a = v/t$	LT^{-2}	m s^{-2}	m s^{-2}
Force, F	$F = m \times a$	MLT^{-2}	kg m s^{-2}	newton, N
Pressure, P	$P = F/A$	$ML^{-1}T^{-2}$	$\text{kg m}^{-1}\,\text{s}^{-2}$	pascal, Pa
Energy, E	$E = F \times l$	ML^2T^{-2}	$\text{kg m}^2\,\text{s}^{-2}$	joule, J

Example—units of R, the gas constant

The equation of state for an ideal gas can be presented in the form

$$PV = nRT \tag{1.175}$$

which can be rearranged to define R.

$$R = \frac{PV}{nT} \tag{1.176}$$

Textbooks commonly give R as $8.314\,\mathrm{J\,K^{-1}\,mol^{-1}}$. Can we prove that these units are correct by dimensional analysis of equation (1.176)?

The method is to reduce each of the variables to its fundamental quantities. n, the number of moles, and temperature T are already in this form, leaving P and V. The SI unit of pressure is the pascal, symbol Pa, and volume has units $\mathrm{m^3}$. Substitution of these leads to

$$\text{unit of } R = \frac{\mathrm{Pa \cdot m^3}}{\mathrm{mol \cdot K}} \tag{1.177}$$

This is quite a curious set of units but as pressure is itself a derived quantity then the analysis must be continued. From Table 1.5 we see that the definition of pressure is that it is the ratio of a force to the area on which it acts,

$$P = \frac{\text{force}}{\text{area}} \tag{1.178}$$

Area is defined as L^2 with units of $\mathrm{m^2}$. Force is still a derived quantity—the product of mass and acceleration. Acceleration is yet another derived quantity, the rate of change of velocity with time,

$$\text{acceleration, } a = \frac{v}{t} \tag{1.179}$$

Velocity is itself the ratio of distance to time.

$$\text{dimensions of velocity } v = \frac{L}{T} = LT^{-1} \tag{1.180}$$

Working back through this breakdown we have

$$\text{dimensions of acceleration } a = \frac{L}{T^2} = LT^{-2} \tag{1.181}$$

$$\text{dimensions of force } F = MLT^{-2} \tag{1.182}$$

$$\text{dimensions of pressure } P = \frac{MLT^{-2}}{L^2} = ML^{-1}T^{-2} \tag{1.183}$$

We can now replace these dimensions with the standard SI units.

$$\text{unit of } R = \frac{ML^{-1}T^{-2}L^3}{mol \cdot K} = \frac{ML^2T^{-2}}{mol \cdot K} \tag{1.184}$$

$$\text{unit of } R = \frac{kg \cdot m^2 \cdot s^{-2}}{mol \cdot K} \tag{1.185}$$

The textbook definition of the gas constant, R, has the unit of energy, the joule involved. We therefore need to look at the dimensions of energy and see if they can be reconciled with this result. Energy is the product of a force and the distance through which it acts. Using the dimensions of force given above we have

$$\text{dimensions of energy} = MLT^{-2}L = ML^2T^{-2} \tag{1.186}$$

$$\therefore \text{ units of energy} = \text{kg m}^2\,\text{s}^{-2} \tag{1.187}$$

This is the unit of the joule. Comparison of equations (1.185) and (1.187) shows that the numerator is indeed the unit of energy and that the unit of R is correct.

1.10.3 Quantity Calculus

Quantity calculus is a technique which allows us to change equations to accommodate the use of different units. Although the SI system of units gives us a framework for uniformity many scientists still use those units which are most convenient to their purpose. It just so happens that volumes in the region of the cubic decimetre (dm^3) are more convenient to use than cubic metres and so it is more convenient to think in these terms. The SI system is broad enough such that it allows certain prefixes to define fractions or multiples of the basic unit. Thus we have n for nano (10^{-9}), m for milli (10^{-3}), c for centi (10^{-2}), d for deci (10^{-1}) and k for kilo (10^3). In measuring length we therefore have the nanometre, millimetre, decimetre, metre and kilometre. Visible-light spectroscopists will use the nanometre because wavelengths are conveniently measured in hundreds of nanometres, most chemists make solutions with concentrations of moles per litre (mol dm^{-3}, moles per cubic decimetre). On the other hand electrochemists often use small volumes and therefore work in moles per cubic centimetre and so on. Many even use non-SI units either because of historical precedence or, once again, convenience. Infra-red spectroscopists use the reciprocal centimetre as a unit of frequency and not the hertz (symbol Hz) or s^{-1}. A whole range of units are used in the measurement of pressure, especially if one refers to old textbooks or research papers. One might find, instead of the pascal (Pa) the *bar*, *torr*, *millimetre of mercury* or the *atmosphere*. It is therefore quite common to have to convert somebody else's most convenient units to those which you, your tutors or at least the International Union of Pure and Applied Chemistry (IUPAC) approve of. One must also keep abreast of the latest versions of these. Periodically the accepted system is modified and what you once thought of as 'correct' may well be 'wrong'.

The method works by deriving an expression, taking into account all of the units involved and rearranging it to produce a conversion factor which can be used for the conversion of particular examples. For example, in the late twentieth century practically the whole world with the exception of the United Kingdom measures distances in kilometres. The British typically use the statute mile. It is relatively simple to find (from an atlas, book of scientific data etc.) that one mile is the same distance as 1.609 kilometres. This can be written as an equation,

$$1 \text{ mile} = 1.609\,\text{km} \tag{1.188}$$

We now rearrange this to find a value of unity.

$$\frac{1 \text{ mile}}{1.609\,\text{km}} = 1 \tag{1.189}$$

This can be used to change another expression—because anything may be multiplied by one without changing its magnitude.

Example—miles vs. kilometres

A British tourist travelling to Paris passes a sign indicating that he has 30 km to travel to his destination. He is naturally curious as to how far he has to go in miles. What is the answer?

We have our conversion factor in equation (1.189). We therefore multiply the quantity given by this factor.

$$30\,\text{km} \times 1 = 30\,\text{km} \times \frac{1\ \text{mile}}{1.609\,\text{km}} \tag{1.190}$$

Note that the units 'km' on the right cancel and we have

$$\begin{aligned} 30\,\text{km} &= \frac{30}{1.609}\ \text{miles} \\ &= 18.65\ \text{miles} \end{aligned} \tag{1.191}$$

It is also possible to use a number of these conversion factors to change whole sets of derived units. In each case we define the conversion factor by an equation and rearrange this to obtain a factor which is equal to unity. In practising this technique it is better to write down the whole equation to make sure that the conversion factor is being used in the correct way.

Example—converting several units

The same tourist breaks down and is able to hire an old car from the local garage, whilst his is being repaired. The mechanic tells him that the old car will do about seven kilometres to the litre. The tourist wonders how this compares to the forty miles per gallon that his new (but not so reliable) fuel-injected motor can achieve. There are 4.55 litres (dm^3) per UK gallon.

We can help define the equation by looking at what we want to achieve. We will call the unknown value x and note that it needs to have units of miles per gallon or mile/gallon. We can write an equation,

$$7\frac{\text{km}}{\text{dm}^3} = x\frac{\text{mile}}{\text{gallon}} \tag{1.192}$$

To change the unit 'km' we must have this on the bottom half of the fraction in the conversion factor. To change the 'dm^3' unit we must have this on the top half of the conversion factor for volume. Taking equation (1.189) we can write the conversion factor for distance,

$$1 = \frac{1\ \text{mile}}{1.609\,\text{km}} \tag{1.189}$$

Note that we could invert this but we need km on the bottom. For volume we note that

$$1\ \text{gallon} = 4.55\,\text{dm}^3 \tag{1.193}$$

Therefore

$$1 = \frac{4.55\,\text{dm}^3}{1\ \text{gallon}} \tag{1.194}$$

We can now apply both of our conversion factors.

$$\begin{aligned} 7\frac{\text{km}}{\text{dm}^3} \times 1 \times 1 &= 7\frac{\text{km}}{\text{dm}^3} \times \frac{1\ \text{mile}}{1.609\,\text{km}} \times \frac{4.55\,\text{dm}^3}{1\ \text{gallon}} \\ &= \frac{7 \times 1 \times 4.55}{1.609 \times 1} \times \frac{\text{km}}{\text{dm}^3} \times \frac{\text{mile}}{\text{km}} \times \frac{\text{dm}^3}{\text{gallon}} \\ &= 19.79\frac{\text{mile}}{\text{gallon}} \end{aligned} \tag{1.195}$$

or 19.79 miles per gallon. In this we also see why we need to decide in advance where the units we are converting are to go—so that they will cancel in the final equation.

Example—the gas constant

The gas constant has been defined, above, as $8.314\,\text{J K}^{-1}\,\text{mol}^{-1}$. It is sometimes given as $0.082\,05\,\text{atm dm}^3\,\text{K}^{-1}\,\text{mol}^{-1}$. Prove that this is also a correct representation.

Initially we need to define a number of relationships:

One atmosphere (atm) pressure is equal to 101 325 Pa, so

$$1 = \frac{1\,\text{atm}}{101\,325\,\text{Pa}} \tag{1.196}$$

One joule is the same as the newton metre (N m, inferred from Table 1.5).

$$1 = \frac{1\,\text{N m}}{1\,\text{J}} \tag{1.197}$$

One pascal is the same as one newton force per square metre.

$$1 = \frac{1\,\text{Pa}}{1\,\text{N m}^{-2}} = \frac{1\,\text{Pa m}^2}{1\,\text{N}} \tag{1.198}$$

Finally there are 1000 litres (dm^3) in one cubic metre.

$$1 = \frac{1\,\text{m}^3}{1000\,\text{dm}^3} \tag{1.199}$$

Therefore we may calculate R from

$$\begin{aligned} R &= 8.314\frac{\text{J}}{\text{K mol}} \times \frac{1\,\text{Nm}}{1\,\text{J}} \times \frac{1\,\text{Pa m}^2}{\text{N}} \times \frac{1\,\text{atm}}{101\,325\,\text{Pa}} \times \frac{1000\,\text{dm}^3}{1\,\text{m}^3} \\ &= \frac{8.314 \times 1000}{101\,325} \times \frac{\text{J}}{\text{J}} \times \frac{\text{N}}{\text{N}} \times \frac{\text{Pa}}{\text{Pa}} \times \frac{\text{m}^3}{\text{m}^3} \times \frac{\text{atm dm}^3}{\text{K mol}} \\ &= 0.082\,05\,\text{atm dm}^3\,\text{K}^{-1}\,\text{mol}^{-1} \end{aligned} \tag{1.200}$$

Example—wavelength and wavenumber

The wavenumber, symbol $\tilde{\nu}$, is a unit equivalent to frequency which is used to classify electromagnetic radiation. It is defined as the reciprocal of the wavelength of radiation and is usually measured in centimetres.

$$\tilde{\nu}/\text{cm}^{-1} = \frac{1}{\lambda/\text{cm}} \tag{1.201}$$

The right hand side of this may be written as

$$\tilde{\nu}/\text{cm} = \frac{\text{cm}}{\lambda} \tag{1.202}$$

The wavelength of visible light is usually measured in hundreds of nanometres, symbol nm, where one nanometre is 10^{-9} of one metre. The nanometre and centimetre are therefore related by

$$1\,\text{cm} = 10^7\,\text{nm} \tag{1.203}$$

Consequently

$$1 = \frac{10^7\,\text{nm}}{1\,\text{cm}} \tag{1.204}$$

Wavelength may therefore be converted to wavenumbers by the equation

$$\begin{aligned} \nu/\text{cm}^{-1} &= \frac{\text{cm}}{\lambda} \times \frac{10^7\,\text{nm}}{\text{cm}} \\ &= \frac{10^7\,\text{nm}}{\lambda} \end{aligned} \tag{1.205}$$

which is the same as

$$\tilde{\nu}/\text{cm}^{-1} = \frac{10^7}{\lambda/\text{nm}} \tag{1.206}$$

That is, we obtain the frequency in wavenumbers by dividing 10^7 by the wavelength in nanometres. For example, 500 nm is converted by

$$\tilde{\nu}/\text{cm}^{-1} = \frac{10^7}{500} = 20\,000 \tag{1.207}$$

or 500 nm is equivalent to 20 000 cm^{-1}.

1.11 The Slope of a Curve

Early in this chapter we stated that the slope of a straight line is one of its most important properties. In chemistry this may be a very important parameter, e.g., a rate of reaction or thermodynamic quantity. This is also true for functions which give rise to curved plots. The most important difference is that the slope changes as the curve progresses. We may be able to estimate a line of best fit for linear data, guessing that the line fits neatly

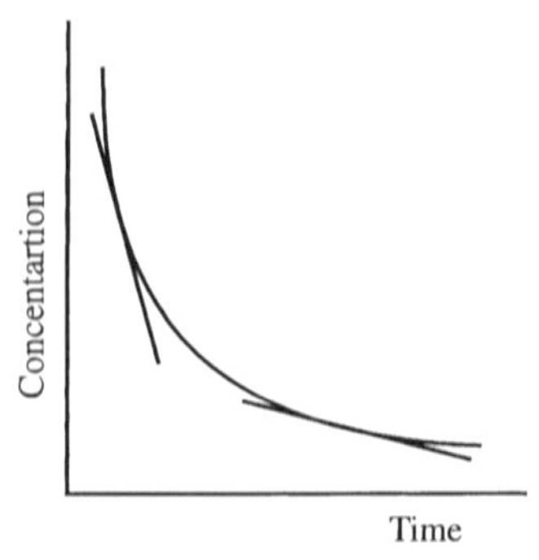

Figure 1.17 The slope of a curve may be estimated by drawing a tangent to the curve at the point of interest. In this kinetic plot the slope, equal to the rate of reaction, decreases in magnitude as the reaction continues. The rate decreases as the concentration decreases and we see directly that there is a link between concentration and rate

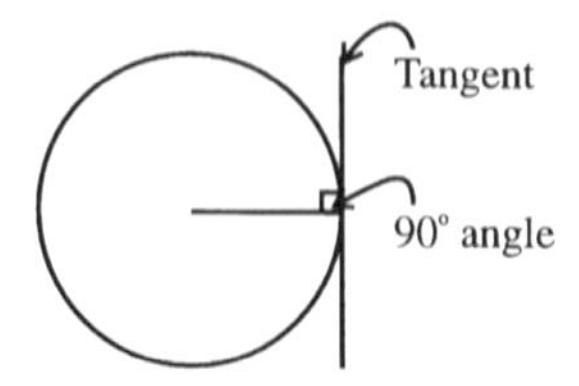

Figure 1.18 The definition of a tangent as a line perpendicular (at right angles) to the radius of a circle

in-between, to the best of our judgement, the scatter of data points. How can we assimilate this for a curve?

This problem will be addressed in some detail in Chapter 4, Differential Calculus. It is possible, however, to estimate a slope directly from a graph. Figure 1.17 shows part of a plot of concentration versus time for a simple chemical reaction. Two straight lines have been drawn at two different points along the curve. These are called *tangent* lines.

The tangent to a circle, Figure 1.18, is the line which is perpendicular, at an angle of 90°, to a line drawn from the centre of the circle to its circumference. The slope of a tangent line is the same as the slope of the curve at the point at which they intersect. By drawing a tangent (which is a straight line) to a slope (which is curved) we may therefore estimate the gradient of the slope at that point.

Estimating the correct position and slope of a tangent line takes some judgement, especially if the data is experimentally produced and therefore subject to error and scatter of the data points. The method is, nevertheless, sometimes used for estimating the rates of reactions. Figure 1.17 shows the change in concentration of a reactant with time—which is indeed the definition of a rate of reaction. The slope (concentration/time) is the very rate of reaction in which we are interested.

It is possible to infer several things from a plot like this. Not least is the fact that as time progresses, so the concentration falls and so does the slope of the curve. We can say that the rate of reaction falls as the concentration of the reactant falls. Turning this on its head we can say that the rate rises as the concentration rises or that rate is proportional to concentration. This is useful in itself but if we could derive some figures, so as to suggest a neat mathematical relationship between the rate and the concentration we are well on the way to finding a complete mathematical model for the reaction. To estimate the rate all

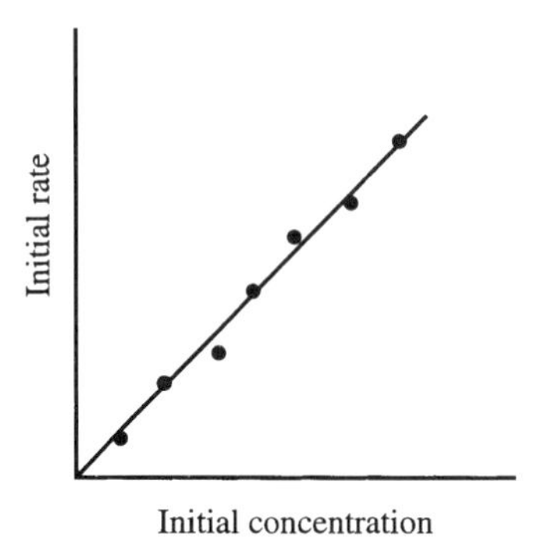

Figure 1.19 Plot of initial rate versus initial concentration. This shows more explicitly the relationship between rate and concentration than the experimental concentration time plot

we need to do is to draw a tangent to the curve at given points corresponding to given concentrations and estimate the slope of the tangent.

A related experiment involves monitoring a reaction several times, each time with a different concentration of a particular reactant. From a plot of concentration versus time we estimate the rate at the very beginning by drawing a tangent at time equals zero. We then plot this initial rate against the initial concentration of the reactant (Figure 1.19). It is very often found that this rate is proportional to the concentration raised to a simple power (0, 1, 2 etc.) and that this can be related directly to the mechanism of the reaction.

These ideas will be discussed in much greater detail in Chapters 4, 5 and 6 showing how they might be investigated in a more rational manner. The tangent method is one which may be used as part of preliminary investigations in the laboratory. Even if a computer is available to plot the data, unless it has *curve fitting* ability, and you understand how to use it, drawing a plot by hand and estimating the tangent might be just as quick and easier to use.

1.12 Concluding Remarks

In this chapter we have presented some basic ideas which underline the use of mathematics by scientists, and particularly chemists. We must be able to use, rearrange and manipulate equations, turning them around to get the information we require. We must be able to plot data in the manner an equation indicates is correct or if necessary change the equation so that it is in an appropriate form. As experimentalists we must be able to use data, plot and correctly label graphs, derive parameters from the slope or the intercept and present the result, with the correct units. We must always check that not only is an equation balanced properly but that the units also balance. Finally we must be able to convert units to those which are more useful or required by a tutor, examiner or scientific institute.

Problems

1.1 Prove that

$$\frac{1}{1/x+2}=\frac{x}{1+2x} \tag{P1.1}$$

1.2 If

$$\frac{x-y}{y}=\frac{a}{b} \tag{P1.2}$$

write this equation explicitly as a function of y, i.e., in the form $y = \ldots$ with y only on the left hand side.

1.3 In studying a reversible reaction in which reactant A (with concentration A) forms product B (with concentration B) the following equation was derived.

$$\text{rate} = -k_1A + k_2B \tag{P1.3}$$

where k_1 and k_2 are constants.

(a) If the rate is zero, write an equation for the ratio B/A.
(b) If the initial concentration of A is A_0 and no B is present then we can say that the concentration of B at any time is the difference between the initial and current concentration of A,

$$B = A_0 - A \tag{P1.4}$$

Substitute this in your answer to (a) and then rearrange this to find an equation for A in terms of A_0 and the constant terms.

1.4 A reactant A can form two products B (concentration B) and C (concentration C) in separate equilibria. The equilibrium constants K_1 and K_2 are written in the usual way,

$$K_1=\frac{B}{A},\ K_2=\frac{C}{A} \tag{P1.5}$$

As in Question 3, $A = A_0$ at the start of the reaction ($t = 0$).

(a) Write B as a function of A and K_1 and C as a function of A and K_2.
(b) The sum of all concentrations is equal to A_0. Write this statement as an equation.
(c) Substitute values for B and C from your answer above in section (a).
(d) Rearrange this to give A as a function of A_0 and other constant terms.
(e) Use your answer to (d) to write equations for B and C.

1.5 $C_1 = N_1/V_1$ and $C_2 = N_2/V_2$. If $N_1 = N_2$ write an equation for C_1 in terms of the other variables.

1.6 In surface chemistry the following equation arises,

$$\frac{\theta}{1-\theta}=\frac{k_2P}{k_1} \tag{P1.6}$$

Rearrange this equation to give θ explicitly as a function of P.

1.7 Factorise the following functions.

(a) $3x + 3y$ (b) $x^2 + x$ (c) $x^2 + 2x + 1$
(d) $x + 1$ (e) $x^2 - y^2$ (f) $x^2 - 4$

1.8 Simplify the following equations

(a) $y = x^6/x^5$
(b) $y = x^{1/2} \times x^2$
(c) $y = x^{1/3}/x^5$
(d) $y = 2x^{-1} \div x^{0.5}$
(e) $y = x^2 + z^2$
(f) $y = \dfrac{x^6}{x^2 \times x^{3/2}}$

1.9 Which of the following have real roots and what are they?

(a) $x^2 + 2x + 1 = 0$
(b) $2x - x^2 - 2 = 0$
(c) $x^2 + 2x + 2 = 0$
(d) $5x^2 + 6x + 1 = 0$
(e) $6x - 3x^2 - 3 = 0$
(f) $3x^2 + x - 4 = 0$

1.10 Find the intersection of the following pairs of straight lines.

(a) $2y = 2x + 4$ and $y = 2x + 1$
(b) $2y = 3x + 1$ and $y = 1 - 2x$
(c) $y = x + 3$ and $y = 2x - 1$

1.11 Plot the following data to determine the equation which relates x and y values.

(a) (1, 1), (3, 2), (5, 3), (7, 4), (9, 5), (10, 5.5), (12, 6.5).
(b) (−3, −5), (−2, −3), (−1, −1), (−0.5, 0), (0, 1), (0.5, 2), (1, 3), (3.5, 8), (4.5, 10), (5, 11).
(c) (−10, −20), (−8, −14), (−6, −8), (−4, −2), (−2, 4), (4, 22), (6, 28), (8, 34).

1.12 Estimate the equations which describe the following data by plotting it and drawing the line of best fit.

(a) (2.0, 5.5), (4.0, 9.3), (6, 12.6), (8.0, 17.1), (10.0, 20.0) (12.0, 23.0), (14, 30.5), (16.0, 32.4), (18.0, 37.0)
(b) (100, 350), (500, 1540), (800, 2800), (1700, 5200), (2500, 7650), (2900, 8800), (3300, 10 000), (3700, 11 100), (4500, 13 600), (5300, 15 900)
(c) (0.2, 0.5), (0.4, 1.5), (0.8, 1.8), (0.9, 2.6)

1.13 (a) ΔG and ΔH both have units of kJ mol^{-1}. Temperature has the unit kelvin (K). Given the equation

$$\Delta G = \Delta H - T\Delta S \tag{P1.7}$$

what is the unit of change in entropy, ΔS?

(b) The rate of a reaction is the change in concentration with time or

$$\text{rate} = \frac{c_2 - c_1}{t} \tag{P1.8}$$

If concentration has units mol dm^{-3} and time has units seconds (s) then what is the unit of rate of reaction?

(c) A first order rate equation may be written as follows,

$$\text{rate} = kc \tag{P1.9}$$

where c is concentration and has units mol dm^{-3}; what is the unit of the first order rate constant k?

(d) A second order rate equation may be written as

$$\text{rate} = kc^2 \quad \text{(P1.10)}$$

What are the units of the second order rate constant k?

(e) What are the units of a concentration gradient ?

$$\text{gradient} = \frac{\text{change in concentration}}{\text{distance}} = \frac{c}{x} \quad \text{(P1.11)}$$

(f) Flux is a term used in transport in solution and is the number of moles of a solute passing through an area (unit m^2) per second, i.e., it has units of $\text{mol m}^{-2}\,\text{s}^{-1}$. Fick's first law relates the flux to the concentration gradient through the diffusion coefficient, D.

$$\text{flux} = -D\frac{\Delta C}{\Delta x} \quad \text{(P1.12)}$$

What are the units of D?

(g) In an electrochemical cell the current at an electrode, $i/\text{C s}^{-1}$, is given by

$$i = nFAj \quad \text{(P1.13)}$$

where j is the flux, n is the number of electrons (and is therefore unitless), F is the Faraday constant ($96\,500\,\text{C mol}^{-1}$) and A is the area of the electrode surface. Show that this equation is dimensionally correct.

1.14 (a) A chemist makes up a solution of concentration $5\,\text{mmol cm}^{-3}$. What is the concentration of the solution in mol dm^{-3}.

(b) Analytical chemists often work in units of parts per million (ppm) which is approximately $1\,\mu\text{g/cm}^3$. What is the molar concentration of a 5 ppm solution of sodium ions?

(c) A chemical reaction has an enthalpy of reaction of 52 kilocalories per mole (kcal mol^{-1}). If one calorie is equivalent to 4.19 joules what is the enthalpy of reaction in kJ mol^{-1}?

(d) An organic compound absorbs radiation of $1715\,\text{cm}^{-1}$. What is the wavelength of the radiation in nm and m?

(e) A reaction of a compound of molecular mass 72 is measured as 5 g per hour. What is the first order rate constant ($\text{mol dm}^{-3}\,\text{s}^{-1}$) if the reaction takes place in a fixed volume of $50\,\text{cm}^3$?

2 Special Functions

2.1 Introduction

In this chapter we introduce some special functions. These are logarithms, exponential functions and the trigonometric functions, sine, cosine and tangent. Related to the sine we also introduce the concept of the wave and some of its important properties.

2.2 Logarithms

2.2.1 Definition

If we were to write

$$y = 10^x \tag{2.1}$$

then we would say that x is the exponent of 10 and that

$$x = \log_{10} y \tag{2.2}$$

which reads 'x is the log to base 10 of y', log being the common abbreviation of *logarithm*. Therefore, a logarithm of a number is the power to which a base is raised to produce that number. The base is written as a subscript to the word 'log' and it is possible to use logarithms to any base. For example, $\log_2 8$ is 3 because 2 raised to the power 3 is equal to 8. A more common base used is 10—the normal base in which we count: $\log_{10} 100$ is 2 because 10^2 is equal to 100. Equation (2.1) is an exponential function, x is related to y as the exponent of 10. Logs and exponential functions are inverse functions. When using logarithms, raising the base to some power may be referred to as taking the *antilogarithm*. Raising 10 to the power of three is called taking the antilogarithm to base 10 of the number three.

The inverse relationship between exponents and logarithms can be demonstrated by plotting graphs of $y = \log_{10} x$ and $y = 10^x$ as shown in Figure 2.1. The two curves are identical in shape but it is as if the axes have been exchanged.

Logarithms have the same properties as other forms of exponents described in Chapter 1. This includes being a higher priority function than multiplication or division. Table 2.1 lists some of these properties, written in the notation of the logarithm. Note that the base

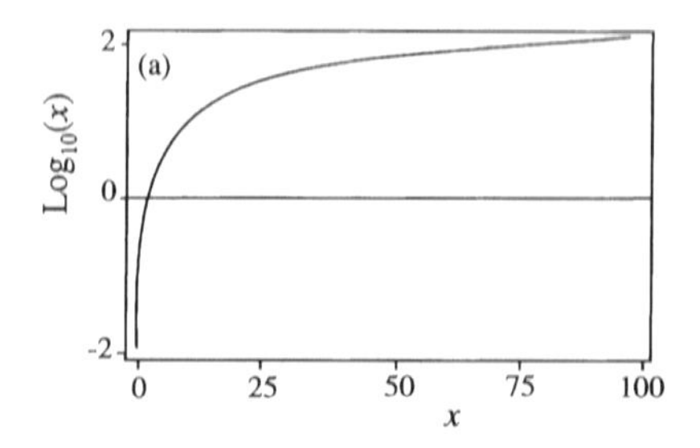

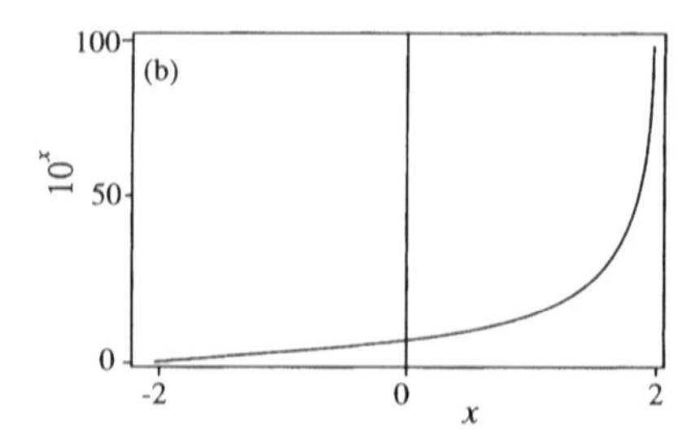

Figure 2.1 (a) $y = \log(x)$; (b) $y = 10^x$

Table 2.1 Properties of logarithms

$\log(xy) = \log(x) + \log(y)$
$\log(x/y) = \log(x) - \log(y)$
$\log(x^y) = y \cdot \log(x)$
$\log(1/x) = -\log(x)$
$\log(1) = 0$
$\log_e(x) = \log_e(10) \cdot \log_{10}(x)$*

*Proof given in Section 2.3

has been omitted because these properties apply to all logarithms. We should also note that we can only take logarithms of numbers, not quantities, and so we should always divide the quantity by its unit when writing equations.

2.2.2 Uses of Logarithms

Before the invention of cheap pocket calculators and the widespread use of computers logarithms were used to facilitate multiplication and division. Although this is particularly useful for very large or small numbers the following trivial example illustrates this use.

Example—multiply 8 × 16

If we were to use logarithms to the base 2, $\log_2 8$ would be 3 since 2^3 is equal to 8 and $\log_2 16$ would be 4 since 2^4 is equal to 16. To multiply 8 by 16 we could represent these numbers as powers of 2, that is logarithms to the base 2.

$$8 \times 16 = 2^3 \times 2^4 = 2^{3+4} = 2^7 \tag{2.3}$$

To get the result, 128, we raise 2 to the power 7. In terms of logarithms this equation is

$$\begin{aligned}\log_2(8 \times 16) &= \log_2(8) + \log_2(16) \\ &= 3 + 4 \\ &= 7\end{aligned} \tag{2.4}$$

The final step is to take the antilogarithm of 7. Antilogarithms were listed along with logarithms in books of mathematical tables and in the base 2 this is 128. You can check this on your calculator using the x^y button. Should you wish to try this using logarithms to base 10 then the antilogarithm is either provided by a special 10^x button or by an *inverse* button which returns the inverse of any following function—in this case the logarithm.

The pocket scientific calculator has made this use of the logarithm completely redundant. Even the largest and smallest numbers can be manipulated at the push of a few buttons. Why then is the logarithm still important to chemists?

One reason is that in chemistry we deal with very large and very small ranges of data. For example, the concentration of protons (hydronium ions) in aqueous solution may vary between several moles per litre and less than 10^{-14} moles per litre. If we tried to represent this huge range graphically we would have an impossible task. Inevitably some of the data would be squashed up at the ends with just a few points stretched out over vast distances. If, however, we take logarithms to base ten this range shrinks to a manageable zero to -14. We can make this even more convenient by taking the log to base 10 and then multiplying by minus one, so that the range is zero to $+14$. This is, of course, the pH range, discussed in Section 2.2.3.

The other reason for the importance of logarithms is that many naturally occurring phenomena can be described using logarithmic functions. Although logarithms to base 10 are most commonly used for performing mathematical manipulations another base arises from these natural phenomena. This is the number 2.718 281 8... which is usually referred to as 'e'. e is an irrational number, one which cannot be represented by a simple fraction (such as $\frac{1}{2}$ or $\frac{3}{4}$). This is why the number is written with several dots at the end, to imply that the decimal representation should carry on indefinitely. Logarithms to base e, $\log_e$ are also called *natural* logarithms or *Naperian* logarithms, after the mathematician Napier. Instead of writing $\log_e(x)$ the notation $\ln(x)$ is used. This is to distinguish between logs to base e and logs to base 10, for which the notation $\log(x)$ is often used, omitting the base. If the base is omitted it is safe to assume that the base 10 is implied.

Figure 2.2 shows plots of $y = \ln(x)$ and $y = e^x$. The curves are essentially the same as the curves in Figure 2.1 and likewise show the inverse relationship of these two functions.

2.2.3 Uses of Logarithms in Chemistry—the 'p' Scale

One of the most common uses of logarithms is the 'p' scale. A general definition is

$$pX = -\log_{10}(X) \tag{2.5}$$

where X is some property of the chemical system. In the case of pH or pOH it is the molar concentration of the proton (more properly the hydronium ion, H_3O^+) and hydroxyl ion respectively. pK refers to an equilibrium constant, usually the dissociation constant of an

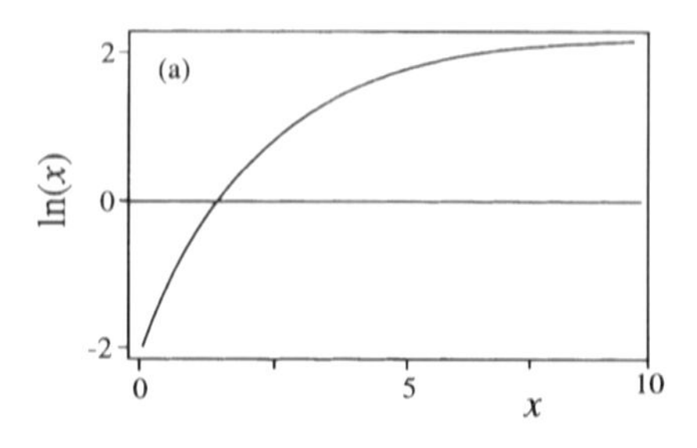

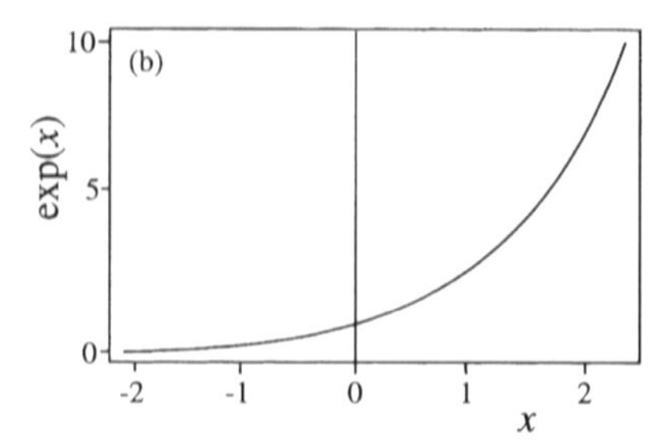

Figure 2.2 (a) $y = \ln(x)$; (b) $y = e^x$

acid, K_a, base, K_b, or water, K_w. As it happens these pK values are all interrelated and all involve pH.

Example—relationship between pH, pOH and pK_w

These dissociation constants are defined in the following way. For an acid, HA, in the presence of water*

$$HA + H_2O \rightleftharpoons H_3O^+ + A^- \tag{2.6}$$

$$K_a = \frac{[H_3O^+][A^-]}{[HA]} \tag{2.7}$$

and for a base B

$$B + H_2O \rightleftharpoons BH^+ + OH^- \tag{2.8}$$

$$K_b = \frac{[BH^+][OH^-]}{[B]} \tag{2.9}$$

For the autodissociation of water

$$H_2O + H_2O \rightleftharpoons H_3O^+ + OH^- \tag{2.10}$$

$$K_w = [H_3O^+][OH^-] \tag{2.11}$$

Equation (2.11) is very important because it is always true. The value of K_w is known at

*It will be noted that the concentration of water, $[H_2O]$ is omitted from these equations. The reason is given on p. 57.

various temperatures and at 25 °C (298 K) K_w has a value of 1×10^{-14}. Thus if we know either the concentration of hydronium ion or hydroxide ion we can always calculate the other.

Taking logarithms of equation (2.11) we get,

$$\log_{10}K_w = \log_{10}([H_3O^+][OH^-]) \tag{2.12}$$

Since $\log(xy) = \log(x) + \log(y)$ we can write

$$\log_{10}K_w = \log_{10}[H_3O^+] + \log_{10}[OH^-] \tag{2.13}$$

and multiplying through by -1 we get

$$-\log_{10}K_w = -\log_{10}[H_3O^+] - \log_{10}[OH^-] \tag{2.14}$$

By definition of pX this is the same as

$$pK_w = pH + pOH \tag{2.15}$$

From the value of K_w we know that (at 298 K) $pK_w = 14$. Thus the sum of pH and pOH is always 14, even if the equilibrium is perturbed by the addition of acid or base to the solution under investigation.

We can perform similar operations on equations (2.7) and (2.9) to get

$$pK_a = pH - \log_{10}\left(\frac{[A^-]}{[HA]}\right) \tag{2.16}$$

for the acid and

$$pK_b = pOH - \log_{10}\left(\frac{[BH^+]}{[B]}\right) \tag{2.17}$$

for the base.

Equation (2.16) is most often used in the rearranged form

$$pH = pK_a + \log_{10}\left(\frac{[A^-]}{[HA]}\right) \tag{2.18}$$

This is known as the Henderson–Hasselbalch equation and is used to calculate the pH of solutions containing acids and salts of these acids. When a salt is added to the acid the $[A^-]$ term is the total for the acid and the salt. If the acid is weak then the amount of A^- produced by the acid is small compared to that produced by the salt and the total concentration can be approximated by the concentration of the salt. The relatively large amount from the salt means that this term is large and dominates equation (2.19).

$$[A^-]_{total} = [A^-]_{acid} + [A^-]_{salt} \tag{2.19}$$

If we say that

$$[A^-]_{acid} \ll [A^-]_{salt} \tag{2.20}$$

we can make the approximation

$$[A^-]_{total} \approx [A^-]_{salt} \tag{2.21}$$

Solutions of weak acids and salts of these acids are known as buffer solutions. Additions of small amounts of acid to these solutions do not affect the total amount of A^- very much and this term still dominates equation (2.18). Thus the pH remains constant even when acid (or base) is added. Buffer solutions are particularly important in biochemical systems where pH must be maintained in order for living systems to operate. If we are using the base version of equation (2.18) then pH may still be calculated by calculating pOH and then using the relationship between pK_w, pOH and pH.

It should not go unnoticed that $[H_2O]$, the concentration of water, does not appear in any of these equations. If we calculate the concentration of pure water as 1000/18 (1000 g being the mass of one litre of water and 18 being the relative molecular mass) which is equal to 55.55 mol dm^{-3}, this appears too big to ignore.

This is considered by many students to be something of a fraud. One reason given for its omission is that it is effectively constant—even at molar solute concentrations the concentration of water is hardly affected. Chemistry is about changes and since the change is negligible the concentration is effectively a constant term in the equation. What then is the rationale in converting a very large 55.55 mol dm^{-3} into an ineffective 1 mol dm^{-3}?

All concentrations are referred to what is called a *standard state*. This is a concept used widely in thermodynamics. We always refer our measurements to some standard reference condition. In measurements of pressure it is 1 atmosphere pressure. In measurements of concentration it is 1 mol dm^{-3}. There are two advantages in this. The act of referring to a standard in mathematical terms is defined by the equation

$$\text{concentration} = \frac{\text{actual concentration/mol dm}^{-3}}{\text{reference concentration/mol dm}^{-3}} \tag{2.22}$$

Because of the division the units associated with concentration are eliminated. Thus we are free to do any mathematical manipulations (taking logarithms for instance) without worrying about the units. The second advantage is that because the numerical value of the reference quantity is unity we can simply use the value of concentration that we are given.

In the case of water, because it is usually the solvent and present in excess a different reference condition, or standard state, is used. This is the state of pure water. In most solutions the concentration is very nearly that of pure water. The concentration as defined by equation (2.22) is therefore very conveniently approximated to one.

2.2.4 Logarithms in Thermodynamic Equations

Many equations used in thermodynamics include a logarithmic term and most of these have the general form

$$X = X^0 + \text{constant} \cdot \ln(\text{concentration term}) \tag{2.23}$$

where X is some property and X^0 is the property under standard conditions. The standard condition or standard state arises when the concentration term is equal to one. Since $\ln(1) = 0$ the ln term becomes zero and the equation reduces to $X = X^0$.

An important example is the equation describing the dependence of the Gibbs function (G) on the pressure of a gas, P (which for a gas is analogous to concentration).

$$G = G^0 + RT \cdot \ln\left(\frac{P}{P^0}\right) \tag{2.24}$$

The derivation of these equations leads to the use of natural logarithms but in many cases common logarithms are used. This is because tables of common logarithms used to be more readily available than tables for natural logarithms. Even though we no longer use mathematical tables this is still useful when the hydronium ion is involved. When this happens we can introduce pH (an easily measurable quantity) into the equation.

$$X = X^0 + RT \cdot \ln[H_3O^+] \tag{2.25}$$

becomes

$$X = X^0 + 2.303RT \cdot \log[H_3O^+] \tag{2.26}$$

which is the same as

$$X = X^0 - 2.303RT \cdot \mathrm{pH} \tag{2.27}$$

The number 2.303 appears because it is equal to $\log_e 10$. The proof of the relation $\log_e x = \log_e 10 \cdot \log_{10} x$ is as follows.

Proof of $\log_e x = \log_e 10 \cdot \log_{10} x$

Suppose we write

$$y = \log_{10} x \tag{2.28}$$

then by definition

$$x = 10^y \tag{2.29}$$

or from equation (2.28)

$$x = 10^{\log_{10} x} \tag{2.30}$$

If we now take logarithms to base e of both sides we get

$$\log_e x = \log_e(10^{\log_{10} x}) \tag{2.31}$$

which from the relation $\log(x^y) = y \cdot \log(x)$ becomes

$$\log_e x = \log_{10} x \cdot \log_e 10 \tag{2.32}$$

An equation which describes the chemical potential μ of a solution is

$$\mu = \mu^0 + RT \cdot \ln(a) \tag{2.33}$$

where a is the activity or effective concentration. Activity is used rather than concentration because many solutions behave as if the concentration is different to that expected from the amount of solute added. The activity is related to the expected concentration m (for molality, the number of moles of solute per kilogram of solvent; for dilute aqueous solutions this is approximately equal to the molarity, moles per litre) by the equation

$$a = m\gamma \tag{2.34}$$

where γ is called the activity coefficient. In ionic solutions it is impossible to separate the effects of the positive ion from those of the negative ion and so the mean ionic activity coefficient is used. This has the symbol $\gamma_\pm$ and is defined in equation (2.42).

Using relation (2.34), equation (2.33) becomes

$$\mu = \mu^0 + RT \cdot \ln(m\gamma) \tag{2.35}$$

which can be expanded to become

$$\mu = \mu^0 + RT \cdot \ln(m) + RT \cdot \ln(\gamma) \tag{2.36}$$

In solution thermodynamics this is very useful. The equation can be thought of as having two parts. The first two terms on the right hand side of the equals sign form the ideal part and the third term the non-ideal part. In the ideal case the solution would behave as if the effective concentration was the same as the amount of solute added, $\gamma = 1$ and $a = m$. In this case the third term ($\ln(\gamma)$) is the logarithm of one, which is zero. Thus the equation reduces to being the same as equation (2.33). If the solution is non-ideal then γ changes from unity and the chemical potential changes accordingly. These changes are, however, totally encapsulated in this $\ln(\gamma)$ term.

Peter Debye and Erich Hückel attempted to model these non-idealities for ionic solutions by deriving an expression for $\gamma_\pm$. The Debye–Hückel equation is

$$\log_{10}\gamma_\pm = -A|z_+z_-|\sqrt{I} \tag{2.37}$$

where A is a constant which depends on the density and permittivity of the solvent. For water it is approximately equal to 0.509. z_+ and z_- are the charges on the positive and negative ions respectively and I is the ionic strength (a measurement of concentration, emphasising the electronic charges). The vertical lines on either side of the product of the charges indicate that we take only the magnitude of the product of charges and not the sign which would be negative for the product of positive and negative charges. This notation may be used for any number and is called taking the *modulus* of the number inside the lines.

Equation (2.37) is called the Debye–Hückel limiting law and is found to work well for solutions up to millimolar concentration. A more complete expression, and one which works for higher concentrations, is the complete Debye–Hückel equation,

$$\log_{10}\gamma_\pm = \frac{-A|z_+z_-|\sqrt{I}}{1 + Ba\sqrt{I}} \tag{2.38}$$

where B is another constant (related to A) and a is the distance of closest approach of two ions in solution. This parameter can usually only be determined by experiment. In most cases the product Ba is equal to one and the equation is

$$\log_{10}\gamma_\pm = \frac{-A|z_+z_-|\sqrt{I}}{1 + \sqrt{I}} \tag{2.39}$$

For most cases in undergraduate chemistry the limiting law is sufficient and can be applied to a number of problems.

Example—the dissolution of a sparingly soluble salt

Silver chloride dissolves in water to form silver cations and chloride anions.

$$AgCl \rightleftharpoons Ag^+ + Cl^- \tag{2.40}$$

The thermodynamic dissociation constant K_T is defined as

$$K_T = \frac{[Ag^+][Cl^-]}{[AgCl]} \cdot \frac{\gamma_+\gamma_-}{\gamma_{AgCl}} \tag{2.41}$$

The activity of the solid AgCl ($[AgCl] \cdot \gamma_{AgCl}$) is by convention taken to be unity and the product of the ion activity coefficients is redefined as the square of the mean ionic activity coefficient $\gamma_\pm$

$$\gamma_+\gamma_- = \gamma_\pm^2 \tag{2.42}$$

$\gamma_\pm$ is called the *geometric mean* of the two activity coefficients. Equation (2.41) now becomes

$$K_T = [Ag^+][Cl^-]\gamma_\pm^2 \tag{2.43}$$

We take logarithms to base 10 of both sides,

$$\log_{10}K_T = \log_{10}([Ag^+][Cl^-]) + 2\log_{10}\gamma_\pm \tag{2.44}$$

and we insert the expression for $\log \gamma_\pm$ from the Debye–Hückel limiting law.

$$\log_{10}K_T = \log_{10}([Ag^+][Cl^-]) - 2A|z_+z_-|\sqrt{I} \tag{2.45}$$

Then rearrange to give

$$\log_{10}([Ag^+][Cl^-]) = \log_{10}K_T + 2A|z_+z_-|\sqrt{I} \tag{2.46}$$

Thus, if it is possible to measure the concentration of either silver or chloride ions in solution (we only need one since they must be equal to each other) and the overall ionic strength we can calculate the thermodynamic equilibrium constant. This is usually achieved using a range of ionic strengths (controlled by adding an inert but soluble salt) and by plotting the left hand side of equation (2.46) against $I^{1/2}$.

Another equation having the same form as equation (2.23) is the Nernst equation. This describes the dependence of electrochemical potential on the concentration of reacting species. For a simple electron transfer

$$\text{ox} + ne^- \rightleftharpoons \text{red} \tag{2.47}$$

where ox is an oxidised species which gains n electrons (e^-) to form a reduced species, red. The Nernst equation is

$$E = E^0 + \frac{RT}{nF}\ln\left(\frac{[\text{ox}]}{[\text{red}]}\right) \tag{2.48}$$

This equation may be derived directly from the equation for the Gibbs function and the nF term is used to account for the energy change due to transfer of charge ($\Delta G = -nFE$). A more fundamental derivation starts with the equation for chemical potential and the nF term again arises from the transfer of charge. As before when the concentration term equals unity the electrode potential is the standard electrode potential. The concentration terms should strictly be considered as activities but for many cases the activity coefficient can be assumed to be equal to one. Alternatively it may be possible to use electrochemical measurements to calculate the activity coefficient, by separating out the log term as shown above for the Debye–Hückel treatment of activities.

Equation (2.48) is the simplest version of the Nernst equation. It assumes only one species is involved on each side of the equilibrium sign. A more complete equation acknowledges that there may be more.

$$a \cdot \text{ox}_1 + b \cdot \text{ox}_2 + ne^- \rightleftharpoons c \cdot \text{red}_1 + d \cdot \text{red}_2 \tag{2.49}$$

This is accounted for by raising each concentration to the power of the number of each species involved.

$$E = E^0 + \frac{RT}{nF} \ln\left(\frac{[\text{ox}_1]^a[\text{ox}_2]^b}{[\text{red}_1]^c[\text{red}_2]^d}\right) \tag{2.50}$$

The reason for this is that the 'ln' term is really the inverse of the equilibrium constant for the reaction. This is defined as the ratio of the product of concentrations of products to the product of the concentrations of reactants.

$$K = \frac{[\text{product 1}][\text{product 2}]}{[\text{reactant 1}][\text{reactant 2}]} \tag{2.51}$$

If more than one mole of a particular species is involved then the product of concentrations becomes the concentration raised to the power of the number involved (because $a \cdot a = a^2$ etc.). In most of the derivations below it will be assumed that only one of each species is involved.

It is possible now to derive a complete Nernst equation for an electrochemical cell which consists of two redox reactions similar to equation (2.47). These normally use subscripts referring to the reaction at the right hand electrode (R) and the left hand electrode (L).

$$\text{ox}_R + ne^- \rightleftharpoons \text{red}_R \tag{2.52}$$

$$\text{ox}_L + ne^- \rightleftharpoons \text{red}_L \tag{2.53}$$

Each of these 'couples' has its own electrode potential and the cell potential difference is the mathematical difference between the two.

$$E_{\text{cell}} = E_R - E_L \tag{2.54}$$

Performing the same subtraction using the chemical equations leads to

$$(\text{ox}_R + ne^-) - (\text{ox}_L + ne^-) \rightleftharpoons \text{red}_R - \text{red}_L \tag{2.55}$$

This can be rearranged in the same way we rearrange a mathematical equation, assuming the equilibrium sign performs the same function as the equals sign:

$$\text{ox}_R + \text{red}_L + ne^- \rightleftharpoons \text{red}_R + \text{ox}_L + ne^- \tag{2.56}$$

The chemical equation is the complete cell reaction. Note that ne^- appears on both sides and may be cancelled. Thus the two half-cell reactions may be combined into a full cell reaction. The values of half-cell potentials are listed in tables of thermodynamic data. Subtraction of the half-cell potentials as shown in equation (2.54) results in a theoretical full cell potential difference. Because this is related to the Gibbs energy for the complete cell reaction (by $\Delta G = -nFE$) we can predict whether this reaction is thermodynamically favourable. A positive potential difference means a favourable reaction because ΔG is always negative for spontaneous reactions.

Returning to equation (2.54) and using the Nernst equation for each term we get

$$E_{\text{cell}} = E_{\text{R}}^0 + \frac{RT}{nF} \cdot \left(\frac{[\text{ox}_\text{R}]}{[\text{red}_\text{R}]}\right) - E_{\text{L}}^0 - \frac{RT}{nF} \cdot \ln\left(\frac{[\text{ox}_\text{L}]}{[\text{red}_\text{L}]}\right) \tag{2.57}$$

This is simplified by combining the two standard electrode potentials into a standard cell potential difference

$$E_{\text{cell}}^0 = E_{\text{R}}^0 - E_{\text{L}}^0 \tag{2.58}$$

and by factorising the ln terms (both are multiplied by RT/nF). Recalling that the difference between two logarithms is given by $\log(a) - \log(b) = \log(a/b)$ we get

$$\begin{aligned} \ln\left(\frac{\text{ox}_\text{R}}{\text{red}_\text{R}}\right) - \ln\left(\frac{\text{ox}_\text{L}}{\text{red}_\text{L}}\right) &= \ln\left(\frac{\text{ox}_\text{R}}{\text{red}_\text{R}} \div \frac{\text{ox}_\text{L}}{\text{red}_\text{L}}\right) \\ &= \ln\left(\frac{\text{ox}_\text{R} \cdot \text{red}_\text{L}}{\text{ox}_\text{L} \cdot \text{red}_\text{R}}\right) \end{aligned} \tag{2.59}$$

and so the complete equation is

$$E_{\text{cell}} = E_{\text{cell}}^0 + \frac{RT}{nF} \cdot \ln\left(\frac{[\text{ox}_\text{R}][\text{red}_\text{L}]}{[\text{red}_\text{R}][\text{ox}_\text{L}]}\right) \tag{2.60}$$

All electrode potentials are measured with reference to the hydrogen half-cell which is given the arbitrary value of $E^0 = 0$ volts. Standard electrode potentials are measured under standard conditions, that is with the concentrations of all species equal to unity. In this case all ln terms become zero ($RT\ln(1)/nF = 0$). We find the standard electrochemical cell potential difference by subtraction of the half-cell potentials. Most real problems involve reactions not under standard conditions. In this case we must balance the half-cell reactions, so that the correct number of electrons are transferred from one couple to the other and use the complete Nernst equation.

Once again if one of the species involved is a proton then we can use $\log_{10}$ rather than ln and introduce pH. This is not just a mathematical exercise. pH is measured in the laboratory using a hydrogen-sensitive glass membrane and we measure the potential difference between this and a reference electrode (Figure 2.3). As the hydrogen ion

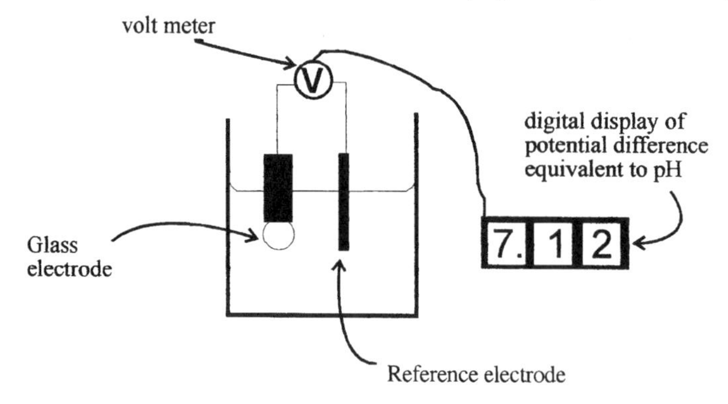

Figure 2.3 The glass hydrogen electrode. The potential difference between the glass electrode and a reference electrode is a measure of pH as defined by the Nernst equation

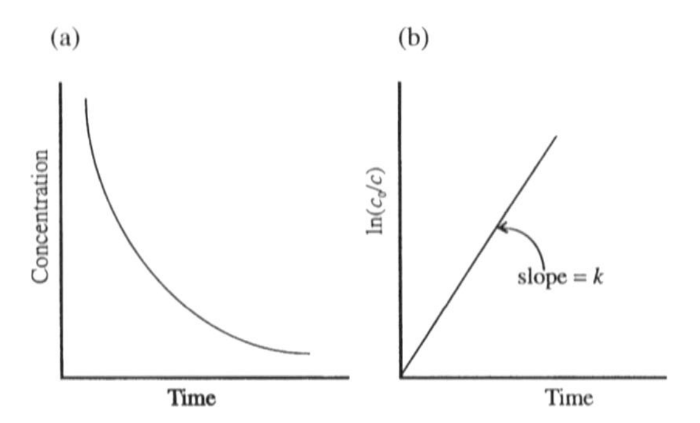

Figure 2.4 (a) Experimental concentration time plot for a first order reaction. (b) The logarithmic relationship is confirmed by plotting $\ln(c_0/c)$ versus time

concentration changes then the term in $[H^+]$ changes and the measured cell potential difference changes giving a direct reading of pH.

2.2.5 Logarithms in Kinetic Equations

Logarithms arise whenever a change occurs which depends on the amount of a substance present. A typical example is a first order chemical reaction. 'First order' in chemical kinetics means that the rate of a reaction is directly proportional to the concentration of the substance reacting. Figure 2.4(a) shows the concentration of x as a function of time for a first order reaction. The rate of reaction is the slope of this plot—the rate of change of concentration with time. Biochemists often call this the velocity of the reaction because it is analogous to velocity as the rate of change of distance with time. In this example the rate is negative because we are observing the loss of a reactant. We could easily monitor the increase in concentration of a product but if we follow the reactant we see directly the effect it has on the rate.

Considering Figure 2.4(a) we see that at the start of the experiment (time = zero) the concentration (c_0) is high because none has reacted yet. Near this point the slope is steep. As the reaction progresses then the gradient gets less and less, as the concentration (c) gets less and less. That is, the rate is proportional to the concentration. Comparison of this graph with that of the ln function shows a striking similarity and indeed if we plot $\ln(c)$ as a function of time, the plot is linear for this kind of reaction. The relation is usually plotted as

$$\ln\left(\frac{c_0}{c}\right) = kt \tag{2.61}$$

where k is a constant called the rate constant and is found from this plot. This equation will be derived in Chapter 6 but one advantage of using the ratio c_0/c is that the term inside the brackets is unitless. We can therefore use any units for concentration we choose. In practice we may use any measurement of concentration, such as the absorption of radiation or the volume measured by titration. Plotting the left hand side of this equation against time gives a linear plot with a positive slope equal to k, the rate constant. The slope is positive because as concentration falls the ratio c_0/c gets larger and the logarithm therefore gets larger. Because the logarithm is unitless the product kt must be unitless and therefore k must have units of reciprocal time, t^{-1}.

The rate of radioactive decay depends on the number of radioactive nuclei present in a sample. As the observed kinetics are first order an equation similar to (2.61) can be derived. The concept of *half-life* is used in connection with radioactive decay but is equally valid in all kinetic studies. The half-life is defined as the time taken for half of the amount of substance to react, or in the case of radioactivity for half of the radioactive nuclei to decay. We can state this mathematically using $t_{\frac{1}{2}}$ as the time taken for the initial concentration c_0 to fall to half of its value, i.e., $c_0/2$. Therefore equation (2.61) becomes

$$\ln\left(\frac{2c_0}{c_0}\right) = kt_{1/2} \tag{2.62}$$

which is the same as

$$\ln(2) = kt_{1/2} \tag{2.63}$$

Note that this equation does not contain a term in concentration. This therefore suggests that when first order kinetics apply the half-life is independent of the concentration of the substance of interest. Also, a measurement of half-life can be used (with equation (2.62) to estimate the value of the rate constant.

Example—calculations involving half-life

A first order reaction has a half-life of 65 seconds. Calculate the time taken or the concentration of the reactant to fall to a quarter of its original value.

As with most problems a systematic approach should be used. Using equation (2.63) we can calculate the rate constant since we know $t_{\frac{1}{2}}$ and ln(2), i.e.,

$$k = \frac{\ln(2)}{t_{1/2}} \tag{2.64}$$

therefore

$$k = \frac{0.693}{65} = 1.07 \times 10^{-2}\,\mathrm{s}^{-1} \tag{2.65}$$

The concentration has fallen to one quarter of its value when $c = c_0/4$. Putting this in equation (2.61) we can calculate t,

$$\ln\left(\frac{4c_0}{c}\right) = 1.07 \times 10^{-2}t \tag{2.66}$$

which rearranges to

$$t = \frac{\ln(4)}{1.07 \times 10^{-2}} = \frac{1.39}{1.07 \times 10^{-2}} \tag{2.67}$$

which is 130 seconds. This is logical since after one half-life of 65 seconds the concentration has halved, then after another 65 seconds it has halved again, to a quarter of its original value.

The main pitfall in solving some problems is in misunderstanding the wording of the

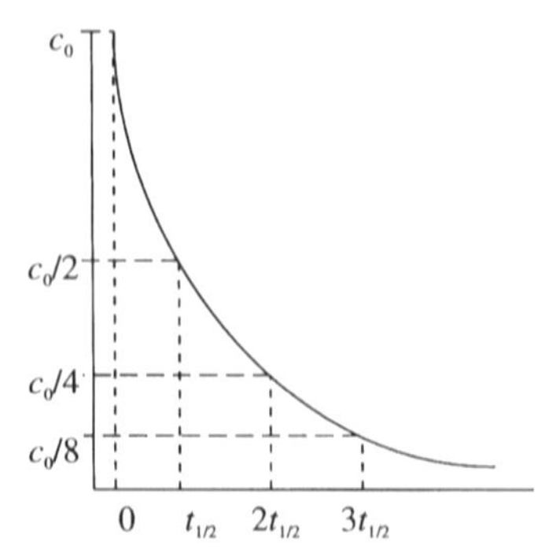

Figure 2.5 For a first order reaction the time taken for the concentration to halve is always the same, regardless of the starting concentration

question. The student must be able to define the ratio c_0/c. For example, if the question asks to find the time for a reaction to go to 90% completion then the final concentration is only 10% of the original or $c_0/10$. Alternatively the student may have to find the final concentration. This can be found as long as we know the initial concentration, but the overall strategy is the same. We find the rate constant from the half-life and then put this in the first order rate equation.

The half-life concept can also be used to determine whether the reaction is first order with respect to the reactant under investigation. A plot is made of concentration versus time (Figure 2.5) and a value of concentration is chosen at an arbitrary time $t = 0$. The time taken for this concentration to fall to half is measured from the graph. Successive initial concentrations are chosen and each half-life is determined. If these half-lives are independent of the initial concentrations chosen then the reaction is first order.

2.2.6 The Dependence of Rate on Temperature

Rates of reaction usually depend on the temperature at which the experiment is performed. The Arrhenius equation relates the rate constant, k, to the temperature T

$$\ln(k) = \ln(A) - \frac{E_a}{RT} \tag{2.68}$$

where R is the gas constant and E_a is called the activation energy and may be considered to be the minimum amount of energy needed for the reaction to take place (an initial input of heat for example). A, called the pre-exponential factor, arises when E_a/RT is equal to zero and can be related to the number of molecular collisions which take place per second. A plot of equation (2.68), $\ln(k)$ versus $1/T$ will have a slope of $-E_a/R$ and an intercept of $\ln(A)$. The Arrhenius equation will be discussed further in Section 2.3 on the exponential function.

2.2.7 Absorption of Light—the Beer–Lambert Law

Although this law is often used in linear form its origin is a logarithmic equation. If light of intensity I_0 travels through a solution of an absorbing species some of the light will be absorbed and the intensity will fall. If over a distance l half of this light is absorbed the intensity will be $I_0/2$, then as the beam travels another distance l the intensity will have halved again so that the intensity will now be $I_0/4$. This is analogous to the half-life concept in kinetics and a full equation may be derived in the same way, substituting the distance travelled by the beam for time and the intensity of light for concentration. As the light travels through the solution the intensity falls in a logarithmic fashion. The law itself is written

$$\log\left(\frac{I_0}{I}\right) = \varepsilon cl \tag{2.69}$$

where I is the measured intensity of light after it has passed through a length l of a sample of concentration c, I_0 is the initial intensity of light and ε is a proportionality constant known as the molar absorption coefficient (in older books it may be referred to as the molar extinction coefficient). ε may be calculated by quantum mechanical means but such calculations are beyond the scope of this book and most undergraduate courses. ε is, however, easily measured by experiment and as most spectrometers measure the parameter $\log(I_0/I)$ directly this is given its own symbol, A, and called the absorbance. The Beer–Lambert law is usually written

$$A = \varepsilon cl \tag{2.70}$$

2.2.8 Using Logarithms to Find Relationships between Variables

If we have a set of experimental results and a plot of two variables measured shows a curve, how do we find out the relationship between the two? One way is to assume that the equation describing this relationship is of the form

$$y = x^n \tag{2.71}$$

This assumption may of course be wrong but it is true of many systems encountered in chemistry so it is worth trying. If we now take logarithms of both sides

$$\begin{aligned} \log y &= \log x^n \\ &= n \cdot \log x \end{aligned} \tag{2.72}$$

A plot of $\log(y)$ versus $\log(x)$ should be linear and the slope is equal to the value of the index n. This works even if the index is negative, fractional or simply a positive integer. Note, also, that it does not matter if we use $\log_{10}$ or ln.

Example—determination of rate law

The rate law is the equation which describes how the rate of a reaction depends on the concentration of the reactants. For a simple reaction this is usually of the form

$$\text{rate} = k \cdot [\text{reactants}]^n \tag{2.73}$$

where the square brackets, as usual, indicate the concentration of the species inside. k is the rate constant and the exponent n is called the order of reaction. The order of reaction is important because it is related to the number of molecules taking part in the rate determining step. (Theory suggests that even if a reaction takes place in several steps only one is important in determining the overall rate of reaction.) If we take logarithms of equation (2.73) we get

$$\log(\text{rate}) = \log(k) + n \cdot \log[\text{reactants}] \tag{2.74}$$

We now have a linear equation (sometimes called a log–log plot) which should yield the value of n from the slope and k from the intercept.

One way of determining the rate law is to measure the initial rate of reaction at various initial concentrations of reactants (Chapter 1, Section 1.11) The initial rate may be measured from a plot of concentration versus time and measuring the tangent to the slope at time equals zero. Data for the initial rate of decomposition of nitrogen (IV) oxide to nitrogen (II) oxide and oxygen at various initial concentrations of the starting material are listed in Table 2.2.

The plot of initial rate versus initial concentration, Figure 2.6(a), is curved, but the plot of log(*initial rate*) versus log(*initial concentration*), Figure 2.6(b), is linear. The accurately measured slope is 1.98. The order of reaction is usually an integer and so it is reasonably safe to say that the order of this reaction is 2. This implies that the rate determining step involves the collision of two NO_2 molecules.

2.3 The Exponential Function

2.3.1 Definition

Although any number raised to a power may be described as an exponential function, what we commonly mean by *the* exponential function is the number e raised to some power. The number e itself may be defined by the equation

$$\mathrm{e} = \left(1 + \frac{1}{n}\right)^n \tag{2.75}$$

where the number n is allowed to increase. As n gets very large e approaches the value 2.718 28.... The exponential function, e^x (also written as $\exp(x)$ in cases where x is a typographically large or unwieldy function), may be defined by the equation

$$\mathrm{e}^x = 1 + x + \frac{x^2}{2!} + \frac{x^3}{3!} + \frac{x^4}{4!} + \dots \tag{2.76}$$

Table 2.2 Data for decomposition of NO_2

Initial conc. $[NO_2]$/mol dm^{-3}	Initial rate/mol dm^{-3} hour^{-1}	log ($[NO_2]$/mol dm^{-3})	log(rate/mol dm^{-3} hour^{-1})
0.0230	1.00	−1.69	0.00
0.0300	1.80	−1.52	0.255
0.0400	3.00	−1.40	0.477
0.0500	5.0	−1.30	0.699
0.0600	7.0	−1.22	0.845
0.0700	9.2	−1.15	0.963
0.0800	11.5	−1.10	1.06
0.0900	15.3	−1.04	1.18
0.1000	19.5	−1.00	1.29

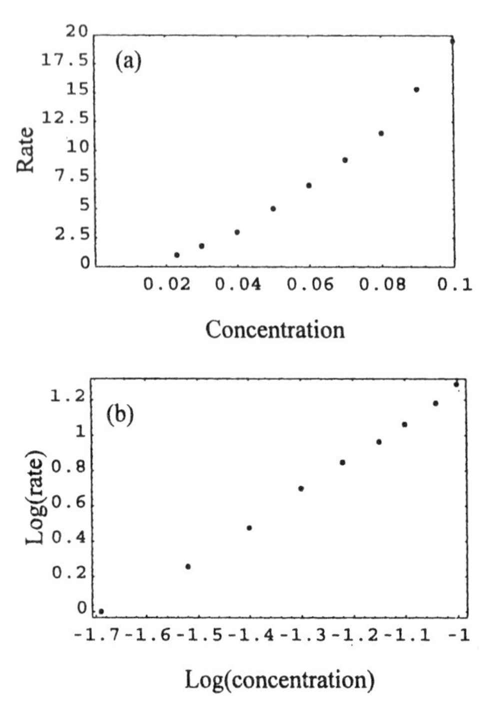

Figure 2.6 Upper figure (a) shows the experimental plot of rate versus concentration which is curved. The lower plot (b) is a log–log plot which shows, as the slope, the value of n, the order of reaction

where $2! = 2 \times 1, 3! = 3 \times 2 \times 1, 4! = 4 \times 3 \times 2 \times 1$, and so on.

The value of e^x may be determined using Equation (2.76). How many terms in the series it is necessary to use depends on the value of x. Table 2.3 lists calculated values for a range of values of x using more and more terms of the equation. The first column of results uses two, i.e., $1 + x$ and consecutive columns one more term per column.

As illustrated, for values of x between -0.25 and $+0.25$ a limiting value of e^x is reached within three terms of this expression. For values of x between -0.8 and $+0.8$ a limiting value is reached within 5 terms. Only for much larger positive or negative values must several more terms be used or the function increase rapidly.

The value of these observations is not in being able to calculate values of the exponential function. This can be easily achieved using a scientific calculator (note that the e^x button should not be confused with the EXP or EE button which are used for floating point numbers). The true value of this knowledge comes in manipulating equations which involve the exponential function. If in the course of deriving some theory we were to find the function $y = e^x$ and we knew that the function defining x was small we could make the substitution

$$e^x = 1 + x \tag{2.77}$$

and get the equation

$$y = 1 + x \tag{2.78}$$

Table 2.3 Calculated values of e^x

x	2 terms	3 terms	4 terms	5 terms	6 terms	7 terms
−100	−99	4901	−161 770	4×10^6	-8×10^7	1.3×10^9
−1	0.000	0.5000	0.333	0.375	0.367	0.368
−0.6	0.400	0.5800	0.544	0.549	0.549	0.549
−0.4	0.600	0.6800	0.669	0.670	0.670	0.670
−0.2	0.800	0.8200	0.819	0.819	0.819	0.819
−0.1	0.900	0.905	0.905	0.905	0.905	0.905
−0.05	0.950	0.951	0.951	0.951	0.951	0.951
0.05	1.050	1.051	1.051	1.051	1.051	1.051
0.1	1.100	1.105	1.105	1.105	1.105	1.105
0.2	1.200	1.220	1.22	1.22	1.22	1.22
0.4	1.400	1.480	1.491	1.492	1.492	1.492
0.6	1.600	1.780	1.816	1.821	1.822	1.822
1	2.000	2.500	2.667	2.708	2.716	2.718
100	101.0	5101	1.72×10^5	4.34×10^6	8.77×10^7	1.47×10^9

In doing this we have successfully turned an exponential equation, which is by definition non-linear, into a linear equation—a process called linearisation. Such an equation is much easier to manipulate and a plot of y versus x should be a straight line. If the simplifying assumption becomes invalid then this linear plot will start to curve. In practice such plots are only linear in certain regions—where the simplification is valid.

2.3.2 Properties of the Exponential Function

The function e^x has one special property, above all others, which accounts for its relevance to chemistry. e^x is the mathematical version of 'to he who has shall be given'. Looking at a plot of this function (Figure 2.2(a) and Figure 2.7) it is seen that the curve rises slowly at first and then increasingly sharply. If we were to estimate the slope at any particular value of x we would find that the slope is actually equal to e^x. Thus as x increases, so does the slope. In other words, the rate of increase of the function is equal to the magnitude of the function itself. This will be proved in Chapter 4, Differential Calculus, but for now it can be seen by inspection of Figure 2.7.

This behaviour is reflected in many systems. The function has even been called the compound interest law. The more money you have in the bank, the more interest you get, consequently you have even more money and more interest next time. It is also seen in biological systems, the growth of colonies of bacteria following an exponential curve. We have already met such chemical systems. The rate of a reaction depends on how much of a chemical is present, the rate of radioactive decay depends on the amount of radioactive material and the amount of light absorbed by a sample depends on the amount of material in the sample. Although we have discussed these as logarithmic functions it is more readily seen that they obey the exponential function. As the two are inverse functions, whatever we have said about $\log_e$ applies equally to the exponential function.

If we have a function, y is proportional to e raised to the power of x, we may write the equation

$$y = a \cdot e^x \tag{2.79}$$

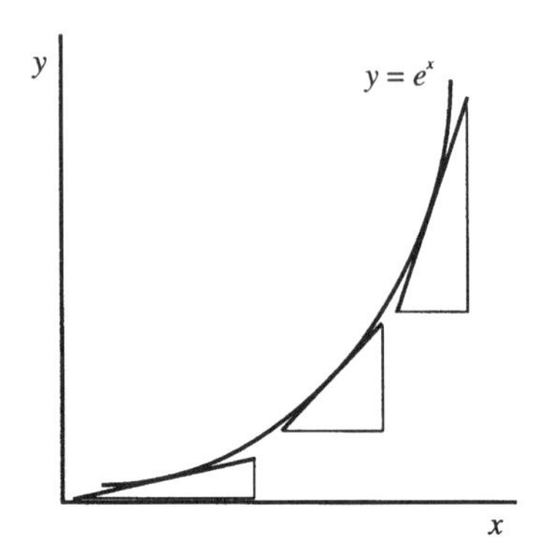

Figure 2.7 Schematic plot showing that the slope of the function $y = e^x$ increases as the value of x increases

in which a is a proportionality constant. If we take this function and calculate y when x is equal to zero, since any number raised to the power zero is equal to one, we get

$$y_{(x=0)} = a \tag{2.80}$$

So we might replace a with the constant y_0 to indicate that this is the value of y when $x = 0$. Thus

$$y = y_0 \cdot e^x \tag{2.81}$$

If we now take logarithms to base e of both sides of this equation we obtain

$$\ln(y) = \ln(y_0) + x \tag{2.82}$$

and rearranging

$$\ln(y) - \ln(y_0) = x \tag{2.83}$$

which is the same as

$$\ln\left(\frac{y}{y_0}\right) = x \tag{2.84}$$

which is now the same form as the equation for first order kinetics and the Beer–Lambert law. In individual cases we have to make sure that the concentration terms are the correct way up. Equation (2.84) has the constant initial value of y on the bottom, and x increases as the ratio y/y_0 gets bigger. In the kinetic equation we were following the fall in concentration. In order that the slope of the plot would be positive we had the ratio as y_0/y, which increases as y falls.

All of the equations highlighted here may be considered as exponential or logarithmic functions. Whichever we choose, there is always a constant involved which accounts for the case when the independent variable is equal to zero. The equations discussed so far are usually considered to be logarithmic equations but there is one special case when the exponential function is used in preference to the logarithmic version, the Boltzmann equation.

2.3.3 The Exponential Function in Chemistry—the Boltzmann Equation

Much of chemistry is concerned with how much energy atoms and molecules have, and how that energy is distributed. For example, a sample of a gas will have a certain amount of energy associated with it. Some of that energy may be attributed to the motion of the molecules (in various forms—translational, rotational, vibrational), some to the motions of electrons in the atoms and bonds which make up the molecules, some to the spinning of the nuclei of the atoms and some to the randomness or disorder of the molecules (the entropy). If we concentrate on the translational motion we would find that the molecules do not all travel at the same speed but there is a range of molecular speeds. Similarly there is a distribution of energy with rotations, vibrations and electronic structure. This distribution must also take account of the observation that in the microscopic world of atoms and molecules energy occurs in discrete packages called quanta. Molecules, atoms, electrons and nuclei can only have discrete amounts of energy, and changes must also occur in these same discrete portions.

Knowledge of this distribution is very important indeed. It is found that when reactions occur there is a minimum amount of energy required before a species can react—the activation energy. Knowing the distribution of energy will tell us what proportion of the sample has this minimum energy and we can estimate the rate of the reaction. Being able to do this may allow us to optimise reaction conditions so as to maximise efficiency.

The distribution of energy may be considered in two senses. One is the amount of energy per particle, for example the electronic energy levels which are used to estimate the electron configuration, or the different vibrational energy levels, Figure 2.8(a). The other is the number of particles per energy unit. In this case the energy levels are considered as universal and quite abstract from the molecules themselves. Rather, a molecule with a particular form of energy is considered to reside in a particular energy level. If the molecule absorbs energy (in the form of electromagnetic radiation perhaps) it is considered to leap into a higher energy level. It may fall into a lower level by losing such energy. This view is particularly relevant to spectroscopic studies. Particles are considered to occupy energy levels as shown in Figure 2.8(b). Absorption of light involves interaction between a particle and a photon of light. This is more likely to occur if there are more particles available to undergo the interaction, just as a reaction is more likely to occur if

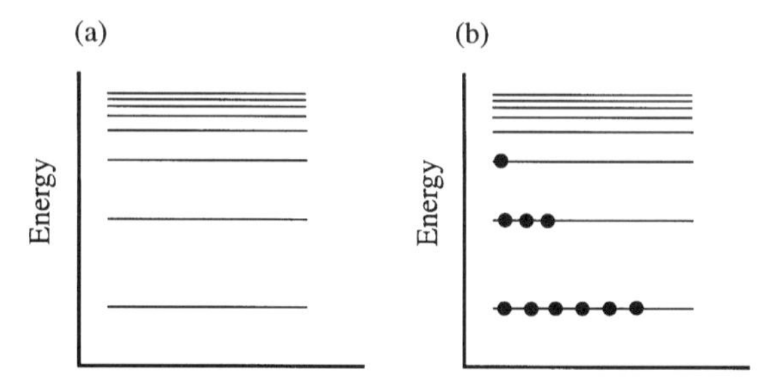

Figure 2.8 (a) Energy level diagrams may be drawn to represent any atomic or molecular species. They are often represented as discrete horizontal lines plotted against a vertical energy axis. In some cases the lines may be curved representing variable energy within that quantum region. (b) Energy levels are often taken as an abstraction to show the distribution of energy within an assembly of particles. In this example most have the lowest energy available with fewer and fewer in the higher levels

there are more molecules with sufficient energy to react. Thus the more heavily populated levels give rise to more intense absorptions. If we know which level is absorbing the light, we have an additional piece of information in mapping out the structure of the atom or molecule.

The all-important equation which is used to calculate these distributions is called the *Boltzmann equation*. If we have a system with two energy states or levels, upper and lower, with corresponding energies E_{upper} and E_{lower} the ratio of the numbers of particles in each state, N_{upper} and N_{lower} is given by this equation

$$\frac{N_{\text{upper}}}{N_{\text{lower}}} = \exp\left(-\frac{E_{\text{upper}} - E_{\text{lower}}}{kT}\right) \tag{2.85}$$

where T is the absolute temperature and k is the universal constant known as the Boltzmann constant. This has a value of $1.38 \times 10^{-23}\,\text{J K}^{-1}$. This is equal to the gas constant divided by the Avagadro constant (the number of particles in a mole). It may therefore be thought of as the gas constant per atom or molecule or alternatively that the gas constant is the molar version of the fundamental Boltzmann constant.

If E_{upper} is greater than E_{lower}, as is logical, then the term in the brackets is negative (due to the presence of the minus sign). The ratio of $N_{\text{upper}}/N_{\text{lower}}$ must be less than one, as $e^{-x} = 1/e^{x}$. This means that there is a greater population in the lower energy state. This is usually found to be the case.

An exception to this rule occurs when the levels are quite close together and at least one of the levels is multiply degenerate. Degeneracy means that more than one state has the same energy. This is like the p orbitals of the hydrogen atom. They are all different, pointing in different directions, but have the same energy. They are triply degenerate. This can occur with any sort of energy distribution. For example, a molecule rotating in space could be considered to be rotating in any of the three normal planes of space. The complete Boltzmann equation, taking account of this phenomenon, is

$$\frac{N_{\text{upper}}/g_{\text{upper}}}{N_{\text{lower}}/g_{\text{lower}}} = \exp\left(-\frac{E_{\text{upper}} - E_{\text{lower}}}{kT}\right) \tag{2.86}$$

where g_n is the degeneracy of state n. If a particular state is, say, doubly degenerate then half the number of particles go into each state. Therefore each number, N, is divided by its degeneracy. For example, the difference in energy between the lowest rotational state and the first excited rotational state of a molecule is of the order of 10^{-23} joules. Putting this energy into equation (2.85), taking $T = 298\,\text{K}$ we get $N_{\text{upper}}/N_{\text{lower}} = 0.98$. The first excited state is, however, triply degenerate and so we get

$$\frac{N_{\text{upper}}/g_{\text{upper}}}{N_{\text{lower}}/g_{\text{lower}}} = 0.98 \tag{2.87}$$

which may also be written as

$$\frac{N_{\text{upper}}}{N_{\text{lower}}} \cdot \frac{g_{\text{lower}}}{g_{\text{upper}}} = 0.98 \tag{2.88}$$

g_{lower} is one and g_{upper} is three, so we get

$$\frac{N_{\text{upper}}}{N_{\text{lower}}} \cdot \frac{1}{3} = 0.98 \tag{2.89}$$

which gives the ratio as 2.94, indicating that the excited level is nearly three times as populated as the lower one.

The importance of the Boltzmann equation in chemistry cannot be over-stressed. It appears wherever there are changes in energy, which is practically everywhere. Some specific areas are discussed below.

2.3.4 Spectroscopy

Spectroscopy is concerned with the absorption of light energy, raising the molecule, atom, nucleus or electron which interacts with light, into a higher energy state. As illustrated above the Boltzmann equation can be used to calculate the populations of these energy states. These populations are very important because the amount of light absorbed will be proportional to the number of species in the lower state. Thus if there is a large excess in the lower state we would expect to see a more intense absorption. This occurs in electronic and vibrational spectroscopy where energy differences are up to the order of 2×10^{-18} joules (equivalent to 1000 kJ mol^{-1}). Putting this value into our equation gives very large excesses in the lower state. For magnetic resonance spectroscopy the spin states of nuclei are of the order of 10^{-26} J apart. This results in almost equal populations. This is reflected in the relatively weak signals that are obtained using these methods.

2.3.5 Kinetics

Taking the exponential form of the Arrhenius equation discussed above gives

$$k = A \cdot \exp\left(-\frac{E_a}{RT}\right) \tag{2.90}$$

This can now be seen to be analogous the Boltzmann equation. R is used rather than k because E_a is usually measured in units of kilojoules per mole, and R has the same units as the Boltzmann constant per mole. It can also be seen why A is called the pre-exponential factor. Here the energy difference is that between some nominal average energy that the reactants might have and the energy needed to begin reacting. The rate will be dependent on the number of molecules which have this additional energy.

2.3.6 Electrochemical Kinetics

This is an area where students often have difficulty coping with the mathematics. This is not surprising when they are faced with the rather imposing Butler–Volmer equation

$$I = I_0\left[\exp\left(\frac{\alpha_A nF}{RT}\eta\right) - \exp\left(\frac{-\alpha_C nF}{RT}\eta\right)\right] \tag{2.91}$$

This equation is important because we can, under different experimental conditions, make a number of approximations which greatly simplify the mathematics used. This arises because of the behaviour of the exponential function.

Equation (2.91) is derived by considering an electron transfer reaction in the same way as normal chemical reactions

$$\text{ox} + ne^- \rightleftharpoons \text{red} \tag{2.92}$$

which is the same as Equation (2.47), only now the electron comes from an electrode rather than another solution species. The overall rate is considered to be the sum of the rate of the forward reaction (left to right) and the reverse reaction (right to left). The current passed through the electrode (I in the above equation, which is strictly speaking the current per unit area of the electrode) is a direct measure of the rate of the reaction. We can show this from an analysis of the units. The units of current are coulombs per second, $C\,s^{-1}$. The coulomb is the unit of charge and, since electrons carry charge, the rate of electron exchange also has units of coulombs per second.

Each reaction is considered to follow an Arrhenius-type equation. So for the forward reaction (the reduction)

$$I = I_0 \exp\left(\frac{-\alpha_C nF}{RT}\eta\right) \tag{2.93}$$

Now it can be seen that I_0 is analogous to the pre-exponential factor A, and is a standard rate or current (usually called the exchange current density) when the overpotential, η, is zero. The activation energy or energy difference is replaced by the term $\alpha_C nF\eta$. Recalling the equation linking electrochemical potential, E, to free energy ($\Delta G = -nFE$), and that the overpotential is literally the potential over or greater than some equilibrium potential (symbol E), we can see that this is a free energy term. Indeed, the Arrhenius equation can be considered in terms of free energy of activation, enthalpy of activation, entropy of activation or any other you might care to think of. All that remains is the variable α_C. This is called the electron transfer coefficient and usually has a value between zero and one. In many cases it is equal to 0.5. The subscript C indicates that it is the cathodic reaction we are discussing. α_C can be considered to be the proportion of the overpotential which contributes to the activation energy for the electron transfer.

The reverse reaction (the oxidation) is described by an analogous equation

$$I = I_0 \exp\left(\frac{\alpha_A nF}{RT}\eta\right) \tag{2.94}$$

The differences between this and equation (2.93) are the sign of the exponent and the transfer coefficient which is labelled α_A (to specify anodic reaction). It is also important to note that in accordance with electrochemical theory the sum of the transfer coefficients is equal to unity. The change of sign arises in the complete derivation of the full equation but it will be seen that the result is logical. Students should be aware of different sign conventions used in the UK and the USA. In the UK a reduction results in negative current. In the USA a reduction results in a positive current. Consequently the Butler–Volmer equation appears different depending on the origin of the author. The above equation corresponds to British convention.

The overall current is the sum of the currents in the two directions. Since the cathodic current is negative and the anodic current is positive, the sum manifests itself as a difference. The exchange current density is common to both equations and is factorised out, to appear outside the large bracket.

The full equation is rarely used in all its glory. Rather, three approximations may be made which rely on the size of the overpotential and how the exponential function itself varies with the value of the exponent.

If η is large and positive, then the second exponential term, having a negative sign in the exponent, becomes very small (for example, exp(− 100) = 3.7 × 10^{-44}). The left hand term becomes very large (exp(100) = 2.3 × 10^{43}). Therefore we can effectively ignore the right hand term and approximate I as equation (2.94). If η is large and negative, then the exponent with the negative sign becomes large and positive and this term then dominates. In this case the equation may be approximated by equation (2.93). Looking at the full equation, this term has a negative sign associated and thus the observed current is negative, in accordance with British convention.

Equations (2.93) and (2.94) are usually written in the logarithmic form

$$\log(-I) = \log(I_0) - \frac{\alpha_C nF\eta}{2.303RT} \tag{2.95}$$

for the cathodic reaction and

$$\log(I) = \log(I_0) + \frac{\alpha_A nF\eta}{2.303RT} \tag{2.96}$$

for the anodic reaction. Equation (2.95) includes a minus sign in the bracket because the current, being cathodic, has a negative sign and we cannot take a logarithm of a negative number. The negative sign turns this into a positive quantity. If we remember which reaction is occurring, it is only the magnitude of the current which matters. Equations (2.95) and (2.96) are known as the *Tafel equations*, after Julius Tafel who first noted the logarithmic dependence of current on potential. This discovery occurred several years before the work of Butler and Volmer whose (independent) work led to the equation which bears their names. Only in hindsight do we see that the Tafel equations are in fact special cases of the Butler–Volmer equation.

In practice, plots of log(current) versus potential (or overpotential) are not entirely linear. This is because the equations are only approximations of the more complete equation. At high and low values of potential, when the assumptions outlined above are no longer valid, they curve. The linear portion must be extrapolated to log(I) equals zero in order to estimate the value of I_0 (Figure 2.9).

The third approximation can be made when η is very small. When this is the case we

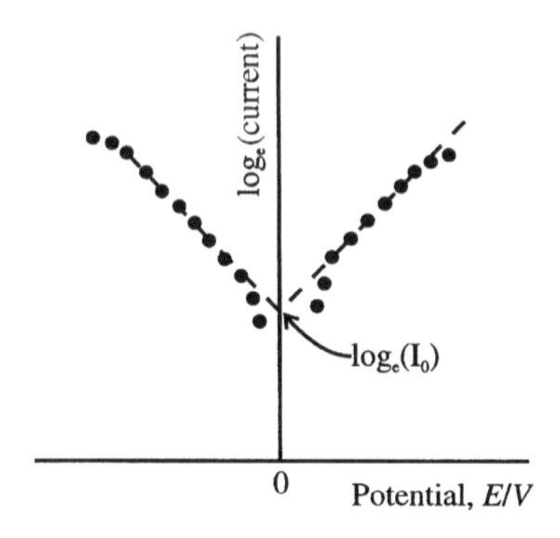

Figure 2.9 A typical Tafel plot has two branches representing anodic and cathodic electron transfers. Both have a limited linear region because the assumption inherent in the Tafel equations breaks down at high and low potentials. This linear region must be extrapolated to estimate the exchange current density I_0

can make the approximation $\exp(x) = 1 + x$. In this case the full equation becomes

$$I = I_0\left[\left(1 + \frac{\alpha_A nF}{RT}\eta\right) - \left(1 - \frac{\alpha_C nF}{RT}\eta\right)\right] \tag{2.97}$$

in which case the 'ones' cancel and the − (minus) of what used to be the second exponential term becomes a +

$$I = I_0\left\{\frac{\alpha_A nF\eta}{RT} + \frac{\alpha_C nF\eta}{RT}\right\} \tag{2.98}$$

Now, the term $nF\eta/RT$ is common to both parts and this equation may be factorised with the sum of the transfer coefficients as the other factor

$$I = I_0\left\{(\alpha_A + \alpha_C)\left(\frac{nF\eta}{RT}\right)\right\} \tag{2.99}$$

If the sum of the transfer coefficients is equal to one, this equation then becomes

$$I = I_0\frac{nF\eta}{RT} \tag{2.100}$$

which is altogether a much simpler equation and shows that current is proportional to the overpotential. In practice this is indeed found to be true when η is very small.

Therefore by making approximations corresponding to different experimental conditions we can simplify the full Butler–Volmer equation and effectively model the real system as it behaves under these different conditions.

2.3.7 Gas Kinetics

Chemistry students usually meet the Maxwell–Boltzmann distribution early on in their careers. This is usually in the form of plots such as those shown in Figure 2.10. The full mathematical treatment is only dealt with in more advanced courses. The distribution describes the range of molecular velocities or speeds in a sample of a gas at a particular temperature. The function of molecular speeds, F(c), is given by $y = \mathrm{F}(c)$ and

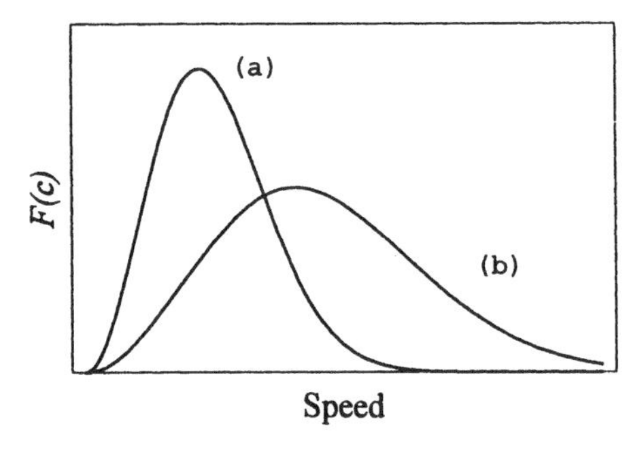

Figure 2.10 Plots of the Maxwell–Boltzmann distribution function. The area under the curve represents the number of particles and the area between a value of 'speed' and the upper limit is the proportion of particles having speed greater than or equal to that value. Curve (b) at a higher temperature has a greater area above any given speed and illustrates the observation that at higher temperatures more particles have higher speeds—resulting in faster rates of reaction

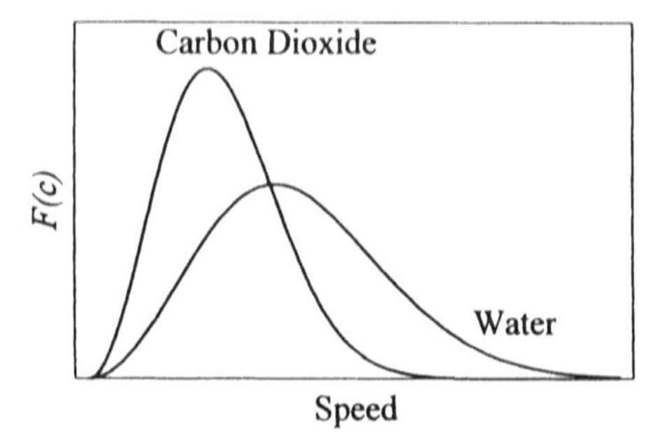

Figure 2.11 Maxwell–Boltzmann plots for two different molecules. The lighter molecule, water, has a curve which peaks at a higher speed (this is the most probable speed) and a greater range of speeds than the heavier molecule, carbon dioxide

$$y = \left(\frac{m}{2\pi kT}\right)^{1/2} c^2 \exp\left(\frac{-mc^2}{2kT}\right) \tag{2.101}$$

This is another equation which at first sight appears rather daunting. On closer inspection it can be seen that the exponential term is no more than the Boltzmann equation with the kinetic energy of a molecule (equal to $mc^2/2$, from classical physics) as the energy term. This therefore represents the number of molecules with a certain kinetic energy. The complete equation is derived using probability theory and the function F(c) is the probability (or chance) that a particle will have a certain speed. Figure 2.10 shows two plots of this equation at two different temperatures. The peak in the plot is the greatest probability and its position on the x axis is the speed most probable or most common in the sample of gas. The plot at higher temperature shows that the value of the most common speed is greater at higher temperatures. If we were to choose a critical speed which might be the speed colliding molecules had to achieve in order to react, at the higher temperature a greater number of molecules have this minimum speed. Therefore we would expect the rate of reaction to increase with temperature, which is usually found to be the case.

Because the distribution is dependent on the kinetic energy, it is also dependent on the mass of the particles of the gas. Figure 2.11 shows how the distribution changes as the mass is increased. Lighter particles have a wider range of velocities and a higher average speed than heavier particles. Care must therefore be taken in interpreting such plots. In terms of reaction rates it is usually the amount of energy which is most important and we must therefore take into account the mass of a gas in calculating the energy due to its translational motion.

2.3.8 Thermodynamics

It is relatively simple to show that the thermodynamic equations described above are another form of a Boltzmann-type equation. In fact the Boltzmann distribution of energies only applies if a system is in equilibrium. Many thermodynamic equations only apply to equilibrium situations.

If we take logarithms to base e of the Boltzmann equation we get

$$\ln\left(\frac{N_{\text{upper}}}{N_{\text{lower}}}\right) = \frac{-(E_{\text{upper}} - E_{\text{lower}})}{kT} \tag{2.102}$$

Multiplying through by kT and taking the energy terms to the left we get

$$E_{\text{upper}} - E_{\text{lower}} = -kT \cdot \ln\left(\frac{N_{\text{upper}}}{N_{\text{lower}}}\right) \tag{2.103}$$

or

$$E_{\text{upper}} = E_{\text{lower}} - kT \cdot \ln\left(\frac{N_{\text{upper}}}{N_{\text{lower}}}\right) \tag{2.104}$$

This is now in the same form as the thermodynamic equations listed and discussed above. We have energy terms (free energy, chemical potential, electrochemical potential), logarithms of concentration terms (concentration must be proportional to the number N of particles present) and the term kT, which is analogous to the RT term (since k is related to R by the Avagadro constant). The purpose of this exercise is merely to show that although the study of thermodynamics involves countless equations, very many of them are of the same form, which is itself related to the most fundamental equation used by chemists.

2.3.9 Statistical Thermodynamics

Statistical thermodynamics uses the known microscopic properties of matter to model the macroscopic, thermodynamic properties. The Boltzmann equation is central to most of its arguments along with the concept of the partition function. The Boltzmann equation can be used to calculate the population distribution between two energy states. In real systems there are usually many more possible energy states to be considered. The partition function is used to add up all the distributions between all the possible pairs of states to give an overall distribution of energy. This topic is largely beyond the scope of this book, but the partition function, q, is given by

$$q = \sum g_i \cdot \exp\left(\frac{E_i - E_0}{kT}\right) \tag{2.105}$$

where i represents a particular energy state and E_0 is the lowest energy state. The capital sigma, Σ, represents a summation of all the terms which follow. In this case g_i and E_i are variable terms and we add up all of the terms which arise when these are given the different values corresponding to a certain assembly of particles.

2.4 Trigonometric Functions—Sine, Cosine and Tangent

2.4.1 Definitions

These functions may be defined by considering the right angled triangle, ABC, in Figure 2.12, with sides of length AB, BC and AC. The sine (usually written ‘sin’), cosine (written ‘cos’) and tangent (written ‘tan’) are defined as follows:

$$\sin(\theta) = \frac{\text{BC}}{\text{AC}} \tag{2.106}$$

$$\cos(\theta) = \frac{\text{AB}}{\text{AC}} \tag{2.107}$$

$$\tan(\theta) = \frac{BC}{AB} \tag{2.108}$$

If we know the value of the angle and the length of the long side (the hypotenuse) the lengths of the two other sides may be calculated from the relationships

$$AC \cdot \sin(\theta) = BC \tag{2.109}$$

from equation (2.106) and

$$AC \cdot \cos(\theta) = AB \tag{2.110}$$

from equation 2.107. Also from these definitions we can redefine the tangent as

$$\tan(\theta) = \frac{AC \cdot \sin(\theta)}{AC \cdot \cos(\theta)} = \frac{\sin(\theta)}{\cos(\theta)} \tag{2.111}$$

In the past values of sines, cosines and tangents would have been found from mathematical tables but once again the pocket calculator can do all that is necessary at the push of a button.

Maths texts usually define these functions in terms of a point with coordinates (x, y) which rotates about the origin $(0, 0)$, as shown in Figure 2.13. The point travels in a circle with radius r and in this case the trigonometric functions may be defined as

$$\sin(\theta) = \frac{y}{r} \tag{2.112}$$

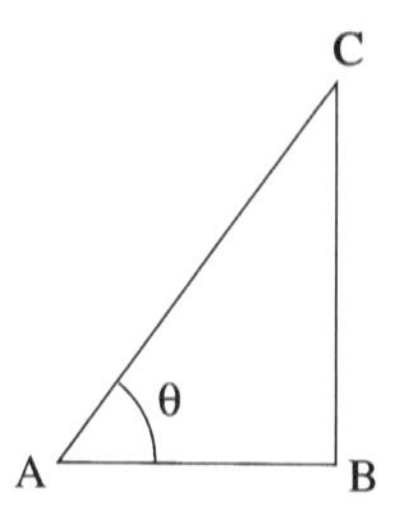

Figure 2.12 Right-angled triangle used to define the functions sine, cosine and tangent

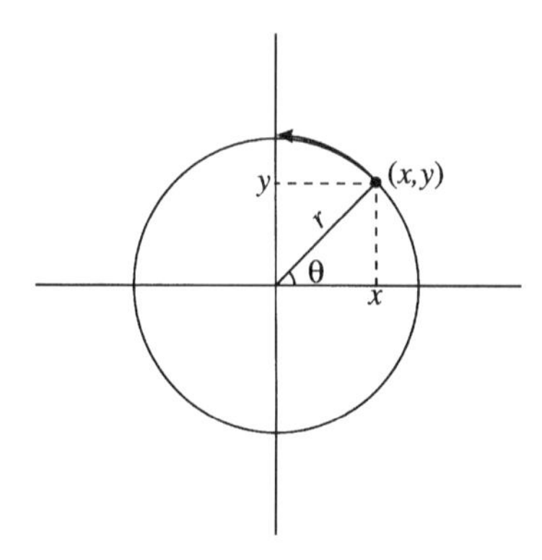

Figure 2.13 Circular motion of the point (x, y) used to define the trigonometric functions. The anti-clockwise direction of motion is taken as positive

$$\cos(\theta) = \frac{x}{r} \tag{2.113}$$

and

$$\tan(\theta) = \frac{y}{x} \tag{2.114}$$

Note that the definition of the tangent is the same as the gradient of the line from $(0, 0)$ to (x, y), as defined in Chapter 1.

This representation also allows us to observe what happens to these functions as the angle increases beyond 90° and eventually undergoes a complete revolution. It also allows the definition of a different measure of the magnitude of an angle. The common unit of angle is the degree, there being 360 degrees in one complete revolution. Consideration of a circle allows the definition of another measure, the radian. A radian is the angle made when the arc of a circle (a portion of the circumference) is the same length as the radius, Figure 2.14. As the circumference of a circle is equal to 2π multiplied by the radius, there are also 2π radians in one complete revolution. There are 360 degrees in one revolution, 180 degrees is equivalent to π radians and 90 degrees is equivalent to $\pi/2$ radians. Radians are much more commonly used in science and students should therefore be familiar with the definition.

As with the exponential function sines and cosines may be represented by power series and we may use these to calculate or approximate the value of a sine or cosine.

$$\sin(\theta/\text{radians}) = \frac{\theta}{1!} - \frac{\theta^3}{3!} + \frac{\theta^5}{5!} - \frac{\theta^7}{7!} + \ldots \tag{2.115}$$

$$\cos(\theta/\text{radians}) = 1 - \frac{\theta^2}{2!} + \frac{\theta^4}{4!} - \frac{\theta^6}{6!} + \ldots \tag{2.116}$$

There are many relationships between sines and cosines. Most of these are not particularly relevant to the undergraduate chemist, but will be found in many science-oriented mathematics textbooks. Some of these are listed in Table 2.4 for reference purposes. They are derived in most elementary maths texts.

2.4.2 Properties of Sine and Cosine—Waves

One of the most interesting properties of the sine and cosine (for the chemist) becomes apparent if we draw plots of these functions. These are shown in Figure 2.15.

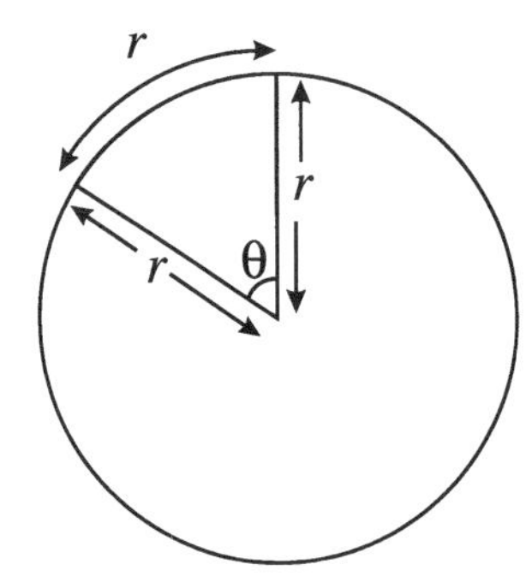

Figure 2.14 The definition of the radian as a unit of angle

Table 2.4 Some important relationships between trigonometric functions

$\sin(-x) = \sin(x)$	$\sin(x + 2\pi) = \sin(x)$
$\cos(-x) = \cos(x)$	$\cos(x + 2\pi) = \cos(x)$
$\tan(-x)\colon = -\tan(x)$	$\tan(x + 2\pi) = \tan(x)$
$\sin(\pi - x) = \sin(x)$	$\sin(\pi/2 - x) = \cos(x)$
$\cos(\pi - x) = -\cos(x)$	$\cos(\pi/2 - x) = \sin(x)$
$\tan(\pi - x) = -\tan(x)$	$\tan(\pi/2 - x) = 1/\tan(x)$
$\sin(x + y) = \sin(x)\cos(y) + \cos(x)\sin(y)$	$\sin^2(x) + \cos^2(x) = 1$
$\cos(x + y) = \cos(x)\cos(y) - \sin(x)\sin(y)$	

Note: $\sin^2(x)$ means $(\sin(x))^2$

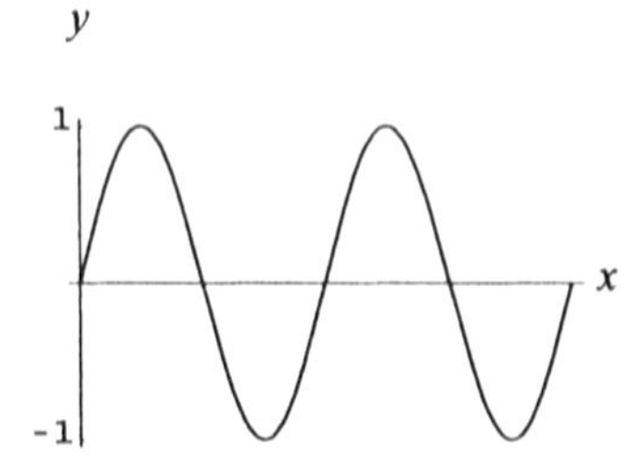

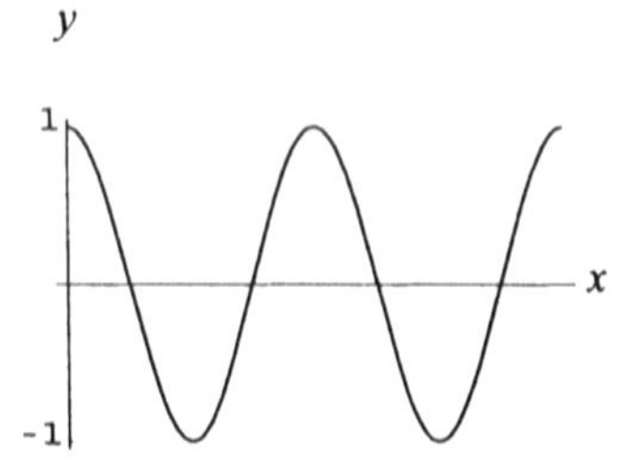

Figure 2.15 Plots of $y = \sin(x)$ (left hand figure) and $y = \cos(x)$ (right hand figure). Note that the shapes are identical but shifted along the x axis

Both plots are identical in shape but the sine function is shifted along the x axis from the cosine function. The difference between the two curves is 90° or $\pi/2$ radians, hence the relationship $\cos(x) = \sin(x + \pi/2)$ in Table 2.4.

The shape of the plot is that of a wave. We are all familiar with the wave motion of the sea or ripples on a pond, but these functions represent many other natural phenomena which are thought of as waves. These include sound waves and the oscillations of electric and magnetic fields which constitute waves of light. Also, since the advent of quantum theory, particles are also considered to have wave-like properties. The important properties are outlined below, with reference to Figure 2.16.

The sine wave is essentially a plot of the vertical distance of a point rotating about a central point versus the angle the line joining the two makes with the horizontal (x axis). Thus the plot is always related to some circular motion.

The horizontal distance from a point on the wave to the position where the function has the same value (e.g., peak to peak) is called the wavelength. This is the same as

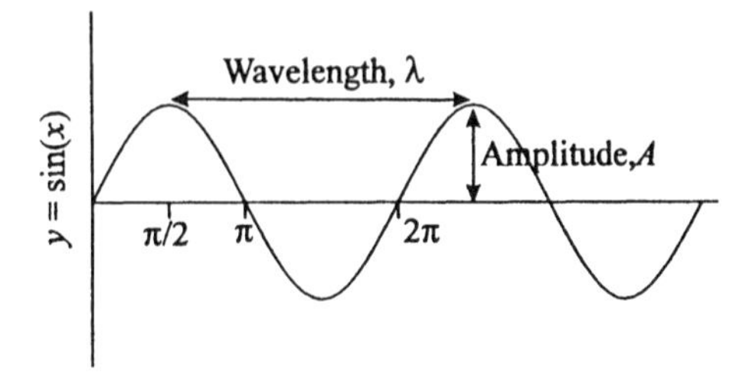

Figure 2.16 Definitions of wavelength and amplitude for the sine wave

completing one cycle of the related circular motion. This is manifested in the relation $\sin(x) = \sin(x + 2\pi)$. The velocity of the wave motion can be derived from the normal definition of velocity as being the distance moved per unit time. A suitable distance to use is the wavelength (symbol λ), and the time to complete one wavelength's motion is called the period (symbol P) of the oscillation. The speed (symbol c) is given as

$$c = \frac{\lambda}{P} \tag{2.117}$$

A more common equation arises from the definition of the frequency (symbol ν) as the number of oscillations per second. That is, $\nu = 1/P$ and

$$c = \nu\lambda \tag{2.118}$$

The frequency is important for two reasons. It defines the colour of visible light. The speed of light is a universal constant and so wavelength and frequency are always related by equation (2.118). It also defines the energy of a photon of light by the equation

$$E = h\nu \tag{2.119}$$

where h is Planck's constant which has a value of 6.626×10^{-34} J s. This equation allows direct correlation between the frequency of light absorbed and the energy level differences in the molecules or atoms which absorb the light.

The units of frequency are cycles per second, commonly called hertz (symbol Hz). Also used is the radian per second (rad s^{-1}). Both can be seen to be related to the angular velocity of the related circular motion.

The general equation which describes a sinusoidal oscillation is

$$y = A \cdot \sin(\theta) \tag{2.120}$$

When $\theta = \pi/2$ (90°), $\sin(\theta) = 1$ and $y = A$, the maximum displacement of the oscillation. This is called the amplitude.

This is very important for two reasons. One is that wave motions can interact with each other. They add together so that two waves starting together, with the same frequency and velocity, reach their peak together and the resultant amplitude is enhanced. This is called being *in phase*, or being *coherent*. If the waves start out at different times, such that one reaches a positive peak as the other reaches a negative peak, they cancel each other out and the resultant is no wave at all, Figure 2.17.

This behaviour can be observed in the diffraction patterns produced by shining light through a diffraction grating, shown in Figure 2.18.

The other reason arises in quantum mechanics and developing a mathematical model for the structure of the atom. For a particle such as an electron which can be described in terms of wave properties, the amplitude of the wave is related to the probability of finding the particle in a given region of space. Thus if we can define a function which describes a wave (a wavefunction) we can also calculate the probability of finding a particle in the region of space defined by the wavefunction.

Another version of equation (2.120) can be derived in terms of the frequency of the wave. Initially we consider a particle travelling in a circle with a speed of ω radians per second. Therefore, after one second the particle will have travelled through ω radians and after t seconds through ωt radians. We can then write

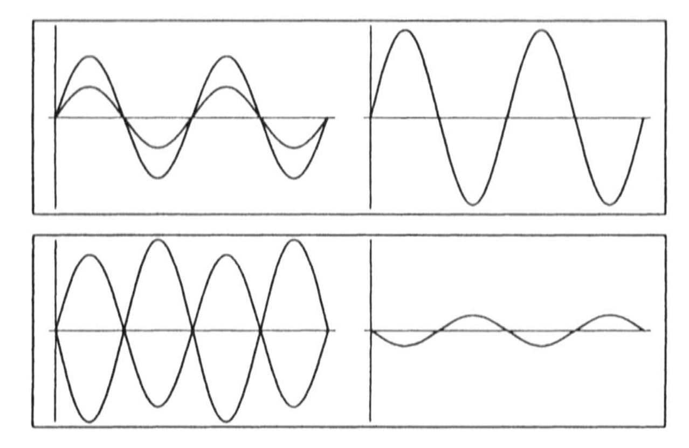

Figure 2.17 Upper figure, left shows two sine waves in phase. The resultant wave, upper figure right, is an enhanced wave. Lower figure, left shows two out-of-phase sine waves. The resultant is a much diminished wave. In the limit the resultant is zero

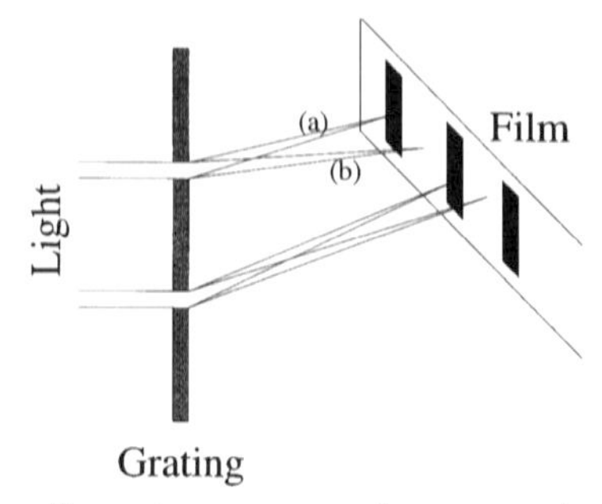

Figure 2.18 A diffraction grating acts as a source of many new beams when light passes through it. When the distance travelled by each of two of these beams is a whole number of wavelengths they are in phase and when they strike a photographic film an image is produced (a). When the distance travelled is different by a number of half-wavelengths they are out of phase and no image results. The overall effect is of a set of light and dark bands

$$y = A \cdot \sin(\omega t) \tag{2.121}$$

and to convert this into units of Hz, since there are 2π radians per cycle, we can write

$$y = A \cdot \sin(2\pi v t) \tag{2.122}$$

We now have an expression which relates the magnitude of the oscillation to both time and distance, because the frequency v is related directly to the wavelength.

2.4.3 The Uses of Sines and Waves in Chemistry

With the exception of the Schrödinger equation there is only one equation which makes use of the sine. This is the *Bragg equation* used in crystallography. This makes use of the sine to calculate the length of a side of a triangle and relies on the definition given above of two waves being coherent.

Figure 2.19 shows two rows of atoms in a crystal lattice. Two parallel waves of X-rays, 1 and 2, strike the lattice at an angle of $\theta°$ to the horizontal. Wave 1 bounces off the upper atom, a, and number 2 bounces off the lower one, c. If the waves are coherent then at the line ab they must be in phase. If the waves are still in phase when beam 2 leaves the lattice then the distance b to c to d (denoted bcd) must be a complete number of wavelengths. Any fractional number would mean that they are no longer in phase and a spot would not appear on the photographic film. The variable n is given to the number of wavelengths and we write

$$\text{bcd} = n\lambda \tag{2.123}$$

Because the sum of the angles of a triangle is always equal to 90°, the angles cad and cab must also be equal to $\theta°$. We have two right-angled triangles with a common hypotenuse ad, whose length, d, is the distance between the planes of atoms. From the formal definition of the sine we may now write

$$\text{bc} = d \cdot \sin(\theta) \tag{2.124}$$

and

$$\text{cd} = d \cdot \sin(\theta) \tag{2.125}$$

Because the light is diffracted at the same angle as it strikes the plane the distances bc and cd must be the same, so

$$\text{bcd} = 2d \cdot \sin(\theta) \tag{2.126}$$

Finally, from equation (2.123) we can write

$$n\lambda = 2d \cdot \sin(\theta) \tag{2.127}$$

This is the Bragg equation. Presuming that we know the value of the wavelength, and can measure the angle of incidence which gives rise to constructive interference (indicated by a spot on a photographic film), we can calculate the distance between layers of atoms in the crystal.

The Bragg equation appears beautifully simple and using it in elementary problems requires no more knowledge than to be able to rearrange the equation and press the correct buttons on your calculator. This, unfortunately, belies the complexity of solving real problems. As shown in Figure 2.20 there are many more parallel planes in a lattice than the horizontal ones in Figure 2.19. Each set of planes will give rise to its own set of spots, and some will be more intense than others. The positions of the spots must be related to the actual diffraction angle and we must also be able to find the value of n, the

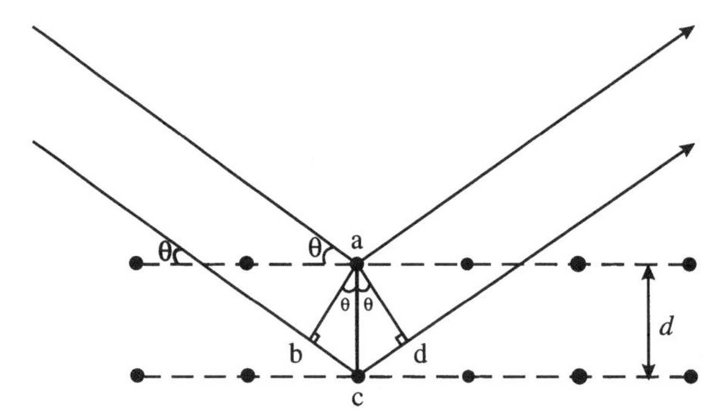

Figure 2.19 The two rows of atoms in a crystal lattice with two beams of X-rays being diffracted. See text for derivation of the Bragg equation

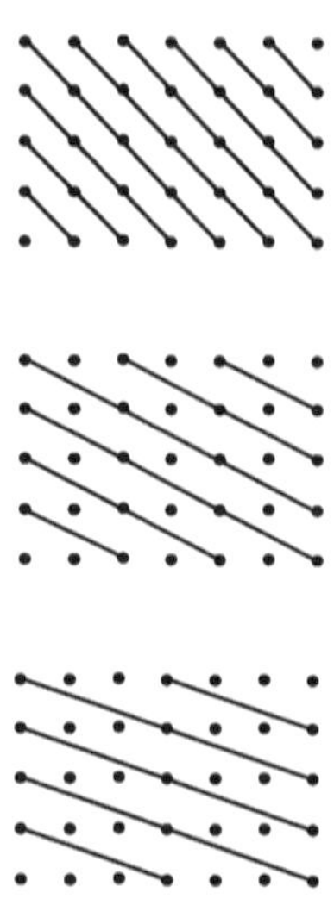

Figure 2.20 Three identical lattices showing that there are many different parallel planes even in one regular crystal

number of wavelengths, and relate the intensity to the position. Having said this, X-ray diffraction is a technique at the centre of modern chemical analysis and is invaluable as a tool to the research chemist.

2.5 Concluding Remarks

In this chapter we met three sets of functions. We called them special functions because they are over and above routine addition and subtraction; other books refer to them as *transcendental* functions. Whatever you wish to call them they are of immense importance to chemists and other scientists. They are defined using simple mathematics as described in Chapter 1, but as one begins to understand the notation they take on a language of their own and become powerful tools. It should be clear that logarithmic and exponential functions are different sides of the same coin and that many equations in chemistry are in essence the same, just written in a different form. In Chapter 8 we will show (using the mathematics of complex numbers) that there is also a relationship between the exponential function and the trigonometric functions. That these functions are all related is not significant just because it brings a kind of unification to the science in which they are used, it also enables us to change equations written in one form into another form which might be more easily solved.

Problems

2.1 Find the values of the following logarithms.

(a) $\log_2(2)$
(b) $\log_2(16)$
(c) $\log_{10}(10)$
(d) $\log_{10}(10\,000)$
(e) $\log_{10}(10^2)$
(f) $\log_{10}(10^{-34})$
(g) $\log_{10}(1)$
(h) $\log_x(1)$

2.2 Calculate the pH of the following solutions.

(a) 1 mol dm^{-3} HCl
(b) 1 mol dm^{-3} H_2SO_4
(c) pure water
(d) 0.1 mol dm^{-3} NaOH

2.3 How would you use a logarithm to calculate $\sqrt{(x)}$?

2.4 The rate of a reaction was estimated for a number of initial concentrations of reactant 'A'. Find the order of reaction with respect to A

initial concentration/ mol dm^{-3}	initial rate/units
0.02	0.000 27
0.025	0.000 432
0.03	0.000 634
0.045	0.001 485
0.07	0.003 756
0.12	0.011 649

2.5 Rearrange the following equation to give x as a function of t.

$$\ln\left(\frac{x}{x_0}\right) = kt$$

2.6 Use the series definition of e^x to find values for the following.

(a) e^x for $x = 0.1$
(b) $\exp(x^2)$ for $x = 0.1$
(c) $\exp(-x^2)$ for $x = 0.01$

2.7 The 'normal distribution' (Chapter 3) is defined as

$$y = \frac{1}{\sigma\sqrt{2\pi}}\exp\left[-\frac{(x-\mu)^2}{2\sigma^2}\right] \tag{P2.1}$$

Estimate the magnitude of y when

(a) $x \gg \mu$
(b) $x \ll \mu$
(c) $x = \mu$

2.8 If the populations of two energy levels are equal ($N_u = N_l$) what can you say about the energy difference between these levels?

2.9 For the following energy level differences calculate the proportion of molecules in the upper state.

(a) 10 000 kJ mol^{-1}
(b) 250 kJ mol^{-1}
(c) 25 kJ mol^{-1}
(d) 0.05 kJ mol^{-1}

2.10 Use the series definitions of sine and cosine functions to calculate the sine and cosine of the following angles.

(a) $\pi/2$ radians
(b) 1 radian
(c) 0.1 radian
(d) 0.05 radians
(e) 0.01 radian
(f) zero radians

2.11 Find expressions for the following functions.

(a) $\sin(2x)$
(b) $\cos(2x)$

2.12 Light of wavelength 154 pm (1 pm $= 1 \times 10^{-12}$ m) is diffracted at an angle of 36.3° from a crystal lattice. What is the separation of the planes of atoms giving rise to this diffraction if $n = 1$?

2.13 Find solutions to the equation

$$\psi = A\sin(kx) + B\cos(kx) \tag{P2.2}$$

when

(a) $\psi = 0$ and $x = 0$
(b) $\psi = 0$ and $x = a$
(c) both (a) and (b) apply

3 Practical Statistics

3.1 Introduction

The mathematics of statistics are used in chemistry as they are elsewhere, for the analysis and representation of experimental data. Importantly they are also used in theoretical aspects of the subject, statistical mechanics, anywhere that uses the Boltzmann equation and even in interpreting the solution to the Schrödinger equation for the hydrogen atom.

In this chapter we illustrate the methods used for summarising data. Finally we introduce the equations used for estimating the line of 'best fit' for pairs of experimental data.

In Chapter 7 we introduce the mathematics of permutations and show how this may be used to derive the all-important Boltzmann equation.

3.2 Statistics for Data Analysis

3.2.1 Summarising Data—Mathematical Approach

In order to define some of the terms used we will initially consider a collection or set of results. Table 3.1 lists the results obtained by seven students for the standardisation of a hydrochloric acid solution using 1.0 mol dm^{-3} sodium hydroxide.

The *range* of the data is the difference between the highest and lowest values. In this example this is

$$\text{Range} = 1.07 - 0.96 = 0.13 \tag{3.1}$$

The *mode* is the most common result. Since two students (numbers 2 and 3) obtained 0.99 mol dm^{-3} this is the mode in Table 3.1.

Table 3.1 Results of titration of HCl with NaOH

Student no.	1	2	3	4	5	6	7
[HCl]/mol dm^{-3}	0.96	0.99	0.99	1.02	1.05	1.06	1.07

If all of the results are listed in ascending order (as they are in this table) the *median* is the middle value. Out of the seven results the fourth is the middle value and this is 1.02 mol dm^{-3}.

The *mean* or *average* of a set of results is the sum of those results divided by their number. Therefore for the data in Table 3.1 the mean concentration is given by the equation

$$\text{Mean} = \frac{0.96 + 0.99 + 0.99 + 1.02 + 1.05 + 1.06 + 1.07}{7} = \frac{7.13}{7} = 1.02 \quad (3.2)$$

A general definition of the mean of any set of n values described by $x_1, x_2, x_3, x_4, \ldots, x_n$, is given by

$$\text{Mean} = \frac{x_1 + x_2 + x_3 + x_4 + \ldots + x_n}{n} \quad (3.3)$$

Conventionally the mean of a series of values of x is given the notation $\bar{x}$, and the sum is signified using capital sigma (Σ) with the limits of the sum indicated above and below,

$$\bar{x} = \frac{\sum_{i=1}^{n} x_i}{n} \quad (3.4)$$

The mean is an estimate of the midpoint of a set of data. In this example it is the same as the median (by definition the midpoint), but this is by no means always the case. It sometimes happens that the calculation of the mean may result in strange or unrealistic numbers. The often quoted average number of children per nuclear family, 2.4, is a good example of this. We know that children come in whole units and therefore this number is clearly absurd. It does, however, give us some idea about the size of a family, usually having two or three children.

The mean is the term most often used as a single representation of a whole data set. There is some justification of this because the calculation (equations (3.3) and (3.4)) involves every value in the data set. By contrast the mode only takes account of the most frequent result, the median relies directly on the number of results obtained and the range on the two extreme values.

To illustrate this let us suppose that an eighth student calculates that the concentration of the acid is 2.02 mol dm^{-3}.

The mode remains the same as 0.99 mol dm^{-3} is still the most common result. It may only change if the new value coincides with one of the original values.

The range increases to the difference between 2.02 and 0.96 mol dm^{-3}, that is 1.06 mol dm^{-3}. However, this does not tell us anything about the values in between. Had the new result fallen in between the current extreme values then the range would not change.

The median is now the middle of eight values. When we have an even number of results we take the mean of the two middle values. For eight, these are the fourth and fifth values.

$$\text{Median} = \frac{\text{4th value} + \text{5th value}}{2} = \frac{1.02 + 1.05}{2} = 1.035 \quad (3.5)$$

Note that we would have obtained the same value for any single additional result greater than or equal to 1.05 mol dm^{-3}.

The new mean may be calculated by adding the new value to the old total and dividing by eight.

$$\bar{x} = \frac{7.14 + 2.02}{8} = \frac{9.16}{8} = 1.15 \tag{3.6}$$

Note that the mean now lies outside the original range of the data. The relatively high value of the eighth student's result has had a large effect on the mean. Unless, however, we could prove that there was something wrong with this new result we cannot reject it just because it changes the apparent profile of the data. That is, we must still take the mean as representative of the whole set.

3.2.2 Summarising Data—Graphical Approach

Graphs and charts can be very useful when summarising large numbers of data. Table 3.2 lists values of the equilibrium constant for a chemical reaction measured by sixty students.

The equilibrium constants are listed in the order that they were recorded. As they stand they are merely a list of numbers and are not particularly informative. We may of course calculate the mean, median, mode and range of these data but we can also present them in a graphical manner.

To do this we may construct a diagram called a *histogram* or *bar chart*. To construct this we begin by rearranging the data into groups. Inspection of Table 3.2 shows that the range of the data is from 10.5 to 15.7. It might be convenient therefore to split the data up

Table 3.2 Results of calculation of equilibrium constant, K

10.5	11.3	11.5
11.4	12.3	12.5
12.3	12.9	12.8
13	13.3	13.5
13	13.7	13.2
13.1	13.9	13.3
14	13.9	13.5
14	14.7	13.2
15.5	14.5	14.8
13.1	15.6	14.9
11.1	11.2	11.7
12.3	12.4	12.8
12.7	13.3	12.7
13.3	13.5	13.1
13.2	13.7	13.3
13.2	13.2	13
13.5	13.8	13
14.2	14.8	13.5
14.7	15.7	14.7
15.3	13.2	14.9

into six equal-sized groups, ten to eleven, eleven to twelve, twelve to thirteen, and so on. We must avoid ambiguity and the possibility of a point falling into two sets. We therefore set definite boundaries which themselves do not overlap. In this example they are 10 to 10.9, 11 to 11.9, 12 to 12.9, 13 to 13.9, 14 to 14.9 and 15 to 15.9. We may now construct another table, Table 3.3, putting the results into these classifications.

On doing this we immediately see a pattern of results emerging. Most results lie in the 13–13.9 region, with fewer in the 12–12.9 and 14–14.9 regions, even less in the 11–11.9 and 15–15.9 regions and just one in the 10–10.9 region. It might be suggested that the pattern is symmetrical with zero in the upper region of 13–13.9, though this is not apparent from the raw data.

To construct the histogram we draw a graph in which the x axis is divided into regions comparable with those in which the data is grouped. Rectangles are then drawn with widths equal to the divisions of the classes and heights proportional to the number, or frequency, of results in that class (these are also called frequency diagrams). The histogram for the data in Table 3.3 is shown in Figure 3.1. The same data may be drawn as a bar chart, Figure 3.2. In this case the bars are drawn without pretence of continuity. The overall shape and impression, nevertheless, remains the same.

As the number of results obtained increases it is possible to use smaller and smaller divisions and the histogram takes on a more defined shape. As shown in Figure 3.3 it is possible to replace the histogram made up of blocks with a smooth curve. This curve is called a *normal* or *Gaussian* distribution curve.

The normal distribution of results is very typical. In this example, as in many others, the mean (13.34) and the median (13.2) fall in the bar in the middle or at the peak of the curve.

Table 3.3 Equilibrium constants, rearranged from Table 3.2

10–10.9	10.5					
	11.1	11.7				
	11.2					
11–11.9	11.3					
	11.4					
	11.5					
	12.3	12.7				
	12.3	12.7				
12–12.9	12.3	12.8				
	12.4	12.8				
	12.5	12.9				
	13	13.1	13.2	13.3	13.5	13.8
	13	13.1	13.2	13.3	13.5	13.9
13–13.9	13	13.1	13.2	13.3	13.5	13.9
	13	13.2	13.2	13.3	13.7	
	13	13.2	13.3	13.5	13.7	
	14	14.7	14.9			
	14	14.7				
14–14.9	14.2	14.8				
	14.5	14.8				
	14.7	14.9				
	15.3					
15–15.9	15.5					
	15.6					
	15.7					

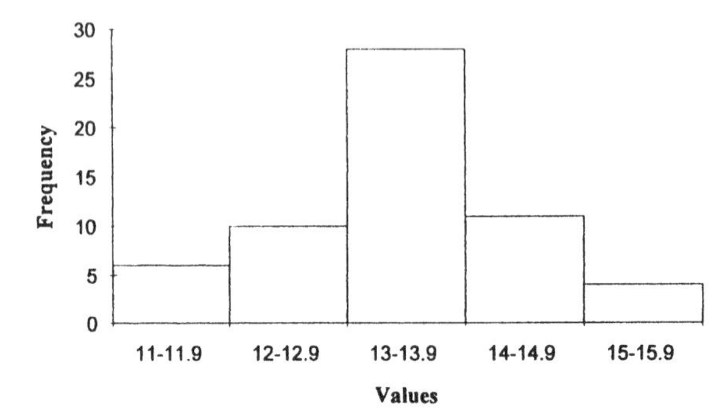

Figure 3.1 Histogram for the data shown in Table 3.3. Note the general shape of the chart which is very common in experimental analyses

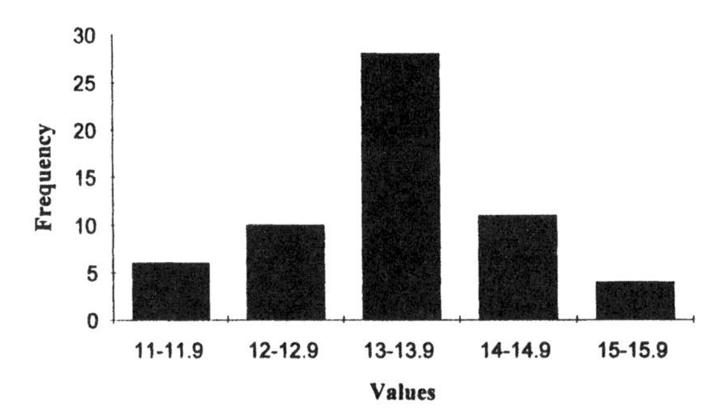

Figure 3.2 Bar chart for the data in Table 3.3. Here there is no pretence of continuity

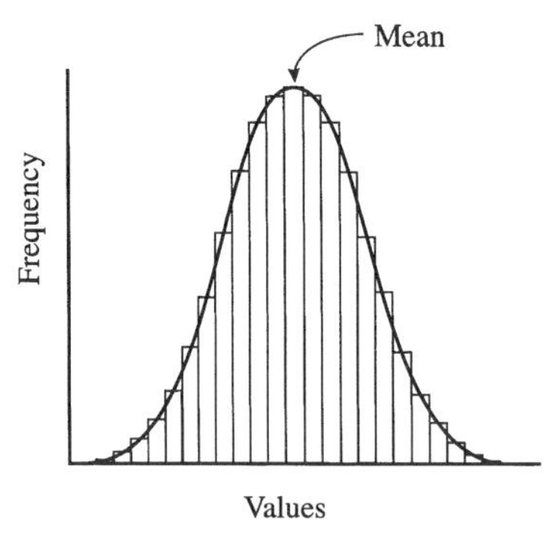

Figure 3.3 When there is a large amount of data smaller divisions may be used. In the limit the bars may be replaced by a smooth curve drawn through the midpoints of the tops of the bars. This curve is known as the normal or Gaussian distribution curve

This type of distribution occurs whenever we measure a variable for a large number of essentially identical objects for which the variation is caused by a number of factors, each exerting a small positive or negative effect. Thus we see normal distributions for such things as analyses of heights or weights of students or measurements of equilibrium constants.

In the ideal world analysts would all perform measurements in an identical fashion. However, each will perform their experiments slightly differently, using slightly different glassware and making slightly different judgements about calibrations and readings.

Mostly, these differences cancel each other out so that most results fall close to the mean. Some differences do accumulate so that there are results far from the mean but this occurrence is less likely and the frequency of these results decreases.

Occasionally an excess of results above or below the mean results in slightly different shapes of distributions, but the overall features are the same, with a large peak with fewer and fewer results on either side.

Such curves are not only useful as visual representations of data but may also be used as a predictive tool. The area under a single block of a histogram is proportional to the number of results it represents. The area under the whole histogram, or curve if enough results are obtained, is proportional to the total number of results.

The *probability* of something occurring is the ratio of the number of ways that this 'something' can occur to the total number of outcomes that may happen. For example, the probability of getting a head on the toss of a coin is 1/2 because there are two possible outcomes (head and tail) but only one way of getting a head. On the roll of a die there is a probability of 1/6 of getting a one because there are six ways the die may land but only one way of getting the one. If we were to specify the result as getting an even number the probability would be 1/2 because there are still six possible results, three of which are even numbers, i.e., three out of six, 3/6, which is equal to one half.*

The ratio of the area of one block to the total area therefore tells us the likelihood or probability that a result gained will fall in that region. If we use a continuous curve the area under a portion of a curve is proportional to the probability that the result will fall in that region. The shape of the normal curve means that the greatest area is concentrated under the peak. That is, the peak is the most likely result to occur.

3.2.3 Accuracy and Precision

Two terms used to describe the validity of a result are *accuracy* and *precision*. Accuracy is the closeness of a result to a true value. This true value may be measured using known standards or alternatively using a different and previously validated technique. It may never be possible to know the true value of anything and much of the discipline of statistics is directed at assessing how good an estimate of the true value a result might be.

Precision is the closeness of a number of results to each other. This is quite different from accuracy and precise results may well be inaccurate. The difference between accuracy and precision is often described using 'target' diagrams like those shown in Figure 3.4. A mathematical measurement of precision is the *standard deviation*. There are two types of standard deviation, the *sample* standard deviation and the *population* standard deviation.

Before defining these we must also define the terms *population* and *sample*. A population has more or less the same meaning as it has in common use. The population of a country is the total number of people inhabiting that country. A population of results is the total number of results it is possible to obtain. It is often not feasible to do the maximum number of experiments, just as in a survey it is not feasible to canvass the entire population of a country. The alternative is to do as many experiments as is reasonable and assume that this smaller sample is representative of the population.

Usually when we calculate the mean of a set of results we are calculating the sample

*Probability may also be written in decimal form, e.g., 1/2 being equal to 0.5 and so on.

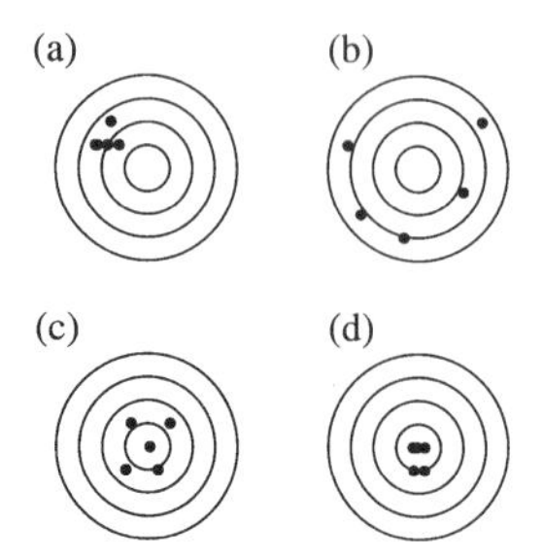

Figure 3.4 Target diagrams used to illustrate the concepts of accuracy and precision. (a) The shots are precise but inaccurate. (b) The shots are imprecise and inaccurate. (c) The shots are accurate but imprecise. (d) The shots are accurate and precise

mean. The true population mean may be different to this and it is given its own symbol, μ. As the size of the sample increases then the value of $\bar{x}$ should approach that of μ.

The definitions of sample and population standard deviations are:

$$s = \sqrt{\frac{\sum_{i=1}^{n} (x_i - \bar{x})^2}{n - 1}} \tag{3.7}$$

$$\sigma = \sqrt{\frac{\sum_{i=1}^{n} (x_i - \mu)^2}{n}} \tag{3.8}$$

In most cases equation (3.7) and the sample mean are used in calculations.

Analysis of these equations shows why they are called deviations and why they are measures of the precision of data. In the numerator inside the square root sign we have a summation of differences or deviations of each result from the mean, $\bar{x}$. We then take an average or mean of these deviations. If the mean deviation is small (compared with the value of the data itself) then the data is precise, if it is large it is imprecise.

This is represented in Figure 3.5 which shows an ideal normal distribution curve. The sample mean is coincident with the peak of the curve and the difference $(x - \bar{x})$ is clearly representative of the spread or precision of the data.

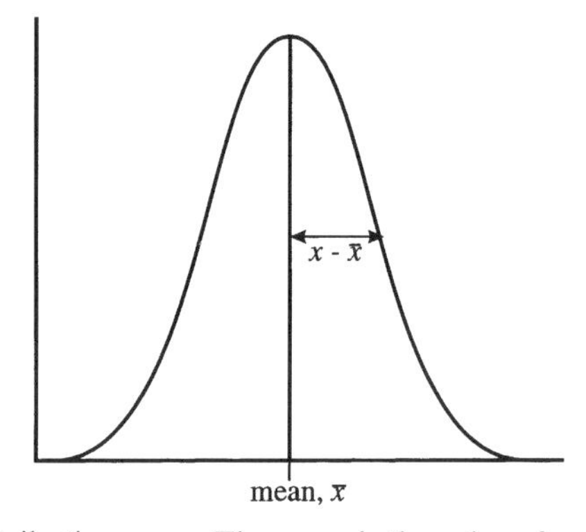

Figure 3.5 The normal distribution curve. The mean is the value of x at the peak of the curve

The reason for taking the squares of these deviations becomes apparent when we note that the sums of all deviations from the mean are equal to zero. That is,

$$\sum(x - \bar{x}) = 0 \tag{3.9}$$

We may test this by investigating the data in Table 3.1. Table 3.4 lists deviations from the mean and the squares of these deviations. The bottom row gives the sums of these deviations and the sum of the squares of the deviations.

Since their sum is zero, it is not possible to calculate a useful mean deviation. That is, the true mean deviation, zero, tells us nothing about the precision. We avoid this by stressing the magnitude of the deviation rather than its sign. This is achieved by taking the square so that they are all positive. As shown in Table 3.4 this has the effect of producing a real sum which can be used in calculating a mean value. The standard deviation is then calculated by taking the square root of the mean.

Using the results in Table 3.4 and equation (3.7) we find the standard deviation to be

$$\begin{aligned} s &= \sqrt{\frac{0.0104}{6}} \\ &= \pm 0.042 \end{aligned} \tag{3.10}$$

Since we are taking a square root the value may be positive or negative and this reflects the fact that deviations or errors from the mean value may be positive or negative. We would probably round the standard deviation to 0.04 as this reflects the number of decimal places in the raw data, and write the result as

$$[\mathrm{HCl}] = 1.02 \pm 0.04\,\mathrm{mol\,dm^{-3}} \tag{3.11}$$

The mean of the squares of deviations is called the *variance* (symbols s^2 and σ^2) and is itself a measure of the precision of a number of results. Scientists usually take the square root of the variance because it has the same units as the original data and is more easily related to the results obtained.

It will have been noted that the only difference in form of equations (3.7) and (3.8) is that (3.7) uses $(n - 1)$ in the denominator while (3.8) uses n (and is therefore a true mean of the deviations). The reason for this is that it is found that for small samples $(n - 1)$ gives a closer estimate to the true population standard deviation. Although this may be more rigorously proved we can use the data in Table 3.2 as an example.

Table 3.4 Calculation of deviations and squared deviations from the mean for data in Table 3.1

	$[\mathrm{HCl}]/\mathrm{mol\,dm^{-3}}$	$x - \bar{x}$	$(x - \bar{x})^2$
	0.96	−0.06	0.0036
	0.99	−0.03	0.0009
	0.99	−0.03	0.0009
	1.02	0.00	0.0000
	1.05	0.03	0.0009
	1.06	0.04	0.0016
	1.07	0.05	0.0025
Sum	7.14	0.00	0.0104

Example—difference between s and σ

Taking the 60 values of K as the population we calculate the population mean, μ, as equal to 13.34. The population standard deviation is calculated to be

$$\sigma = \pm 1.28 \tag{3.12}$$

The data in this table is set out as six subsets. If we calculate the sample standard deviation for each of these sets of ten results and calculate the mean we see that the mean sample standard deviation is equal to

$$s = \pm 1.14 \tag{3.13}$$

This is to be compared with a calculation of σ for each of these small sets. The mean value of σ is

$$\sigma = \pm 1.08 \tag{3.14}$$

Therefore we can see that the use of $(n - 1)$, rather than n results in a closer estimate of the true population standard deviation when using these small samples.

This term $(n - 1)$ is called *the number of degrees of freedom* of the data. This name arises as a consequence of equation (3.9). For any n data points we can calculate a mean and n deviations from this mean. However, we know the deviations add up to zero and therefore we only need to know $(n - 1)$ of the deviations in order to calculate the other one. Therefore, we only need to know $(n - 1)$ deviations in order to characterise the data set.

The use of s and σ will be described in detail in texts on statistics but for our use their importance may be explained as follows. For a population 68.3% (about two thirds) of all data will fall within one standard deviation ($\pm$) of the population mean, 95.5% of the data will fall within two standard deviations of the population mean, and 99.7% of the data will fall within three standard deviations of the population mean. Figure 3.6 shows two hypothetical frequency curves which only differ in the value of the standard deviation for the data. Curve (b) has twice the standard deviation of curve (a). It is evident that the data giving curve (a) is more precise than that for curve (b), the curve being much narrower.

3.2.4 Errors—Definitions

Errors are an inherent part in any chemical analysis and must affect subsequent calculations. This does not mean that a measurement is necessarily incorrect, but reflects that it is

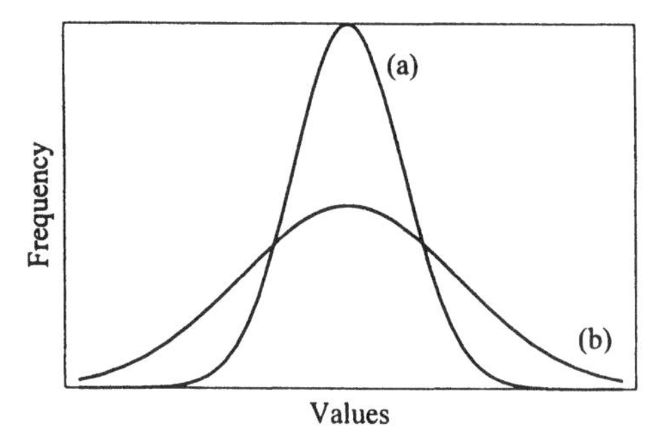

Figure 3.6 Normal distribution curves, (b) has twice the standard deviation of (a)

never really possible to know the true value of any measurement. Whenever two people perform an experiment, a simple titration for example, different results will occur. In doing a titration a number of steps are involved, weighings, making solutions up to the gradation marks correctly, reading of burettes, correct use of pipettes and judging the endpoint of the titration itself. In each of these steps no two people's judgements will be identical and combinations of these differences will result in an overall difference in the result. If no gross mistakes are made, which is the true result? The answer is that we do not know, but we must always be aware that such differences will occur. In any calculation these errors put some limit on the accuracy to which we may quote the final result.

Three types of error are defined. The first are called *random* or *indeterminate* errors. These are those described above such as error in judgement or in calibration of glassware. They are usually small but there may be many sources. Since they largely arise from human judgement they may be reduced by increasing one's skill. They are also reduced to some extent by the increased use of digital displays which eliminates the need for reading analogue scales. One should always, however, take note of manufacturers' specifications for all instrumentation as to how reliable a reading or measurement may be. As one has no real control over random errors and they may be positive or negative these errors affect the precision of a measurement rather than the accuracy.

The second kind of error is called *systematic* or *determinate* error. These can usually be traced to some specific source, either instrument fault or as a result of the method used. Other possible sources of these errors are personal error and prejudice. For example, a person who does not know how to read a vernier scale properly or uses an instrument incorrectly will make consistently wrong readings. Alternatively students often have an idea about what result they are expecting. This idea may be imaginary but in attempting to get the 'right answer' they will try and compensate by judging that a particular reading is slightly higher or lower than it really is.

Systematic errors occur in one direction only (high or low) and affect the accuracy of a result. The detection of systematic errors can be achieved by the use of standard samples, blank samples (with no analyte present) or by use of an alternative method or a different instrument.

The third kind is one nobody likes to admit to but is one which does occur. This is *gross personal* error, simply making a mistake in preparation or performing an experiment. These may be detected by repeated measurements or by comparison with known results. Such errors may not always be owing to incompetence but may also occur as a result of instrument malfunction or temporary power failure. They may possibly be detected by repeated measurement but sometimes only by comparison with an independent analysis.

3.2.5 Presentation of Errors

It is often impractical for students to make repeated measurements and so perform statistical analyses such as calculating means and standard deviations. In doing laboratory experiments as part of a course there may only be time to make each measurement once. Even so, chemists must always be aware of the reliability of their measurements. The accuracy or precision of these should be reflected in the final result which is presented.

Analysis of probable error in each measurement requires some personal judgement and a reliance on manufacturers' specifications for instrumental measurements. A typical case where personal judgement is required is the reading of a burette. A 25 cm^3 burette will have gradations every 0.1 cm^3. It is easy, though, to see if the volume lies between these gradation marks. With practice most people would also be able to judge the level to within 0.02 cm^3, though because there is no mark to this accuracy there is some uncertainty in this final figure. The reading would therefore be given as, for example,

$$\text{Volume} = 15.22 \pm 0.02\,\text{cm}^3 \tag{3.15}$$

This implies that the true reading lies somewhere between 15.20 and 15.24 cm^3. However, volumes measured by burette involve two readings, so how does this uncertainty carry through into the combined reading?

The answer to this lies in the mathematics of error propagation. Every time there is some uncertainty, the error will be carried through all subsequent calculations. If there are a number of uncertainties they will all affect the final, calculated result. We must therefore have a system for passing on these uncertainties through calculations.

Below we list the relevant equations used for this purpose. They are normally used for standard deviations but may also be used for other estimates of errors. For any analysis there may be several sources of errors. There is a rule of thumb to limit the number of errors which have to be accounted for. This is that although there may be many possible sources of errors, one or two will be much larger than the rest and we only need to take account of these. The smaller ones may be regarded as insignificant.

To judge which are important the chemical analysis must be broken down into steps and the accuracy of each step estimated. The insignificant errors are rejected and significant errors noted. The calculation is performed normally with the nominally correct number and then the error is calculated as detailed below according to the type of calculation.

Example—addition and subtraction

The volume calculated from two burette readings results from a subtraction. The error, $\pm y$, involved in a number of additions or subtractions for a function such as $Y = A + B - C$ where errors in A, B and C are a, b and c respectively is

$$y = \sqrt{a^2 + b^2 + c^2} \tag{3.16}$$

Therefore if the two readings are 21.04 cm^3 and 12.24 cm^3 the difference is 8.80 cm^3. If the two readings have an estimated error of 0.02 cm^3 then the overall error is

$$y = \sqrt{0.02^2 + 0.02^2} \tag{3.17}$$

which computes to $\pm 0.03\,\text{cm}^3$.

Example—multiplication and division

In calculating the concentration of a solution which we have made by weighing a sample and making the solution up to a known volume the calculation involves multiplication and division.

$$\text{Concentration, } C = \frac{\text{mass}}{\text{molar mass} \cdot \text{volume}} \tag{3.18}$$

For a function of the type $Y = A/B \cdot C$ where the uncertainties in the variables are y, a, b and c respectively we use the equation

$$\frac{y}{Y} = \sqrt{\left(\frac{a}{A}\right)^2 + \left(\frac{b}{B}\right)^2 + \left(\frac{c}{C}\right)^2} \tag{3.19}$$

where the ratio y/Y is the relative uncertainty or error in Y. In our example of calculation of a concentration suppose 5 g of potassium chloride is weighed with an error of ± 0.01 g. The molar mass of KCl is known to be 74.55 g and the solution is made up to 100 cm^3 with an error of ± 0.2 cm^3. Since we use volume in litres this is 0.1 ± 0.0002 dm^3. We can therefore calculate the relative error in C as*

$$\begin{aligned}\frac{c}{C} &= \sqrt{\left(\frac{0.01}{5.00}\right)^2 + \left(\frac{0.01}{74.55}\right)^2 + \left(\frac{0.002}{0.1}\right)^2} \\ &= \sqrt{4.00 \times 10^{-6} + 1.80 \times 10^{-8} + 4.00 \times 10^{-6}} \\ &= \sqrt{8.10 \times 10^{-6}} \\ &= \pm 2.83 \times 10^{-3}\end{aligned} \tag{3.20}$$

We can now calculate the error, c, by multiplying by C, the calculated concentration. This is 0.671 mol dm^{-3} and so c is given by

$$c = 0.671 \times 2.83 \times 10^{-3} = 1.90 \times 10^{-3} \tag{3.21}$$

We could now write the result as 0.671 (± 0.0019) mol dm^{-3} but this is over-stressing the accuracy to which we know our error. We would, therefore, probably round the 0.0019 up to 0.002 and the result is now more in keeping with the accuracy to which we have given the result, i.e., $C = 0.671$ (± 0.002) mol dm^{-3}.

Example—exponential functions

We now consider equations of the form $Y = A^n$ where there is assumed to be no error in n. The relative error in Y is given by

$$\frac{y}{Y} = n\frac{a}{A} \tag{3.22}$$

Kohlrausch's law relates the molar conductivity of a solution, Λ_m, to the square root of the concentration C by the equation

$$\Lambda_m = \Lambda_m^0 + KC^{1/2} \tag{3.23}$$

where Λ_m^0 is the limiting molar conductivity and K is a constant. If Λ_m^0 is equal to 149.80 S cm^2 mol^{-1} and K is 94.50 S cm^2 mol^{-1} M$^{-1/2}$ (where M represents units of concentration) for KCl and both of these have zero error associated, then the error in Λ_m results solely from the error in C. For a solution of concentration 0.015 (± 0.0005) mol dm^{-3} the uncertainty is

*Note that the relative error in molar mass is much smaller than the other errors. If we omitted this error from the calculation the result would be 2.828×10^{-3}, very close indeed to the result taking in every error. It is therefore possible to ignore small errors as stated in the text

$$\frac{\lambda}{\Lambda} = \frac{1}{2} \cdot \frac{0.0005}{0.015} = 0.017 \quad (3.24)$$

The calculation using equation (3.22) gives Λ_m as 161.37 and the error is consequently ± 2.7. We would probably round the error and the result to 161 (± 3).

Example—logarithms

For an equation of the type

$$Y = \log_{10} A \quad (3.25)$$

the uncertainty y is given by

$$y = 0.434 \frac{a}{A} \quad (3.26)$$

noting that 0.434 is $\log_{10} e$.

In calculating the pH of this solution we take the logarithm of (the numerical value of) the concentration. If the concentration of a hydrochloric acid solution is known to be 0.1 (± 0.02) mol dm^{-3} then the pH would be calculated as $-\log_{10}(0.1) = 1$, plus or minus the error given by

$$\text{Error} = 0.434 \frac{0.02}{0.1} = 0.0868 \quad (3.27)$$

Again we are over-stressing the accuracy of the uncertainty compared with that of the actual result and we would round this error up to ± 0.1, i.e., the pH is calculated as 1 (± 0.1).

Example—antilogarithms

In base 10 this amounts to raising 10 to the power of the number of interest. If we measured the pH of a solution using a pH meter we could calculate the concentration of protons by

$$[H^+] = 10^{-\text{pH}} \quad (3.28)$$

The formula for calculating the error is

$$\frac{y}{Y} = 2.303 \cdot a \quad (3.29)$$

The pH of a solution was measured as 2.65 (± 0.05). The concentration of protons is therefore given by $10^{-2.65}$ which is equal to 0.002 24 mol dm^{-3}. The uncertainty is

$$\frac{y}{0.00224} = 2.303 \times (\pm 0.05) = \pm 0.115 \quad (3.30)$$

or

$$y = 0.00224 \times (\pm 0.115) = \pm 0.000257 \quad (3.31)$$

We would write a result such as this in scientific or floating point notation, i.e., $2.2(\pm 0.3) \times 10^{-3}\,\text{mol dm}^{-3}$.

3.2.6 Significant Figure Notation

A relatively simple way of indicating the accuracy of a measurement is to use the significant figure notation. A measurement is presented by those figures which are known to be accurate plus one more in which some uncertainty is inferred.

Figure 3.7 shows part of the analogue scale of a pH meter. The needle lies between the markings for pH 3.3 and pH 3.4. It is apparent that the needle is just closer to pH 3.3 than 3.4, about 3.34. There is some uncertainty in this last figure and a number of different people might say 3.35, 3.36 or perhaps as low as 3.32. We could, therefore, estimate that the reading was 3.34 ± 0.02.

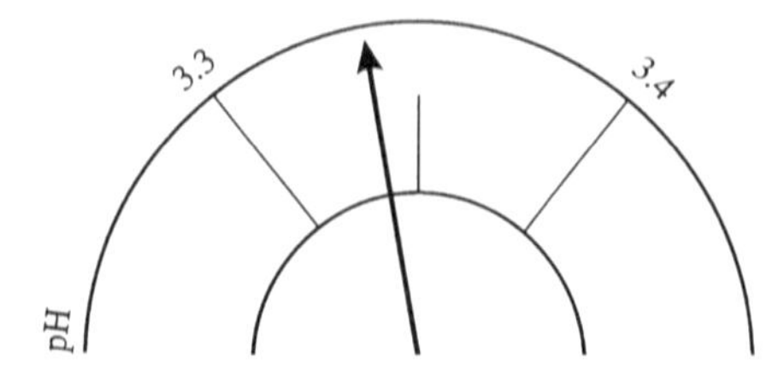

Figure 3.7 Schematic diagram of a pH meter scale. There will be some uncertainty in estimating the value of pH between the scale markings

In many cases data given to a number of significant figures does not carry an estimate of error. In these cases it may be assumed that the error is ± 1 in the final figure. There is certainly ambiguity in significant figure notation and some texts will state that the error in the final figure is ± 0.5 rather than ± 1. Whichever you choose you should avoid ambiguity by stating any relevant definitions when addressing these issues.

When doing calculations on numbers in significant figure notation it is important that the final result is also given to the correct number of significant figures. This is most evident when performing additions or subtractions.

Example—addition and subtraction

We may perform the following calculation,

$$y = 0.234\,3 + 23.262 = 23.496\,3 \tag{3.32}$$

Since there is an uncertainty in 23.262 of ± 0.001, this must carry through into the result. The final figure in 23.496 3 is overridden by this doubt. We could only give the result as 23.496.

When performing multiplications and divisions it is the relative error which must be accounted for.

Example—multiplication and division

We may perform the following calculation using a calculator,

$$y = \frac{1.61 \times 2.434}{0.234\,56} = 16.706\,77 \tag{3.33}$$

The result produced has seven figures, which is in excess of the five significant figures implied by the most accurately defined number, 0.234 56. It is a rule that we can only give results to an accuracy defined by the least accurate of the numbers in the computation. To specify the accuracy in calculations such as these we compare the relative errors for each of the numbers involved. The error in 1.61 is ± 0.01 and so the relative error is

$$\frac{0.01}{1.51} = \frac{1}{161} = 0.006\,21 \tag{3.34}$$

The error in 2.434 is ± 0.001 and the relative error is

$$\frac{0.001}{2.434} = \frac{1}{2434} = 0.000\,41 \tag{3.35}$$

The error in 0.234 56 is $\pm 0.000\,01$ and the relative error is

$$\frac{0.000\,01}{0.234\,56} = \frac{1}{23\,456} = 0.000\,042\,63 \tag{3.36}$$

The error in 1.61 is by far the largest and so it is this which will have most influence in the answer. Remember that the number 0.006 21 is a relative error and we therefore calculate the actual error in the result by multiplication of the result, i.e.,

$$\text{Actual error} = 0.006\,21 \times 16.706\,77 = 0.104 \tag{3.37}$$

That is, there is an error of approximately one in the first decimal place. Therefore, we could only state the result of this calculation as 16.7 ± 0.1 with any confidence.

When doing calculations with logarithms the significant figures are reflected in the number of figures kept after the decimal point. The number on the left of the decimal point in a logarithm (called the *mantissa*) tells us the overall magnitude of the number. For example, in base ten $\log_{10}151$ is equal to 2.179 to three significant figures. The '2' tells us that the magnitude of the number is of the order of 10^2 and nothing more. The logarithm of 15.1 is 1.179, exactly the same after the decimal point but the '1' tells us that the number is of the order of 10^1.

Example—calculating a logarithm

Using a calculator to find ln(5.26), a number given to three significant figures, the result is

$$\ln(5.26) = 1.660\,131 \tag{3.38}$$

Only the numbers after the decimal point are important in representing the accuracy of the number. We therefore give this result as 1.660.

We must also account for this when finding the antilogarithm. The antilogarithm (base 10) of 1.345 is $10^{1.345}$ which is 22.1. There are three figures after the decimal place and therefore only three in the final result.

Example—calculating an antilogarithm

We wish to find the antilogarithm (base e) of 2.3456

$$\text{Antilogarithm (base e) } (2.3456) = 10.439\,535 \qquad (3.39)$$

However, there are only four digits to the right of the decimal point in the logarithm itself and therefore only four significant figures in the result. The result is therefore quoted as 10.44.

There is often confusion about whether the number zero is a significant figure or not. This depends on where the zero appears in the number in question. The rules are:

(1) All zeros between other numbers, e.g., as in 10.44 given above, are significant.
(2) Zeros to the left of the first digit are not significant. For example, both 0.435 and 0.000 654 are three significant figures. We would not typically write 00.461 and so this is also excluded.
(3) Zeros after a number of digits after the decimal point are significant. For example, 0.4560 is a four significant figure number, as it implies that the uncertainty is between 0.4559 and 0.4561. If it were to three figures this would imply an uncertainty of ± 0.001 rather than ± 0001.
(4) Zeros representing tens, hundreds and thousands may or may not be significant. In such cases it is usual to specify the accuracy. For example, 100 may be a one or a three figure number. If it were a one figure number the uncertainty would be between zero and 200 but if it were a three figure number it would be between 99 and 101. Ambiguity may be avoided by using scientific (floating point) notation. 1.00×10^2 is by convention a number given to three significant figures, as in rule (3) above.

In deciding how many significant figures should be used in presenting a result the truncating or rounding of numbers is necessary. One must not fall into the trap of believing that the numbers churned out by a calculator or computer represent the true number of figures to be employed. Unless there is an exact result of less than (say) eight digits a calculator will always give eight-figure numbers.

The common rule is that a number whose digit to be lost is less than five, is truncated by loss of this last digit. Thus 5.64 to two significant figures is truncated to 5.6. If the digit to be lost is five or greater then the number is rounded up, so 5.65 and 5.66 both become 5.7. Some sources suggest that the number five should be rounded towards the nearest even number. This prevents the tendency of always making numbers bigger by rounding. In this case the number 5.65 becomes 5.6 because 6 is the nearest even number. On the other hand, 4.35 becomes 4.4 because this is towards 4, the nearest even number.

Although it is necessary to reflect limits of accuracy in results given it is a mistake to round figures intermediate in calculations. This leads to errors in results and so when using a calculator retain all the digits that the calculator gives you in all stages up to the final result. Only then can you safely round the number to the correct level of accuracy.

3.2.7 The Line of Best Fit

In practical physical chemistry, as compared to analytical chemistry, repeated measurements allowing statistical analysis are not very often made. Experiments involving chemical changes entail repeated measurement but the values are always changing. For example, in a kinetic experiment the concentration of a reactant is repeatedly measured but it changes as a function of time. Although it would be ideal to repeat the experiment several times to obtain a good statistical analysis this is not often permitted in available laboratory time.

Because of the emphasis on measuring quantities which are changing, the significance of mean values and standard deviations is often lost. The question arises, however, can we apply any such statistical analysis to obtain a true value of whatever it is we are trying to calculate? Returning to the example of the kinetic experiment the slope of a plot relating concentration to time provides a value of the rate constant. It is useful, therefore, to be able to estimate the slope of a straight line plot, even though some of the data may exhibit some error.

In Chapter 1 we demonstrated how this can be achieved by estimating the 'best' straight line through the data. We now show how a better estimate of this may be achieved through calculation. The technique is called 'the method of least squares' and is one of a number of techniques called *regression*. These are methods used to find mathematical relationships between data which are generated in some randomised manner. Experimental data comes under this definition because although ideally related to each other (concentration to time etc.) measurements of variables are subject to random error.

In linear regression we assume that there is a linear relationship between the variables, that they are related by an equation of the kind $y = mx + c$. The aim is to find the best or true values of the constants m and c. These may of course be interesting parameters such as the rate constant or equilibrium constant.

For variables which are related by the equation

$$y = mx + c \tag{3.40}$$

where m is the gradient or slope and c is the value of y when x is zero (the y-axis intercept) we may make n measurements of x and y. There will be some error in the measurements and the actual data points will not necessarily fall on a perfect straight line. The method of least squares estimates the equation of a straight line, passing through the data, for which the sums of the squares of the deviations of these points from the line are a minimum. This line is considered the 'best fit' for the data obtained.

With the increased use of microcomputers in laboratories and the increase in availability of dedicated data analysis software it is now possible to perform such calculations without recourse to the mathematics on which they are based. For information, however, we now present the equations used for calculating the slope, m, and the y-axis intercept, c, for this line of best fit.

Initially we must calculate various sums:

$$\text{Sum of } x \text{ values} = \sum x \tag{3.41}$$

$$\text{Sum of values of } x^2 = \sum x^2 \tag{3.42}$$

$$\text{Sum of } y \text{ values} = \sum y \tag{3.43}$$

$$\text{Sum of values of product } xy = \sum xy \tag{3.44}$$

The slope and the intercept for n pairs of x and y coordinates are given by the following equations.

$$m = \frac{\sum x \sum y - n \sum xy}{(\sum x)^2 - n \sum x^2} \tag{3.45}$$

$$c = \frac{\sum x \sum xy}{(\sum x)^2 - n \sum x^2} \tag{3.46}$$

Example—test of Beer–Lambert law

The Beer–Lambert law relates the absorbance A of a solution to the concentration, c, of a species in solution by the equation

$$A = \varepsilon c l \tag{3.47}$$

where l is the path length in centimetres and ε is the molar absorption coefficient. A plot of A versus c should therefore be linear with a slope from which we can calculate ε. Table 3.5 shows measurements of absorbance for a number of concentrations of potassium manganate (VII) ions using a 1 cm cell (i.e., $l = 1$ cm).

The concentrations correspond to values of x and the absorbances correspond to values of y in equations (3.45) and (3.46). The various summations are calculated in Table 3.6.

Therefore the summations (omitting units) to be used in equations (3.45) and (3.46) are

$$\sum x = 5.00 \times 10^{-4} \tag{3.48}$$

$$(\sum x)^2 = 2.50 \times 10^{-7} \tag{3.49}$$

$$\sum x^2 = 5.625 \times 10^{-8} \tag{3.50}$$

$$\sum y = 5.01 \tag{3.51}$$

$$\sum xy = 5.6425 \times 10^{-4} \tag{3.52}$$

Thus the calculation of the slope of the plot of A versus concentration (Figure 3.8) is

Table 3.5 Measurements of absorbance at 522 nm for aqueous potassium manganate (VII) solutions

$10^4[KMnO_4]/mol\,dm^{-3}$	0.50	0.75	1.00	1.25	1.50
Absorbance, A	0.48	0.79	0.95	1.30	1.49

Table 3.6 x and y values and combinations for least squares analysis from Table 3.5

	$10^4 \times x$	y	$10^8 \times x^2$	$10^5 \times xy$
	0.50	0.48	0.25	2.4
	0.75	0.79	0.5625	5.925
	1.00	0.95	1.00	9.5
	1.25	1.3	1.5625	16.25
	1.50	1.49	2.25	22.35
Sums	5.00	5.01	5.625	56.425

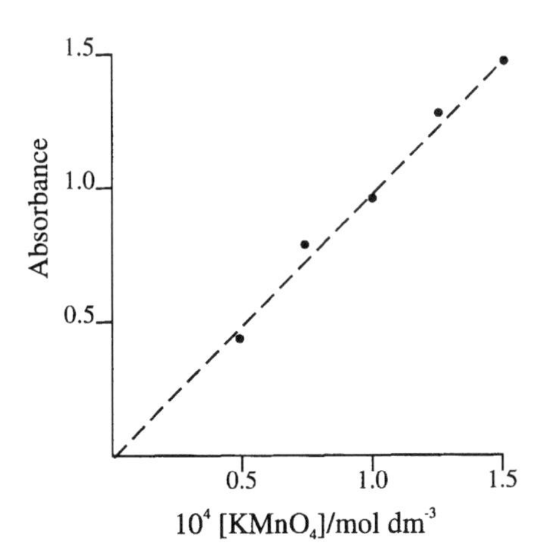

Figure 3.8 Plot of data in Table 3.5

$$m = \frac{(5.00 \times 10^{-4} \times 5.01) - (5 \times 5.6425 \times 10^{-4})}{2.5 \times 10^{-7} - (5 \times 5.625 \times 10^{-8})} = 10\,120 \tag{3.53}$$

$$c = \frac{(5.00 \times 10^{-4} \times 5.6425 \times 10^{-4}) - (5.625 \times 5.01)}{2.5 \times 10^{-7} - (5 \times 5.625 \times 10^{-8})} = -0.01 \tag{3.54}$$

The slope has the units of 1/(concentration). ε may be calculated by dividing this number by the path length, 1 cm. ε has units of concentration^{-1} cm^{-1} or dm^3 mol^{-1} cm^{-1} and has a value of

$$\varepsilon = 10\,120\,\mathrm{dm^3\,mol^{-1}\,cm^{-1}} \tag{3.55}$$

The value of c (the intercept) has units of absorbance. In the example this value is not particularly useful except that we would expect that it would be near to zero.

3.3 In Conclusion

As shown above we may apply statistical ideas to areas of analytical chemistry and to measurements in physical chemistry. Care must always be taken in interpreting results from statistical analyses. For instance, in the example showing the effect of extra results on the mean, median and other parameters, we used the additional value of 2.02 mol dm^{-3} which was quite different from the other values in the sample. When we need to make a judgement concerning the validity of results, one way of doing so is to draw a scatter diagram (Figure 3.9). This immediately shows the difference between the initial seven results and the late addition.

Results such as this, which do not fit an apparent pattern, are called *outliers*. It is tempting to discard them so as to 'improve' estimates of mean or true values and standard deviations. Such spurious results may also occur in laboratory experiments (rather than analyses) and have knock-on effects in calculating the slope of a straight line. However,

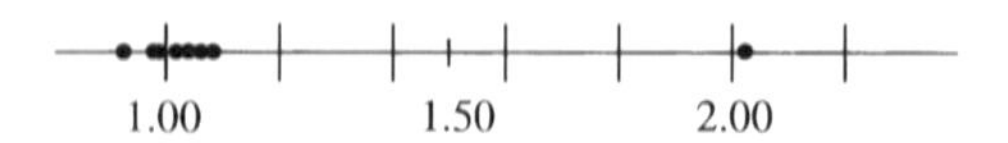

Figure 3.9 A linear scatter diagram of the data in Table 3.1 plus an extra result. This clearly shows that the extra value is substantially different from the bulk of the data

we are not strictly allowed to be selective in which results we use to perform our statistical analyses. We may only reject results if we have a good reason for doing so. For example, if it was found that student 8 had been rather haphazard in making up his standard solution then we might have just cause for rejecting his results.

Finally it should also be noted that we can use the least squares analysis in Section 3.2.7 for non-linear functions. To do this we make them into linear ones. For example, if we expect the variables to be related by the equation

$$y = mx^2 + c \tag{3.56}$$

then we can make the substitution $z = x^2$ and plot

$$y = mz + c \tag{3.57}$$

In general terms for an equation

$$y = m \cdot \mathrm{f}(x) + c \tag{3.58}$$

where $\mathrm{f}(x)$ is a function of x we make the substitution $z = \mathrm{f}(x)$ and plot it as a linear function.

Although there are methods for finding non-linear functions the nature of the linear function is easier to recognise, in that we can all identify a straight line. It is often, therefore, a useful step to linearise a function as shown above.

In closing this chapter we reiterate that whatever methods of mathematical analysis are used the chemist must be aware that the accuracy and precision of measurements are always limited to some extent and this will in turn limit the accuracy or precision of ensuing calculations. This limit should always be reflected in the result which is given.

Problems

3.1 Plot the following points on a graph and draw a smooth curve through the points.

(0, 0.05), (1, 0.15), (2, 0.38), (3, 0.87), (4, 1.8), (5, 3.3), (6, 5.5), (7, 8.1), (8, 10.7), (9, 12.6), (13, 8.1), (14, 5.5), (15, 3.3), (16, 1.8), (17, 0.9), (19, 0.2), (20, 0.1).

(a) Estimate the area under the curve (count the squares on your graph paper).
(b) Draw vertical lines through from the x axis to the curve at the points $x = 5$ and $x = 15$. Estimate the area under the curve between these lines and under the curve. What is the proportion of the total area in this region?

3.2 Find the range, mode, median and standard deviation of the following numbers:

17.7, 27.8, 26.5, 22.4, 23.8, 27.2, 25.2, 34.4, 25.1, 20.2, 29.6, 25.1, 32.3, 21.6, 18.6, 25.0, 24.8, 20.3, 28.5, 25.0, 25.1, 27.5, 26.7, 24.4.

3.3 The following values of a rate constant (units s^{-1}) were measured by a group of students:

12.47	12.54	12.53	12.25	13.09	13.08	12.76
12.56	12.57	12.60	12.61	13.14	13.04	13.04
12.65	12.65	12.65	12.74	12.74	12.69	12.69
12.75	12.75	12.75	12.81	12.82	12.84	12.80
12.94	12.94	12.93	12.92	12.85	12.87	12.88
12.95	12.96	12.96	12.97	12.97	12.98	12.99
13.14	13.14	13.13	13.13	13.14	13.13	13.06
13.17	13.17	13.19	13.18	12.86	21.90	12.76
13.30	13.30	13.28	13.29	13.06	13.09	12.90
13.35	13.35	13.40	13.00	13.01	13.02	13.05
13.50	13.50	13.50	13.20	13.20	13.24	13.22
13.58	13.58	13.50	13.40	12.40	12.40	12.80
13.74	13.31	13.34	13.33	12.87	12.71	13.18
12.86	12.55	13.0				

(a) What is the range of the data?
(b) Arrange the data into sub-ranges and draw a bar chart or a histogram to display the data.
(c) Calculate the sample mean of the data.
(d) Calculate the sample standard deviation for the data.

3.4 The following values of an activation energy (units kJ mol^{-1}) were recorded.

105.9	113.9	106.5	108.2	109.9	112.8	111.6	113.0
106.7	113.30	107.1	108.8	110.1	113.9	109.3	113.4
107.5	111.26	107.5	109.8	111.6	114.7	111.0	114.7
108.3	111.6	108.3	110.2	111.9	114.7	110.5	115.4
109.9	108.2	109.9	111.6	113.0	113.3	112.2	116.7
110.0	108.9	110.1	112.0	113.4	110.5	105.4	109.3
111.6	109.2	111.60	112.9	114.7	104.1	109.3	106.5
111.9	110.2	112.1	110.5	115.4	107.1	111.1	106.8
112.2	113.1	106.6	107.8	108.3	110.8	107.5	107.5

(a) What is the range of the data?
(b) Arrange the data into sub-ranges and draw a bar chart or a histogram to display the data.
(c) Calculate the sample mean of the data.
(d) Calculate the sample standard deviation for the data.

3.5 The normal distribution is described by the equation

$$y = \frac{1}{\sigma\sqrt{2\pi}} \exp\left[-\frac{(x-\mu)^2}{2\sigma^2} \right] \tag{P3.1}$$

where σ is the standard deviation and μ is the population mean. Plot this function for a suitable range of x for

(a) $\mu = 5$ and $\sigma = 3$
(b) $\mu = 10$ and $\sigma = 3$
(c) $\mu = 15$ and $\sigma = 3$
(d) $\mu = 5$ and $\sigma = 5$
(e) $\mu = 5$ and $\sigma = 7$

3.6 Write the following numbers to three significant figures.

(a) 1.2345
(b) 12.345
(c) 123.54
(d) 0.001 234 5
(e) 01.102 35
(f) 101.2
(g) 10.308
(h) 10.38
(i) 1×10^3
(j) -0.0608

3.7 Perform the following calculations and write the result with the cumulative error.

(a) $2(\pm 0.2) + 5(\pm 0.2)$
(b) $10(\pm 1) + 15(\pm 0.2)$
(c) $109(\pm 0.2) + 1(\pm 0.1) - 0.5(\pm 0.001)$
(d) $\sqrt{\dfrac{0.6 \pm 0.2}{10 \pm 0.1}}$
(e) $(10(\pm 0.2)) \times (12(\pm 0.2))/6(\pm 0.02)$
(f) $\log_{10}(21(\pm 2))$
(g) $(7.8 \pm 0.3)^{(2.1)}$
(h) $\log_{10}\left(\dfrac{0.1 \pm 0.001}{0.25 \pm 0.01}\right)$

3.8 Find the slope and the y axis intercept for the following data.

(a) (2.0, 7.3), (6.0, 18.5), (10.0, 31.0), (15.0, 47.0), (30.0, 91.0), (70.0, 200), (101, 306)
(b) (−5.5, −15.0), (2.9, 13.9), (11.3, 42.1), (19.7, 72.5), (28.1, 99.8), (36.5, 131.5), (44.9, 155.6), (53.3, 193.0)
(c) (2, 11.3), (5, 280), (8, 1451), (11, 4400), (14, 10 300), (17, 20 200), (20, 35 800), (23, 58 350), (26, 89 600), (29, 131 400)

4 Differential Calculus

4.1 Differentiation

Differentiation is the process of finding out how a function changes. In Chapter 1 we saw how to calculate the slope of a straight line described by the equation

$$y = mx + c \tag{4.1}$$

The constant m was defined as the slope which could be calculated from the coordinates of two points, (x_1, y_1) and (x_2, y_2) using

$$m = \frac{y_2 - y_1}{x_2 - x_1} \tag{4.2}$$

The slope is defined as the change in y per unit change in x. Equation (4.2) can be written in more general terms by introduction of the symbol Δ to represent the change. When we do this $y_2 - y_1$ becomes Δy which reads as 'delta y'. Similarly $x_2 - x_1$ becomes Δx (delta x) and equation (4.2) can be written as

$$m = \frac{\Delta y}{\Delta x} \tag{4.3}$$

This reads as 'slope equals delta y divided by delta x', the ratio of the change in y with respect to the change in x. It is a rate of change.

Problems arise if we try to apply this method to a function which does not represent a straight line. Equation (4.1) does not apply and therefore we cannot define m in such a way. The nearest we can get is to draw a tangent to the curve (Chapter 1) and estimate this slope using these equations. This method is at best haphazard. Ten people will all give different answers, because everybody's judgement is different. Fortunately there is a much more elegant way of solving this problem. This is the method of differential calculus, first invented by Newton and (independently) Leibniz in the 17th Century.

4.1.1 The Slope of the Function $y = x^2$

Figure 4.1 shows a portion of the curve $y = x^2$. Suppose we wish to find the slope of the

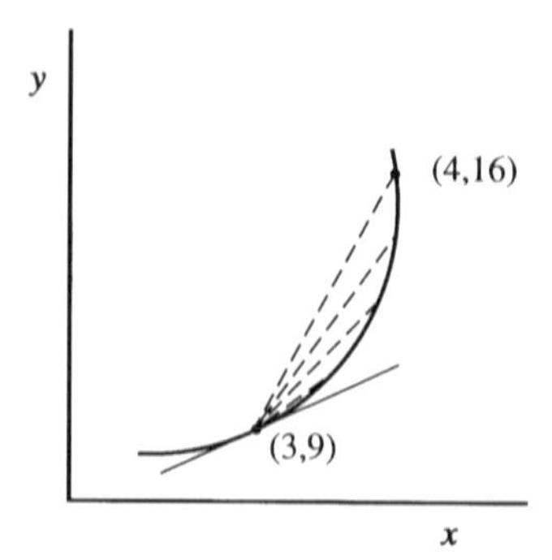

Figure 4.1 Lines drawn between one point and another on a curve are called secant lines (drawn dashed in the figure). It can be seen that as the second point approaches the first the slope of the secant line approaches the slope of the tangent at this point. This is essentially the method of differentiation

curve at a particular point, say (3,9). If we draw a line from this point to another on the line we can, using methods already described, calculate its slope. It can be seen from Figure 4.1 that the slope of this line (called a *secant* line) is somewhat steeper than the actual tangent. We now investigate what happens if we draw successive secant lines from (3,9) to points on the curve that are closer and closer to this starting point.

The first line is from (3,9) to the point (4,16) The slope of this line may be calculated using equation (4.4).

$$m = \frac{\Delta y}{\Delta x} = \frac{y_2 - y_1}{x_2 - x_1} = \frac{16 - 9}{4 - 3} = \frac{7}{1} = 7 \tag{4.4}$$

If we now choose the line which intersects the curve closer to the point of interest, say (3.5, 12.25) the slope may be calculated in the same manner

$$m = \frac{\Delta y}{\Delta x} = \frac{12.25 - 9}{3.5 - 3} = \frac{3.25}{0.5} = 6.5 \tag{4.5}$$

The next point might be (3.1, 9.61)

$$m = \frac{\Delta y}{\Delta x} = \frac{9.61 - 9}{3.1 - 3} = \frac{0.61}{0.1} = 6.1 \tag{4.6}$$

Even closer, the next point is (3.01, 9.0601)

$$m = \frac{\Delta y}{\Delta x} = \frac{9.0601 - 9}{3.01 - 3} = \frac{0.0601}{0.01} = 6.01 \tag{4.7}$$

Using the point (3.001, 9.006 001) we get

$$m = \frac{\Delta y}{\Delta x} = \frac{9.006\,001 - 9}{3.001 - 3} = \frac{0.006\,001}{0.001} = 6.001 \tag{4.8}$$

The value is approaching six. If the new point was chosen to be infinitesimaly close to (3, 9), we would find that the value of the slope would be infinitesimally close to six. So close that we could, for all practical purposes, assume that it was equal to six.

We should now introduce the concept of a *limit*. The process we have observed is that of the second point moving closer and closer to (3, 9). The absolute limit is when the second point falls on top of (3,9). At the limit, the changes in y and x become equal to zero. If this happens the ratio of these changes becomes impossible to define.* What we do in mathematical terms is to take the result as the changes approach the limit of being equal to zero. So long as they do not actually reach the limit we can avoid not being able to define the result. In mathematical jargon we say that *in the limit of* Δx approaching zero, the slope of this line at the point (3, 9) is equal to six.

What is significant about the result obtained is that the value of the slope is equal to twice the value of x at the point of interest. If we had chosen the point (1, 1) the slope would be two, at the point (2, 4) it would be four, at the point (4, 16) it would have been eight. But what about other curves? We need a more general approach—and this is described in the next section.

4.1.2 The Slope of the Line $y = x^2 + 4x + 3$. A General Approach

We now consider an equation which has terms in x^2, x^1 and x^0 (remember that any number raised to the power 0 is equal to 1, and any number, such as 3 in the above equation, can also be considered to be equal to $3x^0 = 3 \times 1 = 3$). In doing so we hope to get a more general view of the process of differentiation—finding the slope of a curve for which we know the equation.

The process is the same as in the previous example. This is, choose a point on the curve and find the slope of a straight line to another point on the curve. The point is then moved a little closer to the first, and the slope of this secant line calculated. The process is continued until the second point is so close to the first as to be practically on top of it. In the limit of the difference tending to zero the slope will be estimated.

In keeping the process as general as possible we will not use real numbers, but the variables x and y, related by the equation

$$y = x^2 + 4x + 3 \tag{4.9}$$

As shown in Figure 4.2, the first point will be given the coordinates (x, y), which are related by equation (4.9). In order to keep the process general, the second point will be given the coordinates $(x + \Delta x, y + \Delta y)$. By choosing these coordinates we can see what happens when the change, Δx, is minimised, i.e., Δx decreases and approaches zero.

It is important to realise that the point $y + \Delta y$ is not randomly chosen but must be related to the point $x + \Delta x$ by equation (4.9). That is, equation (4.9) becomes

$$y + \Delta y = (x + \Delta x)^2 + 4(x + \Delta x) + 3 \tag{4.10}$$

which, expanding the bracketed terms, is the same as

$$y + \Delta y = x^2 + 2x\Delta x + (\Delta x)^2 + 4x + 4\Delta x + 3 \tag{4.11}$$

Remembering the rules about finding the slope of a straight line between two points from Chapter 1, we know that this is given by $m = \Delta y/\Delta x$. What we need to do is to find the value of Δy for a corresponding value of Δx. Hence we need to dispose of the 'y' terms in equation (4.11).

If we subtract y from the left hand side of equation (4.11), we must also subtract y from the right hand side. We only have terms in x on the right hand side, but from equation

*Try it!

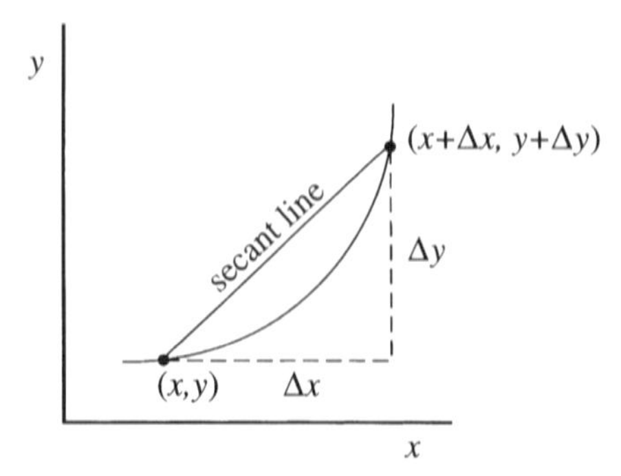

Figure 4.2 The general approach to finding the slope of a curve. The slope of the secant lines between points (x, y) and $(x + \Delta x, y + \Delta y)$ are calculated and the result defined as Δx approaches zero

(4.9) we know the terms in x which are equal to y. We can therefore subtract these, without breaking the rule on manipulating equations given in Chapter 1.

The whole process is the same as performing the subtraction of one equation from another.

$$y + \Delta y - y = \text{eq (4.11)} - \text{eq (4.9)} \tag{4.12}$$

This is

$$\begin{array}{r} y + \Delta y = x^2 + 2x\Delta x + (\Delta x)^2 + 4x + 4\Delta x + 3 \\ -(y = x^2 + 4x + 3) \\ \hline \Delta y = 2x\Delta x + (\Delta x)^2 + 4\Delta x \end{array} \tag{4.13}$$

Now that we have an expression for Δy we can find the value for the slope simply by dividing through by Δx

$$\begin{aligned} \frac{\Delta y}{\Delta x} &= \frac{2x\Delta x}{\Delta x} + \frac{(\Delta x)^2}{\Delta x} + \frac{4\Delta x}{\Delta x} \\ &= 2x + \Delta x + 4 \end{aligned} \tag{4.14}$$

This is a general expression for the slope of a line intersecting any two points on the curve described by equation (4.9). The next step is to see what happens if we make the change, Δx, and the corresponding change in y, very small, so that the secant line approaches the tangent.

It is appropriate at this point to introduce the traditional mathematical notation. The symbol 'Δ' is usually used to signify large changes (in chemistry as well as in mathematics). The small delta, δ, is used to signify small changes. As we have specified that the change becomes small equation (4.14) becomes

$$\frac{\delta y}{\delta x} = 2x + \delta x + 4 \tag{4.15}$$

It is now evident what will to happen as δx becomes very small. The slope will become equal to $2x + 4$. Before we allow this to happen, however, we must introduce some more notation.

If δx becomes equal to zero, then δy also becomes equal to zero and the ratio $\delta y/\delta x$

becomes equal to 0/0 which is impossible to define. To avoid this we allow δx to approach zero, without ever actually reaching this value. We say that δx tends to zero, or that in the limit of δx tending to zero the ratio $\delta y/\delta x$ becomes

$$\lim_{\delta x \to 0} \frac{\delta y}{\delta x} = 2x + 4 \tag{4.16}$$

which reads as 'in the limit of δx tending to zero, $\delta y/\delta x$ is equal to $2x + 4$'. We can replace this notation further,

$$\lim_{\delta x \to 0} \frac{\delta y}{\delta x} = \frac{\mathrm{d}y}{\mathrm{d}x} \tag{4.17}$$

and equation (4.16) becomes

$$\frac{\mathrm{d}y}{\mathrm{d}x} = 2x + 4 \tag{4.18}$$

which reads as 'dy by dx is equal to $2x$ plus 4', or 'the *derivative* of y with respect to x is equal to $2x$ plus 4'. In plain English this reads as 'the gradient of the curve $y = x^2 + 4x + 3$ at any point (x, y) is equal to $2x$ plus 4'.

The result, $2x + 4$, is called the *derivative* of the original equation and it is produced by the process of *differentiation.* dy and dx are called *differentials* and are small changes in these variables. They are changes so small that they are in the limit of tending to, but not being equal to zero. If these are differentials then equation (4.18) is a *differential equation*, because it involves differentials.

4.1.3 Notation

Although this book will continue to utilise the notation described above, there are several other forms of notation used in differential calculus. You may come across these from time to time, but they all mean the same thing: find the derivative of the function involved.

$$\frac{\mathrm{d}y}{\mathrm{d}x}, \frac{\mathrm{d}x^2}{\mathrm{d}x}, \frac{\mathrm{d}\mathrm{f}}{\mathrm{d}x}, \frac{\mathrm{d}}{\mathrm{d}x}\mathrm{f}(x), y', \mathrm{f}'(x), D_x y$$

The version with x^2 explicitly defined refers to any function of x which may be defined. The term 'd/dx' is called an *operator*. This is another term used widely in mathematics. It has the same meaning that it does in everyday use: it implies that we must perform a (mathematical) operation. In this case the operation is to find the derivative of the function involved.

4.1.4 The Short Cut

Section 4.1.2 showed the somewhat lengthy process of differentiating the function given in equation (4.9) from first principles. It would indeed be a lengthy process to do this for every function we might come across. If we examine the result for this function, we find that there is a general approach we can take. Equation (4.9) was

$$y = x^2 + 4x + 3 \tag{4.9}$$

and the derivative was

$$\frac{dy}{dx} = 2x + 4 \qquad (4.18)$$

The term x^2 has become $2x$, the term $4x$ has become 4 and the constant term, 3, has become equal to zero. Examining the result in terms of the powers of x involved we note that the x^2 term has actually become an x^1 term and the x^1 term has become an x^0 term. What has also happened is that the term in x^2 has also been multiplied by the original value of the exponent to become $2x$. The term in x^1 has also been multiplied by the exponent to become $(1 \times 4)x$ ('one times four x'). In this way the constant term 3, which is really $3x^0$, has been multiplied by zero and therefore become zero.

The overall process is as follows

1. Each term is considered separately.
2. Each term has been multiplied by the value of the exponent of x.
3. The value of the exponent has been reduced by one.

If these processes can be applied generally the derivative of a function $y = x^n$ is defined as

$$\frac{dy}{dx} = nx^{n-1} \qquad (4.19)$$

More generally, for a function $y = mx^n$, where m is a constant,

$$\frac{dy}{x} = nmx^{n-1} \qquad (4.20)$$

This formula is quite general. For example, we can see why the slope of a straight line, defined by the equation $y = mx + \text{c}$ is equal to m. Using the above rules, mx reduces to m and c reduces to zero. Therefore the rate of change of y with respect to x, the gradient, is equal to m. Using these rules we can also calculate the derivatives of a number of common functions. These are listed in Table 4.1.

4.2 Proofs of Derivatives of Exponential and Trigonometric Functions

Having declared the importance of the exponential function and the log function in Chapter 2 we can now present some proof. The importance of the exponential function was put down to the fact that the rate of increase of the function was equal to the function itself. To prove that this is the case it is necessary to find the derivative of $y = e^x$. The importance of the derivative of the function $y = \ln(x)$ lies in the inverse process of differentiation, integration. This will be discussed in Chapter 6 (Differential Equations). The proof that the derivative of $y = \ln(x)$ is simply $1/x$ requires more knowledge than has been presented so far, but this will be discussed in Section 4.4. In Chapter 2 the exponential function $y = e^x$ was defined by

$$y = e^x = 1 + x + \frac{x^2}{2 \times 1} + \frac{x^3}{3 \times 2 \times 1} + \frac{x^4}{4 \times 3 \times 2 \times 1} + \ldots \qquad (4.21)$$

Finding the derivative of each term, using the rules given above (equation (4.20)) we get

Table 4.1 Some common derivative formulae

Function	Derivative
$y = \text{constant}$	$\frac{dy}{dx} = 0$
$y = mx^n$	$\frac{dy}{dx} = mnx^{n-1}$
$y = e^x$	$\frac{dy}{dx} = e^x$
$y = \log_e x$	$\frac{dy}{dx} = \frac{1}{x}$
$y = \sin(x)$	$\frac{dy}{dx} = \cos(x)$
$y = \cos(x)$	$\frac{dy}{dx} = -\sin(x)$

$$\frac{dy}{dx} = 0 + 1 + \frac{2x}{2 \times 1} + \frac{3x^2}{3 \times 2 \times 1} + \frac{4x^3}{4 \times 3 \times 2 \times 1} + \ldots \tag{4.22}$$

which is the same as equation (4.21)

$$\frac{dy}{dx} = 1 + x + \frac{x^2}{2 \times 1} + \frac{x^3}{3 \times 2 \times 1} + \frac{x^4}{4 \times 3 \times 2 \times 1} + \ldots \tag{4.23}$$

The derivatives of sine and cosine may also be found by using the series definitions of these functions. These are

$$\sin(x) = \frac{x}{1} - \frac{x^3}{3!} + \frac{x^5}{5!} - \frac{x^7}{7!} + \ldots \tag{4.24}$$

$$\cos(x) = 1 - \frac{x^2}{2!} + \frac{x^4}{4!} - \frac{x^6}{6!} + \ldots \tag{4.25}$$

Here we have used the conventional factorial notation (2!, 3! etc.). Remembering that $2! = 2 \times 1$, $3! = 3 \times 2 \times 1$ etc., and that $2/2! = 1$ and $3/3! = 2 \times 1 = 2!$. In general,

$$\frac{n}{n!} = \frac{1}{(n-1)!} \tag{4.26}$$

The derivatives of these functions are

$$\begin{aligned} \frac{d(\sin(x))}{dx} &= 1 - \frac{3x^2}{3!} + \frac{5x^4}{5!} - \frac{7x^6}{7!} + \ldots \\ &= 1 - \frac{x^2}{2!} + \frac{x^4}{4!} - \frac{x^6}{6!} + \ldots \end{aligned} \tag{4.27}$$

which is the cosine function and

$$\frac{d(\cos(x))}{dx} = -\frac{2x}{2!} + \frac{4x^3}{4!} - \frac{6x^5}{6!} + \ldots$$
$$= -\frac{x}{1} + \frac{x^3}{3!} - \frac{x^5}{5!} + \ldots \tag{4.28}$$
$$= -1\left(\frac{x}{1} - \frac{x^3}{3!} + \frac{x^5}{5!} - \ldots\right)$$
$$= -\sin(x)$$

4.3 More Complex Differentiation

4.3.1 Simplifying Complex Formulae

The simple rules on differentiation are often not sufficient to find derivatives of many functions. It is possible, however, to reduce many functions to combinations of simpler functions. We have already seen this. Without any qualification Rule 1, given above, stated that each term should be treated separately. An equation such as

$$y = x^3 + x^2 \tag{4.29}$$

could be considered as a sum of two functions, u and v

$$y = u + v \tag{4.30}$$

where $u = x^3$ and $v = x^2$.

A function such as

$$y = x^2(x + 1) \tag{4.31}$$

could be considered to be a product of two functions u and v

$$y = uv \tag{4.32}$$

where

$$u = x^2 \tag{4.33}$$

and

$$v = (x + 1) \tag{4.34}$$

Another, more appropriate, example might be

$$y = x^2 e^x \tag{4.35}$$

in which the two constituent functions are $u = x^2$ and $v = e^x$. The Maxwell–Boltzmann equation (Chapter 2) is of a similar form to this, a term in x squared multiplied by a term which is the exponential of x.

In the same way we might find functions which can be expressed as a ratio of simpler functions or possibly as a function of another function. The latter is common and the result is very useful. The function

$$y = e^{kx} \tag{4.36}$$

could be expressed as

$$y = e^{u} \tag{4.37}$$

where k is a constant and u is the function of x

$$u = kx \tag{4.38}$$

Another example is the function

$$y = e^{x^2} \quad (y = \exp(x^2)) \tag{4.39}$$

which can be simplified to

$$y = e^{u} \tag{4.40}$$

where u is defined by

$$u = x^2 \tag{4.41}$$

In all of these examples we have reduced apparently complicated functions to combinations of simpler ones. The derivatives of the simpler functions (Table 4.1) can all be found using the given rules on differentiation. What remains is to define the method for handling these combinations of functions.

Table 4.2 lists formulae, called *reduction formulae*, for finding the derivatives of these combinations of functions. Rather than deriving each of these formulae (these may be found in any textbook on calculus) we now show how some of these formulae are used.

4.3.2 Using Reduction Formulae

A number of examples will serve to illustrate how some of these reduction formulae may be used. The ability to do this lies in the confidence to break down an equation into smaller parts. As with so much of mathematics there are no strict rules just as to what you should do. Much rests on the experience of the mathematician. The following examples, however, may be useful.

Example—the derivative of $y = \ln(x)$

This was given in Table 4.1 as $1/x$. This can be proved using Formula (h) in Table 4.2. If we write the equation

$$y = \ln(x) \tag{4.44}$$

then we can say that

$$x = e^{y} \tag{4.45}$$

This is a function for which we know the derivative.

$$\frac{dx}{dy} = e^{y} \tag{4.46}$$

Using Formula (h) we know that

Table 4.2 Reduction formulae

Function	Derivative	Notes
$y = a \cdot u$	$a \cdot \dfrac{du}{dx}$	(a)
$y = u + v$	$\dfrac{du}{dx} + \dfrac{dv}{dx}$	(b)
$y = \Sigma u$	$\sum \dfrac{du}{dx}$	(c)
$y = uv$	$u\dfrac{dv}{dx} + v\dfrac{du}{dx}$	(d)
$y = uv$	$\dfrac{1}{y}\dfrac{dy}{dx} = \dfrac{1}{u} \cdot \dfrac{du}{dx} + \dfrac{1}{v} \cdot \dfrac{dv}{dx}$	(e)
$y = u/v$	$\dfrac{v\dfrac{du}{dx} - u\dfrac{dv}{dx}}{v^2}$	(f)
$y = f(u)$	$\dfrac{dy}{dx} = \dfrac{dy}{du} \cdot \dfrac{du}{dx}$	(g)
$y = y(x)$	$\dfrac{dx}{dy} = \dfrac{1}{dy/dx}$	(h)

(a) u is a function of x, e.g. $u = x^2$. This is the same as equation (4.20) in that the constant remains unaffected.

(b) This is the same as Rule 1, given above; all terms separated by plus or minus signs are treated separately.

(c) This is a general form of (b), using the summation sign.

(d) u and v are both functions of x.

(e) This is a more general form of (d). This is obtained by dividing the differential equation by $y \cdot y$ is used on the left hand side but uv is used on the right hand side. Thus the equation in (d) becomes

$$\frac{1}{y}\frac{dy}{dx} = \frac{v}{uv} \cdot \frac{du}{dx} + \frac{u}{uv} \cdot \frac{dv}{dx} \tag{4.42}$$

When the common factors are cancelled the equation (e) results. This form is better than (d) because each term contains only one function and therefore might at least be easier to remember. More importantly it can be extended to include any number of constituent functions, e.g., $y = uvw$ and so on.

(g) y is a function of a function of x. This is called the *chain rule*, a chain being another name for a product. Thus a derivative may be expressed as a chain or product of simpler derivatives. This can also be extended to include more functions, e.g.,

$$\frac{dy}{dx} = \frac{dy}{du} \cdot \frac{du}{dv} \cdot \frac{dv}{dw} \cdot \frac{dw}{dx} \tag{4.43}$$

The rule is relatively easy to remember if you think of each derivative as a real fraction. In doing so all of the intermediary functions cancel each other out.

(h) Used for finding derivatives of inverse functions, e.g., $\ln(x)$ from the derivative of e^x.

$$\frac{dy}{dx} = \frac{1}{dx/dy}$$
$$= \frac{1}{e^y} \tag{4.47}$$
$$= \frac{1}{x}$$

Example—the derivative of exp(kx)

This is important because the Boltzmann equation and those related to it are similar to this function. The solution is found using the chain rule and we rewrite the equation as

$$y = \exp(u) \tag{4.48}$$

where

$$u = kx \tag{4.49}$$

Given the chain rule we must find the derivatives of y with respect to u and of u with respect to x.

$$\frac{dy}{du} = \exp(u) \tag{4.50}$$

$$\frac{du}{dx} = k \tag{4.51}$$

The derivative of $y = \exp(k^x)$ is therefore given by the product of these two derivatives.

$$\frac{dy}{du} = \frac{dy}{du} \cdot \frac{du}{dx}$$
$$= k \cdot \exp(u) \tag{4.52}$$
$$= k \cdot \exp(kx)$$

Example—the derivative of $y = \exp(x^2)$

This is another problem related to Boltzmann-type equations. The index '2' could be any constant but the method remains the same. This equation can be written as $y = \exp(u)$ again but this time

$$u = x^2 \tag{4.53}$$

The derivative of y with respect to u is the same as above (equation (4.50)) but the derivative of u with respect to x is

$$\frac{du}{dx} = 2x \tag{4.54}$$

The complete derivative is once again given by the chain rule

$$\begin{aligned}\frac{dy}{du} &= \frac{dy}{du} \cdot \frac{du}{dx} \\ &= 2x \cdot \exp(u) \\ &= 2x \cdot \exp(x^2)\end{aligned} \tag{4.55}$$

4.4 Higher Derivatives

4.4.1 Definitions

Once a function has been differentiated it is possible to differentiate it again. For example, taking the function

$$y = 2x^3 + x^2 + 6x + 8 \tag{4.56}$$

we can define the derivative as

$$\frac{dy}{dx} = 6x^2 + 2x + 6 \tag{4.57}$$

We can differentiate this one more time

$$\frac{d}{dx}\left(\frac{dy}{dx}\right) = 12x + 2 \tag{4.58}$$

If the first derivative is the rate of change of the dependent variable with respect to the independent variable then the second derivative is the rate of change of the first derivative. Velocity is the rate of change of position with time (units of metres per second) whereas acceleration is the rate of change of velocity with time (units of metres per second per second). Acceleration is the second derivative of distance with respect to time.

The significance of this is that the second derivative shows how the function curves. The simplest example is the second derivative of the equation describing a straight line, which by definition does not curve. Using the general equation $y = mx + c$, the first derivative is given by

$$\frac{dy}{dx} = m \tag{4.59}$$

and the second derivative is given by

$$\frac{d}{dx}\left(\frac{dy}{dx}\right) = 0 \tag{4.60}$$

since the derivative of a constant is zero. A straight line shows no curvature and this is reflected in the value of the second derivative. In comparison the first derivative of the curve $y = x^2$ is equal to $2x$, implying that the slope varies as the value of x varies. The second derivative is equal to 2, indicating that it is indeed curved and that the slope changes at a constant rate.

The notation used for the second derivative so far is cumbersome and is replaced by

$$\frac{d}{dx}\left(\frac{dy}{dx}\right) = \frac{d^2y}{dx^2} \tag{4.61}$$

which reads as 'd two y by d x squared' and means that the function which relates y to x is differentiated twice.

Other forms of notation often encountered are analogous to the notations used for the first derivative. The forms D^2y, y'', $f''(x)$ all refer to the second derivative of the function of y with respect to x. This notation can be extended to even higher derivatives though these are unlikely to be used by undergraduate chemists.

4.4.2 Uses of Second Derivatives

Second derivatives are very important in many areas of science, including chemistry. For example, they appear in second order differential equations (Chapter 5) which are used in problems concerning diffusion. In the context of this chapter (the use of differentiation as a process in its own right) they are used to define points on curves called *turning points*. These are points where the curve turns and we have what are called *local maxima*, *local minima* and *local points of inflexion*. These can be defined by referring to Figure 4.3.

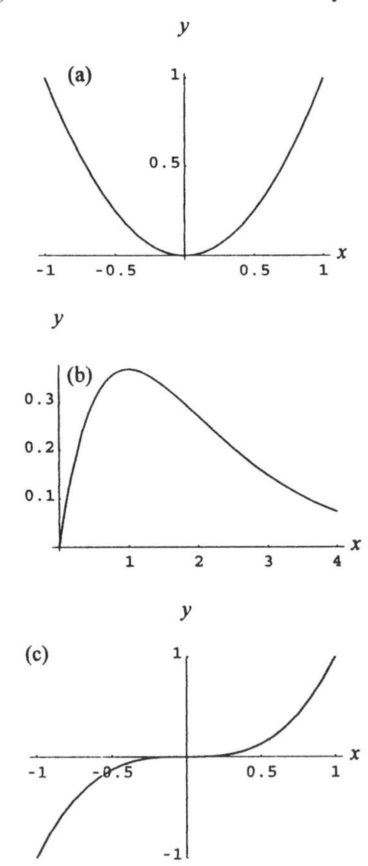

Figure 4.3 (a) $y = x^2$, (b) $y = x.\exp(-x)$, (c) $y = x^3$

Figure 4.3(a) shows a plot of the curve $y = x^2$. At the point (0, 0) the curve has a value of y which is less than any other, equal to zero. The curve appears as a trough or a valley and the lowest point is called a *minimum*. Figure 4.3(b) shows a plot of $y = x.\exp(-x)$, a curve not dissimilar to the Maxwell–Boltzmann distribution curve. This curve appears as a hill in cross-section and it has a peak. The point $(1, e^{-1})$ at which y has a maximum value (before it decreases) is called a *maximum*. Figure 4.3(c) is a plot of $y = x^3$. Near the point (0, 0) the curve starts to level off but then takes off again. This is called a *point of inflexion*. A significant difference between maxima and minima in the mathematical sense and the common experience is that a function may have several maxima and/or minima. Common sense on the other hand suggests that there should only be one maximum and/or one minimum. The sine and cosine functions are two common functions which show many maxima and minima.

What these turning points have in common is that at the point of interest the gradient is equal to zero. Therefore by finding the first derivative as a function of x and finding out at what value of x it is equal to zero (if indeed it is ever equal to zero) we can find out whether the curve has a maximum, minimum or point of inflexion.

Where the second derivative is important is in discovering which it is of these points we have. Although it may be possible to tell simply by plotting the graph of the function this may be a somewhat lengthy process. It may also be subject to errors in drawing the graph. The method of calculus is accurate and rapid.

As a minimum in the curve is approached, x increases and the slope is negative. At the minimum the slope is equal to zero and as x increases further the slope becomes positive. Therefore the rate at which the slope changes is always positive, Figure 4.4(a).

When there is a maximum in the curve the opposite applies. Approaching the maximum the slope is positive. At the maximum the slope is zero and as the maximum is passed the slope becomes negative. In going from positive to zero to negative, the slope is always decreasing. That is, the rate of change of slope, the second derivative, is negative, Figure 4.4(b).

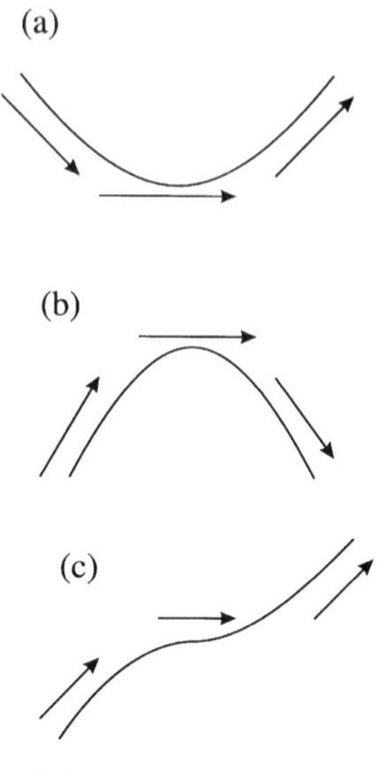

Figure 4.4 Change of slope for (a) minimum, (b) maximum and (c) inflexion in a curve

Table 4.3 Conditions for defining minima, maxima and inflexions

	$\frac{dy}{dx}$	$\frac{d^2y}{dx^2}$
Minimum	Zero	Positive
Maximum	Zero	Negative
Inflexion	Zero	Zero

Before reaching a point of inflexion the slope may be positive or negative. At the point of inflexion it is zero. After the point the curve continues in its original manner. Therefore the rate of change of slope is also zero (Fig. 4.4(c)). This result is summarised in Table 4.3.

4.4.3 Finding Minima, Maxima and Inflexions*

The first three examples, below, are relatively trivial but show the methodology used for these kind of problems.

Example—minimum in a curve

Figure 4.5 shows a plot of $y = x^2$. The first derivative is equal to $2x$ and the second derivative is equal to 2. When the first derivative is equal to zero we must have one of these three points of interest. This can only occur when $x = 0$. This is borne out by the plot in Figure 4.5. The second derivative is equal to 2, whatever the value of x. This is always positive and so by our definitions above this must be a minimum.

Example—maximum in a curve

Figure 4.6 shows a plot of $y = 4x - x^2$. The first derivative is

$$\frac{dy}{dx} = 4 - 2x \tag{4.62}$$

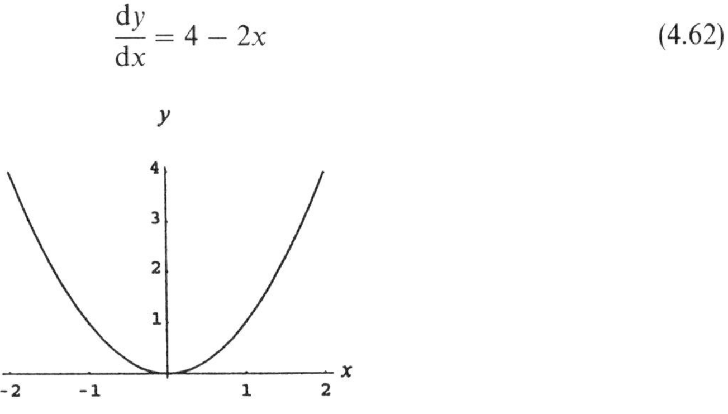

Figure 4.5 $y = x^2$. Minimum occurs at $x = 0$

*These simple rules are adequate for all the examples in this book and most examples you will find in chemistry. They are, however, a simplification of the complete rule which states that the nature of the turning point is decided by the sign of the first non-zero derivative.

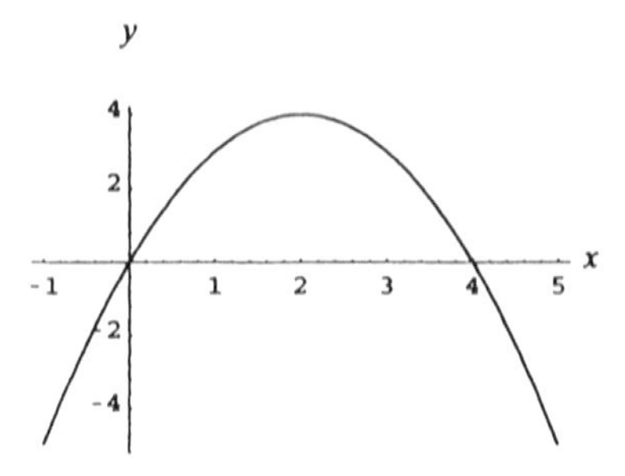

Figure 4.6 $y = 4x - x^2$. Maximum is at (2, 4)

This is set equal to zero

$$0 = 4 - 2x \tag{4.63}$$

Rearranging gives the value of x

$$4 = 2x \tag{4.64}$$

or

$$2 = x \tag{4.65}$$

The second derivative is found by differentiating equation (4.62).

$$\frac{d^2y}{dx^2} = -2 \tag{4.66}$$

This is always negative and therefore we can conclude that this is a maximum. This is shown to be true by inspecting the plot in Figure 4.6.

Example—point of inflexion

Figure 4.7 shows a plot of $y = x^3$ The first derivative of this is given by

$$\frac{dy}{dx} = 3x^2 \tag{4.67}$$

This can only be equal to zero when x is equal to zero. The second derivative is given by

$$\frac{dy}{dx} = 6x \tag{4.68}$$

When x is equal to zero, this is also equal to zero and the point of interest is therefore a point of inflexion as Figure 4.7 indicates.

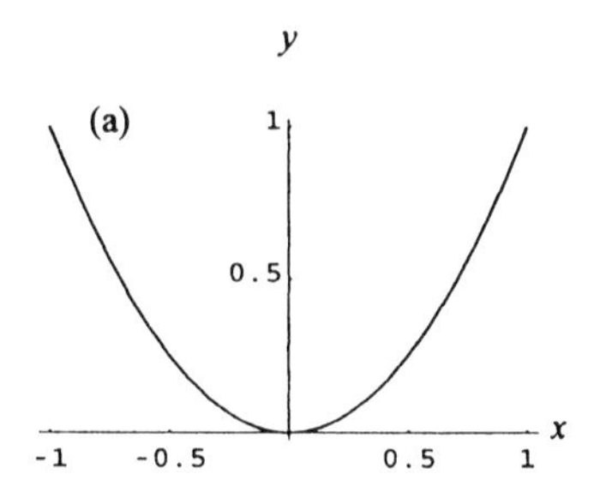

Figure 4.7 $y = x^3$. Inflexion occurs at (0, 0)

Example—the Lennard-Jones potential

This equation is used in problems of molecular interaction. It describes the dependence of the potential energy when two molecules approach each other from a long distance apart. The energy is a play-off between the attractive force between the molecules (proportional to $1/R^6$, where R is the distance apart) and the repulsive force (proportional to $1/R^{12}$). The shape of this function is shown in Figure 4.8. The theory behind this and the resultant equation suggest that when the distance apart is less than σ (an arbitrary distance called the collision diameter) the potential energy (V) increases sharply. The Lennard-Jones potential is

$$V = 4\varepsilon\left\{\left(\frac{\sigma}{R}\right)^{12} - \left(\frac{\sigma}{R}\right)^{6}\right\} \tag{4.69}$$

The parameter ε is the value of V at the minimum of the curve. When R is very small the term $(\sigma/R)^{12}$ becomes very large and dominates the whole equation. When R is very large both terms tend towards zero, implying that when the molecules are very far apart there is little interaction between them (no potential energy). What is interesting about this equation is that it implies that there is a distance apart when the potential energy is a minimum. When energy is at a minimum it implies some sort of stability. We might draw the analogy with running water. Water always runs from high land, where it has a lot of potential energy, to the valley where it has much less potential energy. If the valley is enclosed a lake forms and the water is considered to be in a stable state: it stays where it is. If two molecules or atoms collide and reach some sort of stable state, they may stay together, forming a new molecule.

The question remaining is, can we prove this mathematically? The plot shows a minimum but can we calculate the coordinates of the minimum?

Even without plotting the function we should be able to test for a minimum using the methods already described. First of all we find the first derivative and set it equal to zero. Then we find the second derivative and find its sign. The x coordinate is found from the first derivative result and the y coordinate may be calculated from this from the original equation. The first derivative may be found by expanding equation (4.69) and diferentiating each term in the usual way. To make this example simpler we will set all the constants (including '4') to unity and use the equation

$$V = \left(\frac{1}{R}\right)^{12} - \left(\frac{1}{R}\right)^{6} \tag{4.70}$$

which is the same as

$$V = R^{-12} - R^{-6} \tag{4.71}$$

V may now be differentiated with respect to R,

$$\frac{dV}{dR} = -12R^{-13} + 6R^{-7} \tag{4.72}$$

When this is equal to zero we can rearrange to get

$$12R^{-13} = 6R^{-7} \tag{4.73}$$

This can be rearranged further

$$\frac{12}{6} = \frac{R^{-7}}{R^{-13}} \tag{4.74}$$

Recalling that $x^a/x^b = x^{(a-b)}$ (Chapter 1) this becomes

$$2 = R^6 \tag{4.75}$$

Taking the sixth root of both sides results in

$$2^{1/6} = R \tag{4.76}$$

The y coordinate of this point can be found by putting this value in equation (4.71).

$$V = \left(\frac{1}{2^{1/6}}\right)^{12} - \left(\frac{1}{2^{1/6}}\right)^{6} \tag{4.77}$$

Recalling the rules on indices in Chapter 1 this is given by

$$V = \left(\frac{1}{2^2}\right) - \left(\frac{1}{2}\right) \tag{4.78}$$

since $12 \times 1/6$ is equal to 2 and $6 \times 1/6$ is 1. The coordinate of interest is therefore $(2^{1/6}, -1/4)$.

That this is a minimum can be shown by taking the derivative of equation (4.72).

$$\begin{aligned} \frac{d^2V}{dR^2} &= (-13 \times (-12R^{-14})) + (-7 \times (6R^{-8})) \\ &= 156R^{-14} - 42R^{-8} \end{aligned} \tag{4.79}$$

At the point of interest R is equal to $2^{1/6}$ and we substitute this back into (4.78)

$$\begin{aligned} \frac{d^2V}{dR^2} &= 156 \times (2^{1/6})^{-14} - 42 \times (2^{1/6})^{-8} \\ &= 13.4 \end{aligned} \tag{4.80}$$

which is positive. The point is therefore a minimum. This is verified in Figure 4.8, a plot of the Lennard-Jones potential, equation (4.70).

Example—the Maxwell–Boltzmann distribution

In Chapter 2 we introduced the Maxwell–Boltzman distribution of molecular speed,

$$y = 4\pi\left(\frac{m}{2\pi kT}\right)^{3/2} c^2 \cdot \exp\left(-\frac{mc^2}{2kT}\right) \tag{4.81}$$

This is a curved function with a peak corresponding to the most probable speed. If it were necessary to calculate the most probable speed for a particular gas this could be achieved using the methods already outlined to find the maximum in the curve. Once again the method is to find the first derivative and set it to zero. This (hopefully) produces an expression which relates the speed to the other variables and constants in the equation. That the point of interest is a maximum can then be proved by finding the second derivative and showing that it is a negative quantity.

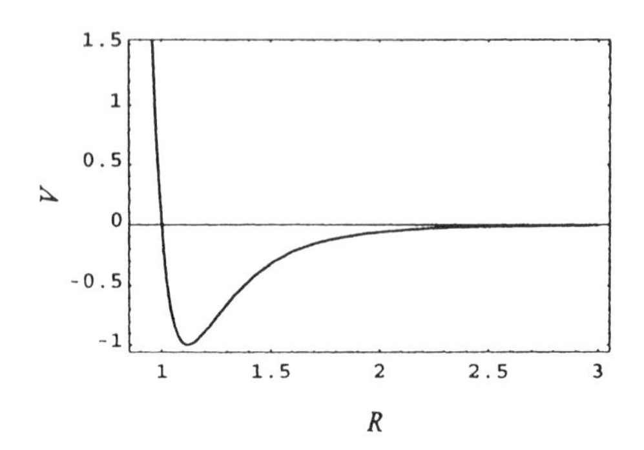

Figure 4.8 Plot of the Lennard-Jones potential, the potential energy, V, between two molecules as they approach each other

We can save time by acknowledging that there are a number of constant terms in this equation. Thus equation (4.81) may be written as

$$y = k_1 c^2 \exp(-k_2 c^2) \tag{4.82}$$

where k_1 and k_2 are defined by

$$k_1 = 4\pi\left(\frac{m}{2\pi kT}\right)^{3/2} \tag{4.83}$$

$$k_2 = \frac{m}{2kT} \tag{4.84}$$

k_1 and k_2 are obviously related as they incorporate the same constants. It will avoid confusing algebra if we keep them as separate terms. We can deal with this at the end if necessary. We acknowledge their relationship merely by giving the same letter different subscripts (1 and 2).

Equation (4.82) can be expressed as a product of two simpler functions, u and v, where

$$u = k_1 c^2 \tag{4.85}$$

$$v = \exp(-k_2 c^2) \tag{4.86}$$

The derivative of y with respect to c can therefore be found in the usual way.

$$\frac{\mathrm{d}y}{\mathrm{d}c} = u\frac{\mathrm{d}v}{\mathrm{d}c} + v\frac{du}{\mathrm{d}c} \tag{4.87}$$

The derivative of u with respect to c is $2k_1c$ and the derivative of v with respect to c is a function of a function found previously in equation (4.55). The only difference is that we now have the constant $-k_2$ involved. The derivative, $\mathrm{d}v/\mathrm{d}c$, is therefore

$$\frac{\mathrm{d}v}{\mathrm{d}c} = -2k_2c\cdot\exp(-k_2c^2) \tag{4.88}$$

The whole derivative is

$$\frac{\mathrm{d}y}{\mathrm{d}c} = k_1c^2\cdot -2k_2c\cdot\exp(-k_2c^2) + \exp(-k_2c^2)\cdot 2k_1c$$

$$= -2k_1k_2c^3 \cdot \exp(-k_2c^2) + 2k_1c \cdot \exp(-k_2c^2) \tag{4.89}$$

This is equal to zero when the the two terms on the right hand side are equal in magnitude:

$$2k_1k_2c^3 \cdot \exp(-k_2c^2) = 2k_1c \cdot \exp(-k_2c^2) \tag{4.90}$$

The factor $2k_1c \cdot \exp(-k_2c^2)$ is common to both sides and may be cancelled. This gives the result

$$k_2c^2 = 1 \tag{4.91}$$

or, on rearrangement,

$$c^2 = \frac{1}{k_2} \tag{4.92}$$

In problems on this distribution we would probably be more interested in the value of the most probable speed, which as this equation tells us is a function of mass and temperature. The value of the probability can be found, however, by substituting this equation back into equation (4.82) and then substituting for the constants k_1 and k_2.

The second derivative is found by differentiating equation (4.89). This may be accomplished by considering this equation as a sum of two functions, each of which is a product. Each term has the $\exp(-k_2c^2)$ function for which we have the derivative and another term for which the derivative is relatively trivial. We can write the full second derivative as

$$\begin{aligned}\frac{d^2y}{dc^2} &= [2k_1c \cdot (-2k_2c) \cdot \exp(-k_2c^2) + 2k_1 \cdot \exp(-k_2c^2)] \\ &\quad - [\exp(-k_2c^2) \cdot 6k_1k_2c^2 + 2k_1k_2c^3 \cdot (-2k_2c) \cdot \exp(-k_2c^2)]\end{aligned} \tag{4.93}$$

Collecting together all the individual terms (taking great care with minus signs) we get

$$\begin{aligned}\frac{d^2y}{dc^2} &= -4k_1k_2c^2 \cdot \exp(-k_2c^2) + 2k_1 \cdot \exp(-k_2c^2) \\ &\quad - 6k_1k_2c^2 \cdot \exp(-k_2c^2) + 4k_1k_2^2c^4 \cdot \exp(-k_2c^2)\end{aligned} \tag{4.94}$$

which can be factorised, removing the exponential function from each term,

$$\frac{d^2y}{dc^2} = \exp(-k_2c^2)[-4k_1k_2c^2 + 2k_1 - 6k_1k_2c^2 + 4k_1k_2^2c^4] \tag{4.95}$$

This appears unwieldy but if we recall from equation (4.88) that the product k_2c^2 is equal to one then it is greatly simplified.

$$\begin{aligned}\frac{d^2y}{dc^2} &= \exp(-k_2c^2)[-4k_1 + 2k_1 - 6k_1 + 4k_1] \\ &= -4k_1 \cdot \exp(-k_2c^2)\end{aligned} \tag{4.96}$$

Since all of the parameters which constitute k_1 are positive then the second derivative must be negative and we have a maximum in the plot.

Example—the van der Waals equation

The assumptions behind the ideal gas equation are, under certain conditions, invalid. At high pressures and low volumes we can no longer assume that the molecules have zero volume or that there are no interactive forces between them. If these assumptions were true, there would be no condensed phases.

The van der Waals gas equation is a modification of the ideal gas equation which allows for these inaccuracies. The parameter a is introduced to account for the attractive forces between molecules and the parameter b allows for the volume of the molecules. The van der Waals equation is

$$P = \frac{RT}{(V-b)} - \frac{a}{V^2} \tag{4.97}$$

where V is now the molar volume of the gas. Examining the equation it can be seen that at low pressures the volume V would be expected to be large and, most importantly, large compared to b and a. In this case the second term on the right becomes very small and the term $(V-b)$ approximates to V. Thus the equation reduces to the ideal gas equation.

Experiments on the behaviour of gases often take the form of measuring the pressure as the volume of the gas is changed. This can be facilitated by compressing the gas in a piston. Figure 4.9(a) shows what the ideal equation predicts, and indeed what occurs at high temperatures for many gases. These plots are called isotherms because the pressure volume relation is plotted at a single temperaure (*iso* means 'the same' or 'constant', *therm* relates to the temperature as in *therm*ometer).

At lower temperatures the experimental isotherms show a horizontal portion. What is happening can be best described by thinking about what happens as the gas is compressed, moving from right to left along the plot. As the volume decreases then the presure exerted by the gas increases. When the pressure reaches a certain value the gas begins to condense and form a liquid. On decreasing the volume further the pressure remains constant as the gas liquefies more and more. Eventually the gas is totally liquefied and as liquids are very difficult to compress the pressure rises rapidly. At higher temperatures the horizontal section of these curves is shorter and shorter. At a temperature known as the critical temperature, T_c, the horizontal part appears as an inflexion in the curve. At the

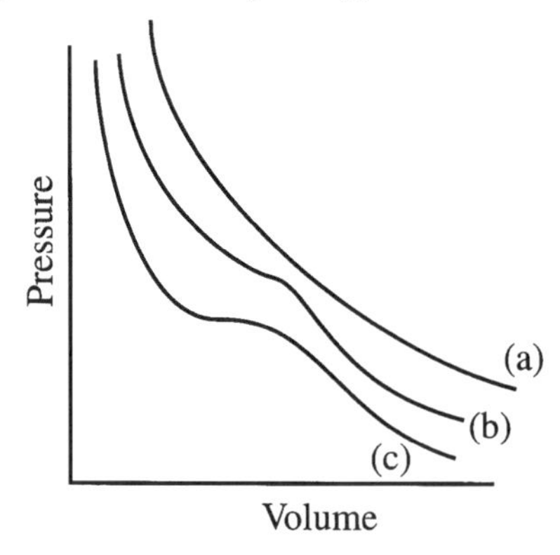

Figure 4.9 Pressure–volume isotherms for gases. (a) Above the critical temperature, (b) at the critical temperature and (c) below the critical temperature

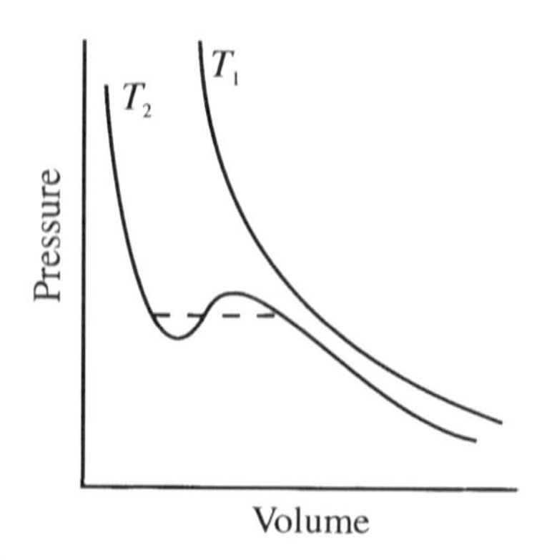

Figure 4.10 Isotherms predicted by the van der Waals equation. (a) Above the critical temperature, (b) below the critical temperature. At the critical temperature an inflexion is predicted

critical temperature the pressure and volume at this inflexion are called the critical pressure, P_c, and the critical volume, V_c. Under these conditions (P_c, V_c and T_c) the gas and the liquid may be thought of as not just being in equilibrium with each other but as being a single phase. The practical significance of the critical temperature is that above this temperature the gas cannot be liquefied just by compressing it. To achieve liquefaction the gas must be cooled below this temperature.

The important question is 'Can this behaviour be predicted from the van der Waals equation?' Figure 4.10 shows theoretical isotherms based on the van der Waals equation. Below the critical temperature the equation predicts a dip in the plot (so that it actually has a maximum and a minimum). At the critical temperature, however, it does predict the presence of an inflexion. The minimum in the curve is nonsense as this implies that the pressure decreases as volume decreases which has never been found in practice. We choose to ignore this aberration and concentrate on what happens at the critical point.

It is possible to calculate the conditions which represent the critical point by the methods of differential calculus. At a point of inflexion both the first and second derivative should be equal to zero. These derivatives are relatively easy to find as shown below.

The second term (a/V^2) can be differentiated directly.

$$\frac{\mathrm{d}}{\mathrm{d}V}\left(\frac{a}{V^2}\right) = \frac{\mathrm{d}}{\mathrm{d}V}(aV^{-2})$$
$$= -2a \cdot V^{-3} \tag{4.98}$$
$$= -\frac{2a}{V^3}$$

The first term can be found as a function of a function

$$y = \frac{RT}{u} \tag{4.99}$$

where y is a function of u and $u = V - b$. The derivative of y with respect to V is given by the chain rule

$$\frac{dy}{dV} = \frac{dy}{du} \cdot \frac{du}{dV}$$

$$= -\frac{RT}{u^2} \cdot 1 \tag{4.100}$$

$$= -\frac{RT}{(V-b)^2}$$

The complete derivative is therefore

$$\frac{dP}{dV} = \frac{-RT}{(V-b)^2} + \frac{2a}{V^3} \tag{4.101}$$

The second derivative is found in exactly the same manner

$$\frac{d^2P}{dV^2} = \frac{2RT}{(V-b)^3} - \frac{6a}{V^4} \tag{4.102}$$

At the point of inflexion both of these are set equal to zero. We can also now give the variables the subscripts which show that they are the critical values. Equations (4.101) and (4.102) become

$$\frac{-RT_c}{(V_c-b)^2} + \frac{2a}{V_c^3} = 0 \tag{4.103}$$

$$\frac{2RT}{(V_c-b)^3} - \frac{6a}{V_c^4} = 0 \tag{4.104}$$

We now have two equations which may be solved simultaneously to get expressions for the critical constants V_c and T_c in terms of the constants a and b. P_c may be found from these results, which are

$$V_c = 3b \tag{4.105}$$

$$T_c = \frac{8a}{27Rb} \tag{4.106}$$

$$P_c = \frac{a}{27b^2} \tag{4.107}$$

4.5 Differentials

In the proofs and definitions used in this chapter we have considered the derivative dy/dx as a complete entity. It is, however, possible to consider it as the mathematical ratio of two separate quantities, dy and dx. These are the infinitesimally small changes in the variables y and x and they are called *differentials*. Being distinct quantities they can be treated as such. For example, for the equation $y = x^2$ we can write

$$\frac{dy}{dx} = 2x \tag{4.108}$$

and this can be rearranged to

$$dy = 2x \cdot dx \tag{4.109}$$

If we know a value for dx then we can calculate a value for dy.

There are two areas where differentials are used in their own right. One is in finding solutions to differential equations, which usually involves integration (Chapters 5 and 6). The other is in thermodynamics.

The notation of the differential is used constantly in thermodynamics. For this reason, if no other, students must at least be aware of the notation and what it means. In many thermodynamic equations differential quantities, dU, dG, dH, dq, dw etc., are related to other differential quantities by purely algebraic equations, e.g.,

$$dU = dq + dw \tag{4.110}$$

which relates the change in internal energy, U, to the changes in heat, q, and work, w, in a thermodynamic system. Thus the change in internal energy can be calculated by adding changes in heat and work done on or by a system.

The term 'reversible' is used widely in thermodynamics. This is not to be confused with the everyday meaning of the word, that something can go backwards or forwards. Rather it means that a change occurs when a system is at equilibrium. Such a change may be reversed by an infinitesimal change in the system. A change such as this would be so small that it could not be observed. Such an infinitesimally small change can be considered to be a differential quantity.

The question might be raised 'What is the use of a quantity that is too small to be measured and does not apparently change anything?' The answer is in the mathematics of the differential. We can add up all the infinitesimally small changes using calculus to model the large changes which can be observed. This act of summation is known as integration. In Chapter 5, Integration, we will return to this discussion of the implication of the differential notation. For the present it is sufficient to note that differentials may be used as distinct quantities.

4.6 Partial Differentiation

4.6.1 Definitions and Notation

In this chapter we have only considered equations which involve two variables. In practice there may be many more. For example, the ideal gas equation links four variables: pressure, volume, temperature and amount of gas. In kinetic measurements we usually follow the concentration of a substance with the passing of time. We may also have to consider the concentrations of other reactants, the temperature, or variables such as the viscosity of the solution. In thermodynamic equations there may be many more variables. It is important therefore that there is a well established mathematical method for dealing with relationships involving several variables.

This method is known as *partial differentiation.* The rules of differentiation apply but the technique relies on keeping all but one of the independent variables constant and differentiating the dependent variable with respect to the remaining one. For example, we can write the ideal gas equation presenting P as a function of T and V, with n, the number of moles, constant.

$$P = \frac{nRT}{V} \tag{4.111}$$

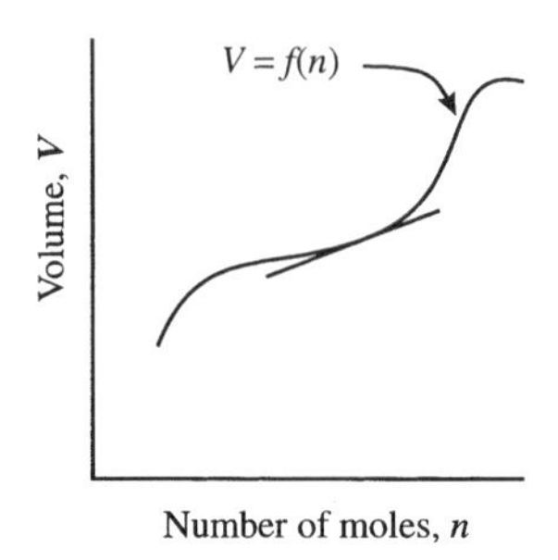

Figure 4.11 The partial molar volume is effectively the slope of a plot of volume versus amount of substance with all other conditions (temperature, amount of other substances etc.) constant

This can be differentiated with respect to T in the normal manner, assuming that the volume remains constant. The derivative will be nR/V as expected but the notation used is

$$\left(\frac{\partial P}{\partial T}\right)_V = \frac{nR}{V} \tag{4.112}$$

The curly ∂ operates in the same way as the usual d (for differential) but indicates that the derivative is only with respect to one of several variables. The other variables, held constant, are listed in the subscript outside the bracket.

Example—partial molar quantities

Partial derivatives are seen in equations concerning partial molar quantities. The partial molar volume of a substance ($\bar{V}$) is defined as the change in volume of a mixture when an infinitesimally small amount of the substance of interest is added.

$$\bar{V}_A = \left(\frac{\partial V}{\partial n_A}\right)_{T,P,n_B...} \tag{4.113}$$

which reads as 'the partial molar volume of substance A is the partial derivative of the the volume of the mixture with respect to the number of moles of A (n_A) at constant temperature T, pressure P and amount of B, n_B'. The dots indicate that the list may continue, that is, that all other possible variables are held constant.

Figure 4.11 shows the significance of such partial molar quantities. If we were to plot the volume of a mixture as a function of a particular component the partial molar volume is the slope of the plot at a point of interest.

One of the most important partial molar quantities is the partial molar Gibbs free energy. This is defined as the chemical potential, μ:

$$\mu = \left(\frac{\partial G_A}{\partial n_A}\right)_{T,P,n_B,...} \tag{4.114}$$

4.6.2 Graphical Representation of Functions Involving More than Two Variables

In order to represent an equation which involves three variables we must have a three-dimensional graph, i.e., one with three axes. There is an obvious difficulty in doing

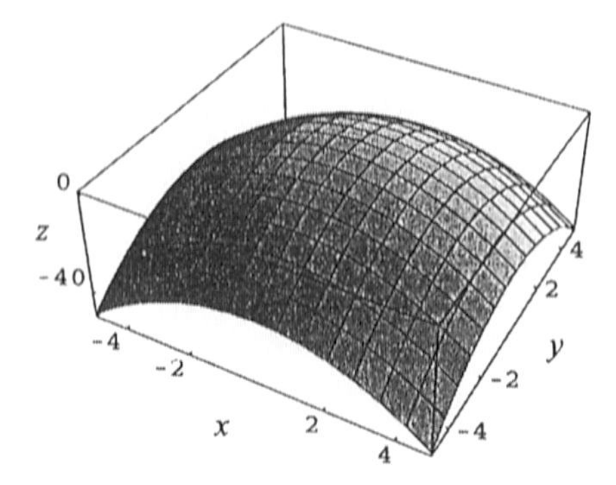

Figure 4.12 Three-dimensional surface plot of z as a function of x and y. This is a surface and the total differential is a plane tangential to the surface at the point of interest

this on two-dimensional paper but Figure 4.12 shows a typical example. The function now appears as a surface rather than a line. The total derivative with respect to both independent variables is now a plane tangential to the point of interest. The partial derivative is a line along this plane, parallel to the variable axis of interest.

The relationship between the two possible partial derivatives and the complete derivative is given by a fundamental theorem of partial differentiation.

4.6.3 Fundamental Theorem of Partial Differentiation

If we consider the rectangle in Figure 4.13, then the area, A, of the rectangle is a function of the lengths of the two sides, x and y.

$$A = x \cdot y \tag{4.115}$$

where x and y are both independent of each other. If the side x changes by Δx and y is held constant then the change in A is given by

$$\Delta A = y \cdot \Delta x \tag{4.116}$$

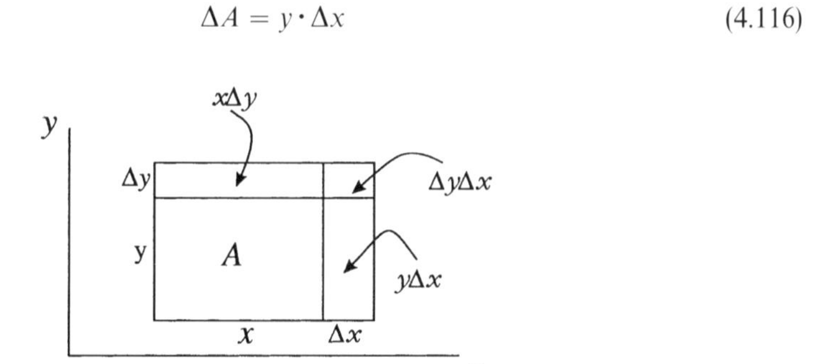

Figure 4.13 The area of the rectangle is a function of the lengths of its sides which are independent of each other. The change in the area is the sum of the areas indicated. These are functions of x, y, Δx and Δy

and consequently

$$\left(\frac{\Delta A}{\Delta x}\right)_y = y \tag{4.117}$$

where the subscript is used to indicate that y is held constant. By the same argument, if y changes by Δy and x is held constant, then

$$\Delta A = \Delta y \cdot x \tag{4.118}$$

and

$$\left(\frac{\Delta A}{\Delta y}\right)_x = x \tag{4.119}$$

If both variables change then the change in the area is the sum of these quantities plus the extra in the corner (Figure 4.13)

$$\Delta A = y\Delta x + x\Delta y + \Delta x\Delta y \tag{4.120}$$

We can then substitute for x and y from equations (4.117) and (4.119)

$$\Delta A = \left(\frac{\Delta A}{\Delta x}\right)_y \Delta x + \left(\frac{\Delta A}{\Delta y}\right)_x \Delta y + \Delta x\Delta y \tag{4.121}$$

If the changes are all very small then this notation can be replaced by differential notation

$$\mathrm{d}A = \left(\frac{\partial A}{\partial x}\right)_y \mathrm{d}x + \left(\frac{\partial A}{\partial y}\right)_x \mathrm{d}y + \mathrm{d}x\mathrm{d}y \tag{4.122}$$

Furthermore, since the product of two small changes is an even smaller change, the term $\mathrm{d}x\mathrm{d}y$ can be effectively ignored and so the fundamental equation for the differential $\mathrm{d}A$ is

$$\mathrm{d}A = \left(\frac{\partial A}{\partial x}\right)_y \mathrm{d}x + \left(\frac{\partial A}{\partial y}\right)_x \mathrm{d}y \tag{4.123}$$

This equation shows how the total differential, $\mathrm{d}A$, is related to the independent variables x and y by the partial derivatives $\partial A/\partial x$ and $\partial A/\partial y$.

Although we will not expand greatly here much use is made of this theorem in thermodynamics. A differential which can be written in this form is called an *exact differential.* Thermodynamic functions which can be written as exact differentials, in the form of equation (4.123), are called *state functions*. These are functions such as the internal energy, enthalpy, entropy and free energy which are functions of the present state of a system and not the way in which it was prepared. For example, the first law of thermodynamics is often written in the form of equation (4.110) where U is internal energy, q is heat input and w is the work done on a system. Heat is related to temperature and work is a function of volume. We might therefore conclude that the internal energy of a system is a function of temperature and volume. If these two are independent and U is indeed a state function we can write $\mathrm{d}U$ as an exact differential.

$$\mathrm{d}U = \left(\frac{\partial U}{\partial T}\right)_V \mathrm{d}T + \left(\frac{\partial U}{\partial V}\right)_T \mathrm{d}V \tag{4.124}$$

The partial derivative $(\partial U/\partial T)_V$ is defined as the heat capacity at constant volume, C_V, and is measurable by experiment. $(\partial U/\partial V)_T$ for an ideal gas is zero (because there are no interactions between molecules, moving them further apart does not affect the internal energy—which is a function of these interactions) but for a real gas can be measured by experiment.

Such is the process of theoretical thermodynamics. Without necessarily knowing the mathematical relationship between given functions we use sound mathematical principles to turn ideas into working equations which may be tested by experiment.

4.7 Concluding Remarks

In this chapter we have introduced the process of differentiation, how it may be used to find the rate of change of a function and how this may be used to find points of interest on a curve, e.g., maxima and minima. We have also introduced the concept of the partial derivative and how this relates to the total differential of a function of two variables.

There are a number of important properties of partial derivatives which have been omitted. These are used in deriving some important thermodynamic relationships, such as the 'Maxwell relations'. The purpose of this chapter, however, has been to introduce differential calculus at a fundamental level in preparation for Chapters 4 and 5, not to present a treatise on thermodynamics. The interested student will find derivations of these equations in both thermodynamics and advanced mathematics textbooks.

One of the problems for the chemist is to find the exact mathematical relationship between variables. It is often difficult to do this directly from an experiment without going through a protracted trial-and-error process of guessing equations which may or may not be correct. Derivatives are very important in chemistry because they can be determined by experiment. Working backwards from the derivative, by the process of integration, results in an expression relating the variables of interest to each other. This is the subject of Chapters 5 and 6, 'Integration' and 'Differential Equations'.

Problems

4.1 Find the derivative, $\mathrm{d}y/\mathrm{d}x$ for the following functions:

(a) $y = 6$ (P4.1)
(b) $y = 6x$ (P4.2)
(c) $y = 6x^2 + 2x$ (P4.3)
(d) $= 1/x$ (P4.4)
(e) $y = \sqrt{x}$ (P4.5)

4.2 Find the derivative $\mathrm{d}y/\mathrm{d}x$ for the following functions:

(a) $y = \ln(x)$ (P4.6)
(b) $y = \frac{c}{x}$ where c is a constant (P4.7)
(c) $y = \exp(x)$ (P4.8)
(d) $y = \sin(x)$ (P4.9)
(e) $y = \ln(x)$ (P4.10)

4.3 In the following problems decide how to break down the function into a combination of functions and then use the correct reduction formula to find the derivative.

(a) $y = \sin(x)\cos(x)$ (P4.11)
(b) $y = \sin^2(x)$ (P4.12)
(c) $y = \tan(x)$ (P4.13)
(d) $y = \log_{10}(x)$ (Hint: write the log in terms of $\log_e$) (P4.14)
(e) $y = x \cdot \ln(x)$ (P4.15)

4.4 In the following problems decide how to break down the function into a combination of functions and then use the correct reduction formula to find the derivative.

(a) $y = \ln(x^2 + 2)$ (P4.16)
(b) $y = \sin(x^2)$ (P4.17)
(c) $y = \exp(\sin(x))$ (P4.18)
(d) $y = \exp(-x^2)$ (P4.19)

4.5 Find the second derivative, $\mathrm{d}^2y/\mathrm{d}x^2$, for the following functions:

(a) $y = x^3 + 2x$ (P4.20)
(b) $y = c/x$ where c is a constant (P4.21)
(c) $y = \sin(x)$ (P4.22)
(d) $y = x.\exp(-x)$ (P4.23)
(e) $y = \exp(-kx)/x$ where k is a constant (P4.24)

4.6 Find the first and second derivative of potential V with respect to distance r for the relationship

$$V = \frac{q}{r} \tag{P4.25}$$

thus proving that

$$\frac{\mathrm{d}^2V}{\mathrm{d}r^2} + \frac{2}{r}\frac{\mathrm{d}V}{\mathrm{d}r} = 0 \tag{P4.26}$$

4.7 Find the first and second derivative of the equation

$$x = A \cdot \sin(kt) + B \cdot \cos(kt) \tag{P4.27}$$

Simplify the second derivative as much as possible.

4.8 The probability P of finding an electron at a distance r from a hydrogen nucleus is given by the equation

$$P = \frac{4r^2}{a_0^3}\mathrm{e}^{-2r/a_0} \tag{P4.28}$$

(a) Define this equation as a combination of other functions, i.e., decide which reduction formulae to use.
(b) Find the derivative of P with respect to r.
(c) Find the value of r when P is a maximum, i.e., the most probable value of r.

4.9 This problem concerns two fundamental definitions for the Gibbs function, G, and the internal energy, U:

$$G = H - TS \tag{P4.29}$$

$$H = U + PV \tag{P4.30}$$

(a) Substitute for H in equation (P4.29) to give an expression for G.
(b) Use this expression to derive an equation for dG in terms of P, V, T, S and their respective differentials.
(c) Under conditions where the only mechanical work is due to expansion of a gas and is given by

$$\mathrm{d}U = -P\mathrm{d}V \tag{P4.31}$$

and entropy changes are zero this equation can be simplified. Show that this is the case.
(d) G is a state function and therefore dG can be written as an exact differential. If G is a function of pressure, P, and temperature, T, write an expression for dG in terms of these variables.
(e) Compare the results for (c) and (d) to give expressions for the volume, V, and the entropy, S, of this system.

5 Integral Calculus

5.1 Introduction

Integration may be defined in two ways. The first is as the inverse operation of differentiation, called *indefinite integration.* The second is as a process of summation, called *definite integration.* The reasoning behind these definitions will become apparent as they are described. Definite integration is useful in finding the areas under curves and in statistical analysis. Indefinite integration is useful in finding a function for which we already know the derivative. As will be seen in Chapter 6, this occurs frequently in chemistry.

The method of finding an area is to divide the large area into smaller strips and then add up the areas of these to find the total. This is exactly analogous to the summation process used in thermodynamics. Here the smaller, sometimes hypothetical changes, are added up to produce an expression which describes the large, observable change in which we are interested.

The mechanics of performing an integration is almost identical for both of these processes and the two are linked by a fundamental theorem.

5.2 Indefinite Integration

5.2.1 Definition

Most mathematical texts will define the process of indefinite integration as follows. Suppose that we are given a function $\mathrm{f}(x)$ such that the derivative is another function $\mathrm{F}(x)$

$$\frac{\mathrm{d}(\mathrm{f}(x))}{\mathrm{d}x} = \mathrm{F}(x) \tag{5.1}$$

then the function $\mathrm{f}(x)$ is given by

$$\mathrm{f}(x) = \int \mathrm{F}(x)\mathrm{d}x \tag{5.2}$$

where the sign $\int$ means that we are to integrate the function called the *integrand* which comes between this and the differential which follows. Equation (5.2) reads as 'the function f(x) is equal to the integral of the function F(x) with respect to dx'. The sign $\int$ is an elongated s and refers to the process of summation.

The following example will illustrate equation (5.2). If we are told that the derivative of a function of x is equal to $2x$ then what is the function itself? By equation (5.2) we can write

$$\mathrm{f}(x) = \int 2x\mathrm{d}x \tag{5.3}$$

Having been told that integration is the reverse of differentiation we can solve this to find f(x). By reference to Table 4.1 (Chapter 4) we know that the $2x$ is the derivative of x^2. The solution to equation (5.3) is therefore

$$\int 2x\mathrm{d}x = x^2 \tag{5.4}$$

or that f(x) is equal to x^2. Equation (5.4) is, however, not strictly correct. We know from Chapter 4 that $2x$ is also the derivative of the following functions.

$$\begin{aligned} \mathrm{f}(x) &= x^2 + 2 \\ \mathrm{f}(x) &= x^2 + 10 \\ \mathrm{f}(x) &= x^2 + 265 \end{aligned} \tag{5.5}$$

This is because differentiation tells us the slope of a function. Constant terms are all reduced to zero. We must be aware of this when performing the inverse function and in general we write the equations (5.5) as

$$\mathrm{f}(x) = x^2 + C \tag{5.6}$$

where C is an unknown constant. Because of this equation (5.4) should be written as

$$\int 2x\mathrm{d}x = x^2 + C \tag{5.7}$$

C is called the constant of integration. Because C is unknown we cannot define the integral precisely, it is indefinite and hence the process is called indefinite integration. We can only find the value of C if we are given extra information. For example, if we are also told that f(x) is equal to five when x is equal to two then we can say that C is equal to one because $5 = 2^2 + 1$. The extra information is called the *boundary condition*.

5.2.2 Indefinite Integration—the Method

Having been told that this is the reverse of differentiation we can write the method as the reverse order of Rules 2 and 3 given for differentiation in Chapter 4. These new rules are

(1) The index of the function of x is increased by one.
(2) The function is divided by the new index.

Rule 1 given in Chapter 4 still holds. Each term (separated by a plus or a minus sign) may be considered individually.

We may write these rules in the form of equations.

$$\int x^n = \frac{x^{n+1}}{n+1} + C \tag{5.8}$$

and

$$\int x^n + x^m \mathrm{d}x = \int x^n \mathrm{d}x + \int x^m \mathrm{d}x$$
$$= \frac{x^{n+1}}{n+1} + \frac{x^{m+1}}{m+1} + C \tag{5.9}$$

Following from the definition of the derivative of $y = mx^n$ given in Table 4.1 a constant term in the integrand may be taken outside the integral sign.

$$\int mx^n \mathrm{d}x = m \int x^n \mathrm{d}x$$
$$= \frac{m \cdot x^{n+1}}{n+1} \tag{5.10}$$

This may be combined with the above rule concerning sums and differences of integrals. For example,

$$\int 3x^2 + 4x^4 \mathrm{d}x = 3 \int x^2 \mathrm{d}x + 4 \int x^4 \mathrm{d}x$$
$$= \frac{3x^3}{3} + \frac{4x^5}{5} + C \tag{5.11}$$

Note that in these equations the constants of integration are all combined into one single constant. The sum of a number of constants is always equal to another constant.

5.2.3 Examples of Indefinite Integrals

For simple functions the above rules may be used to calculate integrals. However, for more complex functions it is usually necessary to rely on our knowledge of differentiation. For example, for the function $y = x^{-2}$ the rules still work and the indefinite integral is $-1/x + C$. For the integrand $1/x$, when the index, -1, is increased by 1 it becomes equal to zero and we get x^0 which is then, according to Rule 2 above, divided by zero. This is, of course, infinity and the method does not work. From our knowledge of differentiation, however, we know that $1/x$ is the derivative of $\ln(x)$. Therefore, the integral of $1/x$ is given by $\ln(x) + C$. Table 5.1 lists the integrals of several simple functions. You should note the similarity between this and Table 4.1.

5.3 Definite Integration

5.3.1 Definition as the Area under a Curve

Definite integration is the most useful type of integration. It is used widely in thermodynamic and kinetic analyses. For example, the work (the energy of producing mechanical changes, symbol w) involved in expanding a gas against an external pressure P is given as the product of the pressure and the volume change

$$w = -P\Delta V \tag{5.12}$$

Table 5.1 Some standard indefinite integrals

Function	Integral	
$y = mx^n$	$\int y\mathrm{d}x = \frac{m \cdot x^{n+1}}{n+1} + C$	(a)
$y = \mathrm{e}^x$	$\int y\mathrm{d}x = \mathrm{e}^x + C$	(b)
$y = \mathrm{e}^{mx}$	$\int y\mathrm{d}x = \frac{1}{m}\mathrm{e}^{mx} + C$	(c)
$y = m^x$	$\int y\mathrm{d}x = \frac{1}{\ln(m)}m^x + C$	(d)
$y = \ln(x)$	$\int y\mathrm{d}x = x \cdot \ln(x) - x + C$	(e)
$y = \sin(x)$	$\int y\mathrm{d}x = -\cos(x) + C$	(f)
$y = \cos(x)$	$\int y\mathrm{d}x = \sin(x) + C$	(g)

Notes: Equations (a), (b), (f) and (g) are essentially the inverse of derivatives. Equations (c) and (d) may be found by the method of substitution, Section 5.4.3, and equation (e) is found by the method of integrating by parts, Section 5.4.5.

Figure 5.1(a) shows a plot of pressure against volume for this process where the pressure is constant. Note that the gas will only expand if the external pressure P is less than that of the gas confined to a cylinder. However, the gas may expand against a constant external pressure. It can be seen from this simple plot that the product of P and ΔV is the area under the graph. For more complex plots, Figure 5.1(b), where the pressure changes, the

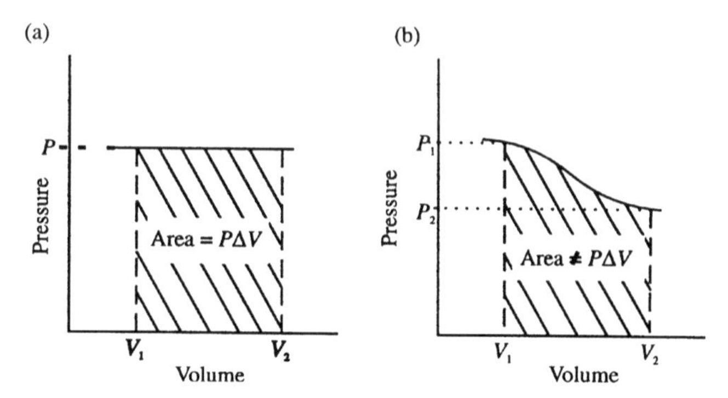

Figure 5.1 (a) The work involved in expanding a gas against a constant pressure is defined as $P\Delta V$, which is the area under the curve between the limits specified. (b) If the pressure also changes then the definition of work done is not so simple. The result still amounts to the area bounded by the curve and the V axis though we must now define P as a function of V and use the method of integral calculus to find the result

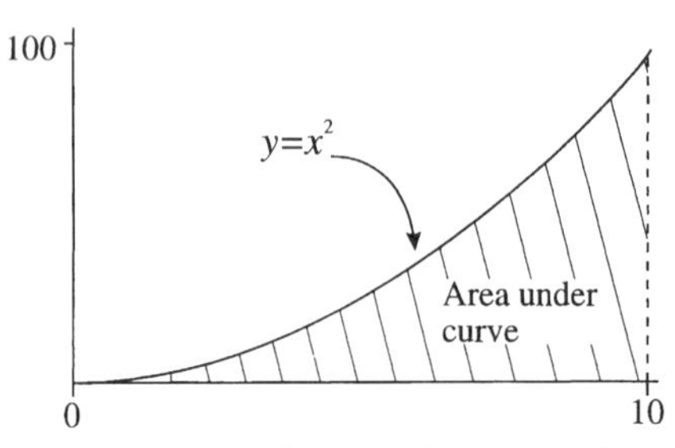

Figure 5.2 The definition of the area under the curve is that area bounded by the curve and the x axis. In definite integration we specify the limits of x

area is still the equivalent of the work done, but the relationship is no longer quite so simple.

We cannot calculate the area as simply the product of the pressure and the volume change. The process of definite integration, however, allows us to calculate the area under a curve just like this.

Before giving specific examples it will be useful to present some evidence that finding an area under a graph is equivalent to the process of integration and that this is the inverse of differentiation.

5.3.2 The Area under the Curve $y = x^2$

A rigorous mathematical proof of the method of integration is beyond the scope of a book of basic mathematics and so we will use a geometric approach. In doing this our aim is to show that we can find an area under a curve as a summation and that this process is indeed equivalent to the inverse of differentiation.

Figure 5.2 shows, once again, part of the curve $y = x^2$. We will estimate the area under the curve between the points $x = 0$ and $x = 10$. In defining the problem in this way we have already made the first steps into definite integration. That is, we have set the limits, or boundary conditions. We do not necessarily have to set the limits of y because we know that we may calculate y from its relationship to x. In more general analyses, where we have not been given the relationship (because we are trying to find it), we may have to set y to given values.

This approach to estimating the area is to divide it into a number of rectangles. This has the advantage that it is easy to calculate the area of a rectangle. The disadvantage is that the estimate is only an approximation. In Figure 5.3(a) there are four rectangles of width two units, all of which fit under the curve. As can be seen from this diagram, the sum of the areas of these rectangles falls short of what the true value of the area must be. In Figure 5.3(b) we have used rectangles of the same width but this time they all rise above the curve. This is the maximum number of rectangles of width two which will fit in the region zero to ten on the x axis. The sum of the areas of these five rectangles will be an overestimate of the area. The true area is, however, somewhere between the two estimates that we have made.

Figure 5.3(c–d) shows what happens if we increase the number of rectangles used to estimate the area. The maximum number of rectangles, width one, is ten. The overestimate (Figure 5.3(d)) uses all ten of these whilst the underestimate (Figure 5.3(c)) uses nine. Again the true value lies between that obtained from these calculations. It can be seen

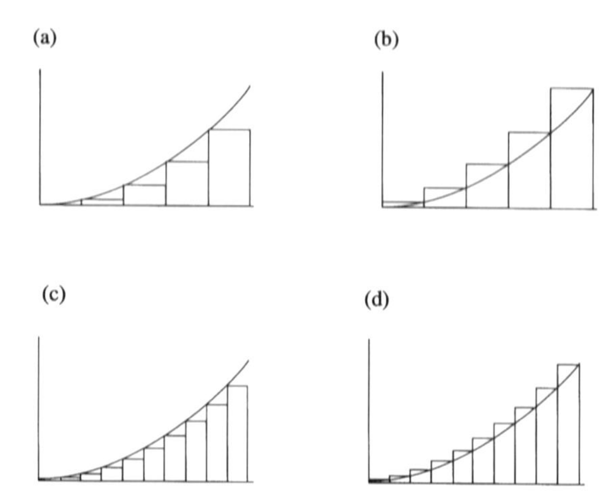

Figure 5.3 We may estimate the area under the curve by drawing rectangles whose area is easy to calculate. We may draw rectangles which lie beneath the curve, (a) and (c). The area estimates will be less than the true area. Alternatively we may draw rectangles which rise above the curve, (b) and (d). The area calculated from these will be an overestimate. Using more and more rectangles leads to greater accuracy, as shown by the difference between (a) and (b), four and five rectangles, and (c) and (d), nine and ten rectangles

from the diagram that this calculation appears to be closer to the true area than the previous one which used only five rectangles.

Table 5.2 shows results of calculations of these areas and compares them to the area estimated using increasing numbers of rectangles. Figure 5.4 shows plots of upper and lower estimates for each of these calculations plotted against $1/n$, where n is the number of rectangles used. As n increases to an infinite number (i.e., $1/n$ tends to zero) then it is seen that the two estimates converge to a single value. This is approximately 333. The area under the curve $y = x^2$, between the limits of x equals zero and ten is therefore estimated to be 333.

The next step is to generalise this procedure and try and produce an expression which may be recognised.

5.3.3 Area under the Curve $y = x^2$. A General Approach

We now calculate the area under this curve between $x = 0$ and some arbitrary point $x = m$. We divide the area into an arbitrary number, n rectangles. This is illustrated in

Table 5.2 Estimates of the area under $y = x^2$

Max. number rectangles, n	Lower estimate	Upper estimate	Difference
5	240	440	200
10	285	385	100
20	309	359	50
40	321	346	25
100	328	338	10

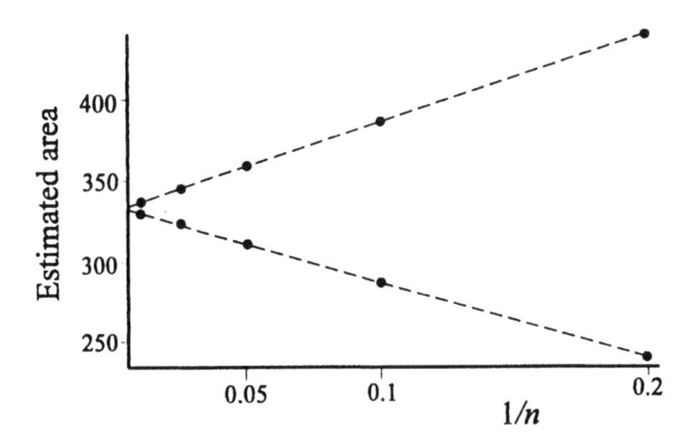

Figure 5.4 Plot of estimated area under the curve versus $1/n$, the number of rectangles used. Both plots are linear and appear to converge at the same intercept, approximately 333

Figure 5.5. The points of intersection of the vertical edges of the rectangles occur at the points $x = m/n$, $2m/n$, $3m/n$, $4m/n$ up to $x = m - 1$ and $x = m$. The underestimate of the area utilises only $n - 1$ of these rectangles (Figure 5.5(a), labelled 'area a') whereas the overestimate uses all of them (Figure 5.5(b), labelled 'area b'). As above we can say that the true area, A, lies between these two areas. In mathematical terms this may be written

$$\text{calculated area a} < A < \text{calculated area b}$$

The heights of the rectangles may be calculated from the relation $y = x^2$. The the areas of the rectangles will be given by the product of m/n and the height of each rectangle.

$$\text{Area} = \frac{m}{n} \cdot \left(\frac{m}{n}\right)^2 \tag{5.13}$$

The total will be given by the sum

$$\text{Area b} = \frac{m}{n}\left(\frac{m}{n}\right)^2 + \frac{m}{n}\left(\frac{2m}{n}\right)^2 + \frac{m}{n}\left(\frac{3m}{n}\right)^2 + \ldots + \frac{m}{n}\left(\frac{nm}{n}\right)^2 \tag{5.14}$$

The factor $(m/n)^3$ is common to each of these terms and so may be removed, giving

$$\text{Area b} = \frac{m^3}{n^3}[1^2 + 2^2 + 3^2 + \ldots + n^2] \tag{5.15}$$

This expression may be validated by checking the results in Table 5.2. There is, however, an expression which may be substituted for the sum in square brackets. This is the sum of the squares of the first n natural numbers,* which is

$$\sum_{r=1}^{n} r^2 = \frac{1}{6}n(n + 1)(2n + 1) \tag{5.16}$$

This may be substituted into equation (5.15) to give

$$\text{Area b} = \frac{m^3}{n^3}\left[\frac{1}{6}n(n + 1)(2n + 1)\right] \tag{5.17}$$

*The proof of this can be found in standard mathematics textbooks.

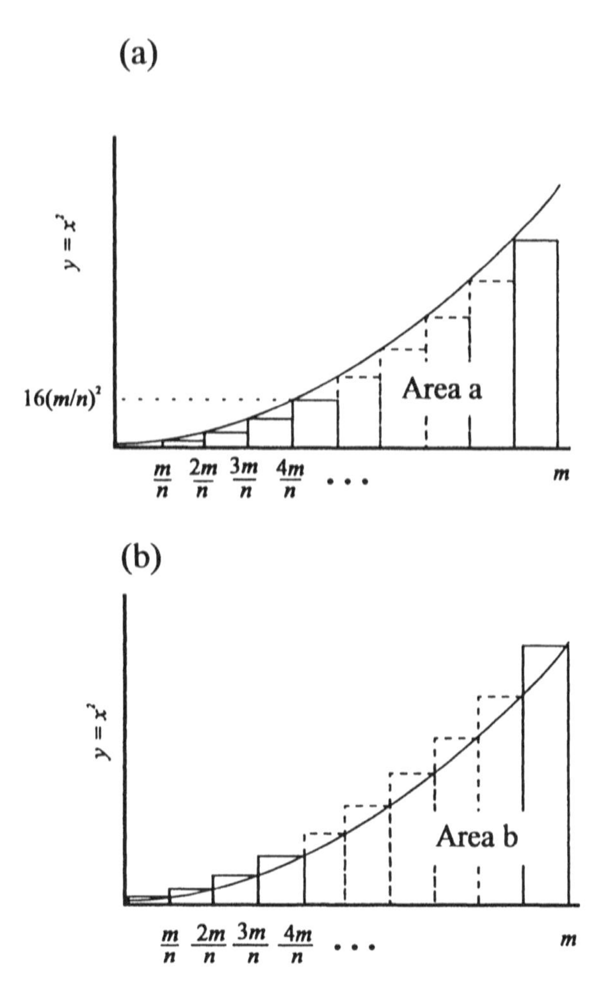

Figure 5.5 The general approach to finding the area under the curve $y = x^2$. The limits are $x = 0$ and $x = m$, divided into n rectangles of width m/n. See text for derivation

Removing the factor 1/6 and expanding the bracketed terms produces

$$\text{Area b} = \frac{m^3}{6n^3}[2n^3 + 3n^2 + n] \tag{5.18}$$

The n terms inside and outside the bracket will cancel to produce

$$\text{Area b} = \frac{m^3}{6}\left[2 + \frac{3}{n} + \frac{1}{n^2}\right] \tag{5.19}$$

This relates the area b to the upper limit m and the number of rectangles used. The area a may be estimated in the same way but this time only $(n - 1)$ rectangles are used. Also, the height of the largest is dictated by the $(n - 1)$th value of x and the analogous equation to (5.15) is

$$\text{Area a} = \frac{m^3}{n^3}[1^2 + 2^2 + 3^2 + \ldots + (n-1)^2] \tag{5.20}$$

Substitution of the expression for the sum of the squares of the first $(n - 1)$ natural numbers

$$\sum_{r=1}^{n-1} r^2 = \frac{1}{6}n(n-1)(2n-1) \tag{5.21}$$

followed by expansion gives

$$\text{Area a} = \frac{m^3}{6n^3}[2n^3 - 3n^2 + n] \tag{5.22}$$

Cancelling the n terms gives

$$\text{Area a} = \frac{m^3}{6}\left[2 - \frac{3}{n} + \frac{1}{n^2}\right] \tag{5.23}$$

We can now write an inequality defining the range for the upper and lower estimates of the true area A. This lies between areas a and b.

$$\frac{m^3}{6}\left[2 - \frac{3}{n} + \frac{1}{n^2}\right] < A < \frac{m^3}{6}\left[2 + \frac{3}{n} + \frac{1}{n^2}\right] \tag{5.24}$$

In the tradition of calculus we will now allow the number of rectangles used to increase to infinity. In doing so the width of each rectangle approaches zero as do the $3/n$ and $1/n^2$ terms in the above inequality.

$$\frac{m^3}{6}[2 - 0 + 0] < A < \frac{m^3}{6}[2 + 0 + 0] \tag{5.25}$$

Both areas have approached the same value and we can write

$$A = \frac{2m^3}{6} = \frac{m^3}{3} \tag{5.26}$$

This is the result we have been looking for. Having concentrated on a point $x = m$, so that $y = m^2$, we have found that the area under the curve (found by a process of summation) is given by $m^3/3$. This number could also be produced by doing the inverse of differentiation, adding one to the index and then dividing by the new index. Therefore we may conclude from this example that integration as defined by a process of summation is also the inverse of differentiation.

We may check this against the result obtained in the previous section. In this example m was taken to be ten. Therefore

$$\frac{m^3}{3} = \frac{10^3}{3} = \frac{1000}{3} = 333.33 \tag{5.27}$$

which is essentially the same result as obtained by estimating the area by counting rectangles.

5.3.4 Area under the Curve $y = x^2$ between Different Limits

The method in Section 5.3.3 is suitable only if we need to find the area under a curve between some number and zero. It would be much more useful to be able to integrate between any limits. In order to define the limits we must introduce the notation used to specify what the limits are. In the above examples we have integrated the curve $y = x^2$ between $x = 0$ and $x = 10$. This is written as

$$\int_{x=0}^{x=10} x^2 \mathrm{d}x \tag{5.28}$$

It is usually assumed that we are intending to integrate between values of the independent variable and therefore these are usually omitted. That is, we write

$$\int_{0}^{10} x^2 \mathrm{d}x \tag{5.29}$$

Now suppose we needed to integrate $y = x^2$ between $x = 3$ and $x = 10$. How this is achieved can be seen by reference to Figure 5.6. Using exactly the same method as above we can find the area between $x = 3$ and $x = 0$. The area under the curve between $x = 3$ and $x = 10$ is merely the area between ten and zero minus the area between three and zero. This may be written as

$$\int_{3}^{10} \mathrm{F}(x)\mathrm{d}x = \int_{0}^{10} \mathrm{F}(x)\mathrm{d}x - \int_{0}^{3} \mathrm{F}(x)\mathrm{d}x \tag{5.30}$$

Recalling our original definition of the indefinite integral (equation (5.2)) we can write this as

$$\int_{3}^{10} \mathrm{F}(x)\mathrm{d}x = \mathrm{f}(x = 10) - \mathrm{f}(x = 3) \tag{5.31}$$

or more generally for integrating between two limits a and b

$$\int_{a}^{b} \mathrm{F}(x)\mathrm{d}x = \mathrm{f}(b) - \mathrm{f}(a) \tag{5.32}$$

where $\mathrm{f}(a)$ and $\mathrm{f}(b)$ are the indefinite integrals of x given the values a and b. This equation is very important as it links the process of definite integration to the process of indefinite integration. As such it is fundamental to integral calculus.

It is also interesting to note that equation (5.32) manages to solve the problem of

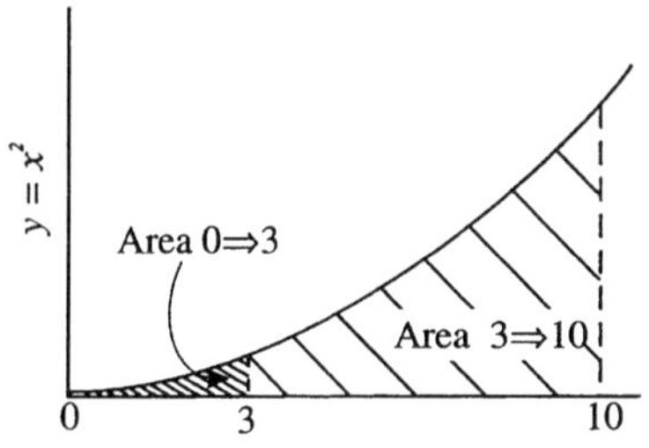

Figure 5.6 The total area under a curve is the sum of the areas 0 to 3 and 3 to 10. The area 3 to 10 is therefore the difference between the total (0 to 10) and 0 to 3

defining the constant of integration. Continuing with our example of integrating $y = x^2$ between the x values of three and ten: the indefinite integral is $x^3/3 + C$, regardless of the specific value of x. Equation (5.32) is therefore written as

$$\begin{aligned}\int_3^{10} x^2 \mathrm{d}x &= \frac{10^3}{3} + C - \left[\frac{3^3}{3} + C\right] \\ &= \frac{10^3}{3} + C - \frac{3^3}{3} - C \\ &= \frac{10^3}{3} - \frac{3^3}{3}\end{aligned} \tag{5.33}$$

The result may then be calculated directly as $333.33 - 9 = 324.33$.

5.3.5 Properties of Definite Integrals

There are a number of important properties of integrals which may need to be considered. These include addition and subtraction of integrals and splitting them up into smaller regions. Taking note of equation (5.30), this could be rearranged to give

$$\int_0^{10} \mathrm{F}(x)\mathrm{d}x = \int_0^{3} \mathrm{F}(x)\mathrm{d}x + \int_3^{10} \mathrm{F}(x)\mathrm{d}x \tag{5.34}$$

Thus we can calculate an integral as the sum of two smaller integrals. A more general equation is

$$\int_a^{b} \mathrm{F}(x)\mathrm{d}x = \int_a^{c} \mathrm{F}(x)\mathrm{d}x + \int_c^{b} \mathrm{F}(x)\mathrm{d}x \tag{5.35}$$

where $a < c < b$. This may be proven by writing the integrals in terms of the function $\mathrm{f}(x)$.

$$\begin{aligned}\int_a^{c} \mathrm{F}(x)\mathrm{d}x + \int_c^{b} \mathrm{F}(x)\mathrm{d}x &= \mathrm{f}(c) - \mathrm{f}(a) + [\mathrm{f}(b) - \mathrm{f}(c)] \\ &= \mathrm{f}(b) - \mathrm{f}(a)\end{aligned} \tag{5.36}$$

The $\mathrm{f}(c)$ terms cancel and the result is the definition of the definite integral between a and b.

Splitting an integral up into smaller parts may be necessary if a function changes sign within the range. For example, the function $y = \sin(x)$ becomes negative at values of x greater than π radians. If we try to integrate between zero and 2π radians we get

$$\begin{aligned}\int_0^{2\pi} \sin(x)\mathrm{d}x &= -\cos(2\pi) - (-\cos(0)) \\ &= -1 - (-1) \\ &= 0\end{aligned} \tag{5.37}$$

This result is clearly incorrect as Figure 5.7 indicates. Our common experience does not normally recognise area as having negative values. Nevertheless, arithmetic manipulation has led to the result of the area below the x axis being given a negative value. In order to

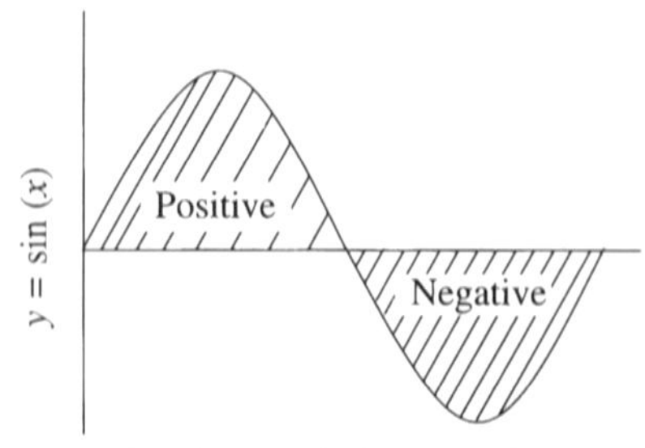

Figure 5.7 A function having a negative sign results in a negative assignment of area. This can lead to errors in calculating the area bounded by the curve and the x axis

calculate the real area which is bounded by the curve and the x axis we must split the curve up into two parts. We must also take account of the negative sign generated by calculating the area under a curve below the x axis. The first part is achieved by noting that this curve crosses the x axis at the point $(\pi, 0)$. This provides a convenient break. The area may then be given by

$$A = \int_0^{\pi} \sin(x)\mathrm{d}x + \int_{\pi}^{2\pi} \sin(x)\mathrm{d}x \tag{5.38}$$

Solving these integrals leads us to

$$\begin{aligned} A &= [-\cos(\pi) - (-\cos(0))] + [-\cos(2\pi) - (-\cos(\pi))] \\ &= [1 + 1] - [-1 - 1] \\ &= 2 - 2 = 0 \end{aligned} \tag{5.39}$$

This is still wrong and so to reverse the effect of the negative area of the second integral we must subtract it. This has the desired effect and equation (5.38) becomes

$$A = \int_0^{\pi} \sin(x)\mathrm{d}x - \int_{\pi}^{2\pi} \sin(x)\mathrm{d}x \tag{5.40}$$

which is

$$A = 2 + 2 = 4 \tag{5.41}$$

The same result may be obtained if we recognise the fact that the function is symmetrical about the point $(\pi, 0)$. In this case equation (5.38) can be interpreted as

$$A = 2\int_0^{\pi} \sin(x)\mathrm{d}x \tag{5.42}$$

Such manipulations are only important in more advanced topics such as quantum mechanics. If an integral is found to be equal to zero it may be used as a measure of the symmetry of a function. Symmetry is a concept which pervades chemistry and quantum mechanics in particular and this use of integration is very important. Even at this level, however, it is important to know something about the function we are investigating in order to obtain the correct solution to the problem in hand.

5.4 Methods of Integration

In this section we will describe some of the methods used to perform an integration. This will include the direct method, recognising integrals from Table 5.1, and some more complex methods used when this is inadequate. The first step is to present the common notation which is used.

5.4.1 Notation

Recalling equation (5.32) describing the definite integral, this is normally written in the form

$$\int_a^b \mathrm{F}(x)\mathrm{d}x = [\mathrm{f}(x)]_a^b \tag{5.43}$$

in which the square brackets are used to abbreviate the subtraction of the integrals. The next step is to use the values of $\mathrm{f}(a)$ and $\mathrm{f}(b)$ and perform this sum

$$[\mathrm{f}(x)]_a^b = \mathrm{f}(b) - \mathrm{f}(a) \tag{5.44}$$

For example, integrating x^2 between limits of $x = 2$ and $x = 3$ we get

$$\left[\frac{x^3}{3}\right]_2^3 = \frac{3^3}{3} - \frac{2^3}{3} = \frac{27}{3} - \frac{8}{3} = \frac{19}{3} \tag{5.45}$$

5.4.2 Direct Integration

Direct integration is used when we can recognise the integrand as the derivative of some function. Table 5.1 lists common functions which may be integrated directly. Some examples are now given.

Example—the integral of one

The simplest integral is that of 'one' and this occurs in many thermodynamic equations. For example

$$\mathrm{d}U = q + w \tag{5.46}$$

which tells us how the change in internal energy ($\mathrm{d}U$) can be calculated using the heat input and the work done on or by a system. This is, however, only an infinitesimal change in U. To calculate the large, observable, change in U we must integrate the term $\mathrm{d}U$ between given states, e.g., a and b (which may be functions of temperature, pressure etc.). The integral is written as

$$\int_a^b \mathrm{d}U \tag{5.47}$$

To solve the integral we recognise that this is the same as

$$\int_a^b 1\,\mathrm{d}U \tag{5.48}$$

which is the same as

$$\int_a^b U^0 \, dU \tag{5.49}$$

Applying the normal rules of integration we get

$$\int_a^b U^0 dU = \left[\frac{U^1}{1}\right]_a^b = U_b - U_a \tag{5.50}$$

where U_b and U_a are the values of U under the conditions specified by b and a. This is usually written using the Δ symbol to represent the large change in a property, i.e.,

$$U_b - U_a = \Delta U \tag{5.51}$$

Consequently equation (5.50) is usually written as

$$\int_a^b U^0 \, dU = \Delta U \tag{5.52}$$

Example—zero order reaction kinetics

In the equation describing the concentration dependence of a zero order reaction a similar integrand to the one above appears. The rate of a zero order reaction is independent of concentration and this is represented as

$$kt = \int_a^c \frac{dc}{c^0} \tag{5.53}$$

where a is the concentration at the start of the experiment at time zero and c is the concentration at time t. The reason for the c^0 term appearing as the denominator in this integrand will become apparent in Chapter 6. However, as c^0 is the same as one this equation is the same as

$$kt = \int_a^c dc \tag{5.54}$$

This gives essentially the same result as for the thermodynamic example above.

$$\begin{aligned} kt &= [c]_a^c \\ &= c - a \end{aligned} \tag{5.55}$$

The zero order reaction shows a linear dependence on concentration of the substance of interest, since a is constant.

Example—first order kinetics

Analysis of a first order mechanism leads to the equation

$$kt = \int_a^c \frac{dc}{c} \tag{5.56}$$

where c and a are the concentrations at time t and time zero as above. The term $1/c$ may be recognised as the derivative of $\ln(c)$ and thus the integral is given as

$$\begin{aligned} kt &= [\ln(c)]_a^c \\ &= \ln(c) - \ln(a) \\ &= \ln\left(\frac{c}{a}\right) \end{aligned} \tag{5.57}$$

Thus the concentration changes in a logarithmic fashion as time progresses for a first order reaction.

Example—second order kinetics

For a simple second order reaction the rate equation may be written as

$$kt = \int_a^c \frac{\mathrm{d}c}{c^2} \tag{5.58}$$

with c and a as above. The integral, with reference to Table 5.1 is

$$kt = \left[\frac{-1}{c}\right]_a^c \tag{5.59}$$

As a result a plot of $1/c$ against time should be linear for a second order reaction.

Example—irreversible expansion of a gas

The work, w, involved in expanding a gas is given by

$$w = -P\mathrm{d}V \tag{5.60}$$

If the gas is expanded against a constant external pressure this equation is solved by writing

$$w = -P\int_{V_1}^{V_2} \mathrm{d}V \tag{5.61}$$

which becomes

$$\begin{aligned} w &= -P[V]_{V_1}^{V_2} \\ &= -P\Delta V \end{aligned} \tag{5.62}$$

This would be the case if the expansion was carried out irreversibly, that is, the changes were large enough that they could not be reversed by an infinitesimal change in the opposite direction.

Example—reversible expansion of a gas

If the change was carried out reversibly then the pressure would have to be altered as the expansion continued so that the external pressure was always infinitesimally different from the internal pressure of the gas. In this case the pressure would have to be considered to be a function of the volume. The relation to be invoked is, of course, the ideal gas equation.

$$P = \frac{nRT}{V} \tag{5.63}$$

This is then substituted in equation (5.61), taking all of the constant terms outside the integral,

$$w = -nRT\int_{V_1}^{V_2} \frac{dV}{V} \tag{5.64}$$

The solution to such is

$$\begin{aligned} w &= -nRT[\ln(V)]_{V_1}^{V_2} \\ &= nRT \cdot \ln\left(\frac{V_2}{V_1}\right) \end{aligned} \tag{5.65}$$

In this we have an equation of the form which occurs time and again in thermodynamic equations. A similar equation links the infinitesimal change in Gibbs function to the change in pressure

$$dG = VdP \tag{5.66}$$

V is again given in terms of P using the ideal gas equation. The left hand side of this equation may be integrated to give ΔG and the right hand side integrated as

$$\Delta G = nRT\int_{P_1}^{P_2} \frac{dP}{P} \tag{5.67}$$

This leads to

$$\Delta G = nRT \cdot \ln\left(\frac{P_2}{P_1}\right) \tag{5.68}$$

This should now be almost recognisable as equation (2.24) in Chapter 2 and the many others that are of the same form. Further manipulation concerning this will be discussed in Chapter 6. However, we can now see the origin of such expressions. That is, in the integration of thermodynamic equations involving reciprocal functions such as P and V related by the ideal gas equation.

Example—the Clausius–Clapeyron equation

The Clausius–Clapeyron equation which relates the temperature dependence of vaporisation to the vapour pressure of a liquid contains the term

$$\ln\frac{P_2}{P_1} = \frac{\Delta H}{R}\int_{T_1}^{T_2} \frac{dT}{T^2} \tag{5.69}$$

where T_1 is the temperature when the pressure is P_1 and T_2 is the temperature at pressure P_2. This is recognisable as the derivative of $-1/T$ and the solution is

$$\ln\frac{P_2}{P_1} = \frac{\Delta H}{R}\left[-\frac{1}{T_2}+\frac{1}{T_1}\right]$$
$$= \frac{\Delta H}{R}\left[\frac{1}{T_1}-\frac{1}{T_2}\right] \quad (5.70)$$

5.4.3 Integration by Substitution

This is used when the integrand is not easily recognisable but can be replaced by a function which can be easily integrated. Initially we will take a somewhat trivial example.

Example—integral of 2x

With reference to Table 5.1 we can integrate this directly,

$$\int 2x\,\mathrm{d}x = \frac{2x^2}{2} + C$$
$$= x^2 + C \quad (5.71)$$

We could, however, substitute $2x$ for a simpler function, say $u = 2x$. Then we have as the integral

$$\int u\,\mathrm{d}x \quad (5.72)$$

This cannot be solved directly but we do have a relationship between u and x. Thus, by differentiating u with respect to x we can find a relationship between $\mathrm{d}u$ and $\mathrm{d}x$.

$$\frac{\mathrm{d}u}{\mathrm{d}x} = 2 \quad (5.73)$$

therefore

$$\mathrm{d}x = \frac{\mathrm{d}u}{2} \quad (5.74)$$

Substituting this in equation (5.72) we now have a solvable integral.

$$\int u\frac{\mathrm{d}u}{2} \quad (5.75)$$

which is the same as

$$\int \frac{u}{2}\,\mathrm{d}u = \frac{u^2}{4} + C \quad (5.76)$$

Replacing u with the function of x it represents we have

$$\frac{u^2}{4} = \frac{(2x)^2}{4} = \frac{4x^2}{4} = x^2 \quad (5.77)$$

We have therefore obtained the same result by integrating a simpler function, remembering that we also have to substitute some suitable term for the differential. This is always found from the relationship which defines the original substitution.

Example—integral of e^{mx}

In this case (see result in Table 5.1) we substitute $u = mx$. Then the derivative is given by

$$\frac{\mathrm{d}u}{\mathrm{d}x} = m$$
$$\therefore \mathrm{d}x = \frac{\mathrm{d}u}{m} \tag{5.78}$$

Putting the complete substitution into the integrand we get

$$\int \exp(mx)\mathrm{d}x = \frac{1}{m}\int \exp(u)\mathrm{d}u$$
$$= \frac{1}{m}\exp(u) + C \tag{5.79}$$
$$= \frac{1}{m}\exp(mx) + C$$

Example—integral of $1/(1-x)$

This integral occurs in derivations of certain kinetic equations. We make the substitution $u = 1 - x$ and note that this time $\mathrm{d}u/\mathrm{d}x$ is equal to -1

$$\int \frac{1}{1-x}\mathrm{d}x = -1 \cdot \int \frac{1}{u}\mathrm{d}u$$
$$= -\ln(u) + C$$
$$= \ln\left(\frac{1}{u}\right) + C \tag{5.80}$$
$$= \ln\left(\frac{1}{1-x}\right) + C$$

A similar result is obtained for the function $1/(1-2x)$ only this time $\mathrm{d}u/\mathrm{d}x = -2$ and therefore $\mathrm{d}x = -\mathrm{d}u/2$. This leads to the result

$$\int \frac{1}{1-2x}\mathrm{d}x = \frac{1}{2}\ln\left(\frac{1}{1-2x}\right) + C \tag{5.81}$$

These integrals will be used further in Chapter 6.

Example—integral of $x \cdot \exp(-x^2)$

This is a very important function which appears in the derivation and manipulation of equations such as the Maxwell–Boltzmann distribution equation. We may find the integral of this function using the method of substitution. The integral is written in the usual form

$$\int x \cdot \exp(-x^2)\mathrm{d}x \tag{5.82}$$

To solve this we make the substitution $u = x^2$, noting that $\mathrm{d}u/\mathrm{d}x$ is equal to $2x$ or that $x \cdot \mathrm{d}x$ is given by $\mathrm{d}u/2$. We may now rewrite the integral as

$$\begin{aligned} \int \exp(-x^2)x\mathrm{d}x &= \int \exp(-u)\frac{\mathrm{d}u}{2} \\ &= \frac{1}{2}\int \exp(-u)\mathrm{d}u \end{aligned} \tag{5.83}$$

From Table 5.1 it can be seen that the substituted integral is solved as

$$\begin{aligned} \frac{1}{2}\int \exp(-u)\mathrm{d}u &= -\frac{1}{2}\cdot\exp(-u) \\ &= -\frac{1}{2}\cdot\exp(-x^2) \end{aligned} \tag{5.84}$$

5.4.4 The Method of Partial Fractions

Sometimes an apparently complex function can be simplified algebraically. Fractional quantities may be separated into simpler fractions which may then be integrated. This is the same as splitting a fraction such as 2/15 into two simpler components, e.g.,

$$\frac{2}{15} = \frac{1}{3} - \frac{1}{5} \tag{5.85}$$

Example—second order kinetics, general approach

Using a general approach to solving kinetic problems (Chapter 6) we may obtain an equation of the form

$$kt = \int \frac{\mathrm{d}x}{(a-x)(b-x)} \tag{5.86}$$

This appears to be much more complicated than other functions we have encountered in this section. As is often the case in mathematics our aim is to simplify the equation so that it is easier to solve.

The first step is to *assume* that the fraction can be rewritten as the sum of two fractions, called partial fractions. This is written as

$$\frac{1}{(a-x)(b-x)} = \frac{A}{a-x} + \frac{B}{b-x} \tag{5.87}$$

A and B are two, as yet undetermined constants. The following steps are aimed at deriving suitable values for them. The first is to multiply equation (5.87) by $(a-x)(b-x)$ to obtain

$$\frac{1\cdot(a-x)(b-x)}{(a-x)(b-x)}=\frac{A(a-x)(b-x)}{(a-x)}+\frac{B(a-x)(b-x)}{(b-x)} \tag{5.88}$$

or

$$1=A(b-x)+B(a-x) \tag{5.89}$$

The value of A may be found by eliminating the term involving B. Assuming that equation (5.89) is true and that we could draw a line to represent this function this is the same as finding the point or points in the line when the second term is zero. This is the case when x is equal to a, so $(a-x)=0$. Then

$$1=A(b-x) \tag{5.90}$$

This can be rearranged to give A directly but we must also remember that we have said that under these conditions $x=a$ and so

$$A=\frac{1}{b-x}=\frac{1}{b-a} \tag{5.91}$$

The value of B is found by eliminating the term in A. This occurs when $x=b$, so that $(b-x)=0$, and

$$1=B(a-x) \tag{5.92}$$

or

$$B=\frac{1}{a-x}=\frac{1}{a-b} \tag{5.93}$$

Substituting for A and B in equation (5.87) we obtain the integral as

$$\int\frac{\mathrm{d}x}{(a-x)(b-x)}=\int\left[\frac{1}{(b-a)}\cdot\frac{1}{(a-x)}+\frac{1}{(a-b)}\cdot\frac{1}{(b-x)}\right]\mathrm{d}x \tag{5.94}$$

Even though both $1/(a-b)$ and $1/(b-a)$ are constants and the integral is solved this equation may be simplified even further. This is achieved by multiplying both the numerator and denominator of the fraction defining B by -1.

$$B=\frac{-1\cdot1}{-1\cdot(a-b)}=\frac{-1}{b-a} \tag{5.95}$$

Equation (5.94) now becomes

$$\int\frac{\mathrm{d}x}{(a-x)(b-x)}=\int\left[\frac{1}{(b-a)}\cdot\frac{1}{(a-x)}-\frac{1}{(b-a)}\cdot\frac{1}{(b-x)}\right]\mathrm{d}x \tag{5.96}$$

The common factor, $1/(b-a)$ may be removed from the integral to give

$$\int\frac{\mathrm{d}x}{(a-x)(b-x)}=\frac{1}{b-a}\int\left[\frac{1}{a-x}-\frac{1}{b-x}\right]\mathrm{d}x \tag{5.97}$$

This may now be integrated as the sum of two terms, each of which may be integrated using the method of substitution described above.

Example—second order kinetics, general approach

Similar to the preceding problem, we have the equation

$$kt = \int \frac{\mathrm{d}x}{(a - x)(b - 2x)} \tag{5.98}$$

The steps in finding the partial fractions equivalent to this integrand are exactly the same as above.

(1) Write the integrand as the sum of two fractions

$$\frac{1}{(a - x)(b - 2x)} = \frac{A}{a - x} + \frac{B}{b - 2x} \tag{5.99}$$

(2) Multiply through by $(a - x)(b - 2x)$ and cancel the common factors

$$1 = A(b - 2x) + B(a - x) \tag{5.100}$$

(3) Find an expression for A by removing the term involving B. In this case this isachieved by setting $x = a$ so $B(a - x) = 0$. Also remember to put $x = a$ in thefinal equation

$$1 = A(b - 2x) = A(b - 2a) \tag{5.101}$$

and

$$A = \frac{1}{b - 2a} \tag{5.102}$$

(4) Find an expression for B by removing the term in A. In this example this is done by setting $x = b/2$ so that $A(b - 2x) = 0$. As above, remember that this substitution must also be made in the final equation.

$$1 = B(a - x) = B(a - b/2) \tag{5.103}$$

which rearranges to

$$B = \frac{1}{a - b/2} \tag{5.104}$$

At this stage we have reduced the expression to two fractions both of which involve a constant term. Before putting this into the integral it is always interesting to see if any further simplifications can be made. As in the first example we might see if the two terms for A and B can be made identical. The term $B = 1/(a - b/2)$ is a little awkward. This can be made more palatable if we multiply both numerator and denominator by 2. So step (5) is, if possible, to simplify the expression.

(5) Simplify!
We can rewrite B as suggested,

$$B = \frac{2}{2(a - b/2)} = \frac{2}{2a - b} \tag{5.105}$$

This is starting to look a little like the expression for A. Multiplying top andbottom by -1 we get

$$B = \frac{-1 \cdot 2}{-1 \cdot (2a - b)} = \frac{-2}{b - 2a} \tag{5.106}$$

This is now ready to be substituted for the original integrand. A final step, though, ought to be to check the result by working out this sum of fractions in longhand. We write the final result as

$$\int \frac{\mathrm{d}x}{(a - x)(b - 2x)} = \frac{1}{b - 2a} \int \left[\frac{1}{a - x} - \frac{2}{b - 2x} \right] \mathrm{d}x \tag{5.107}$$

This again may be solved using the method of substitution described in Section 5.4.3.

5.4.5 Integration by Parts

Mathematics texts will usually inform you that this is the most widely used method of integration. However, it is used only in more advanced mathematical problems associated with chemistry, e.g., quantum mechanics and statistical mechanics. Its utility may, however, be demonstrated by proving some integrals from Table 5.1.

The method is derived initially from the formula for the derivative of a product (Chapter 4), recalling

$$\frac{\mathrm{d}(uv)}{\mathrm{d}x} = u\frac{\mathrm{d}v}{\mathrm{d}x} + v\frac{\mathrm{d}u}{\mathrm{d}x} \tag{5.108}$$

If we now integrate each term with respect to x we obtain

$$\int \frac{\mathrm{d}(uv)}{\mathrm{d}x}\mathrm{d}x = \int \left(u\frac{\mathrm{d}v}{\mathrm{d}x}\right)\mathrm{d}x + \int \left(v\frac{\mathrm{d}u}{\mathrm{d}x}\right)\mathrm{d}x \tag{5.109}$$

On the left hand side we have the integral of a derivative of the function (uv), the result of which is merely the function itself. On the right hand side the '$\mathrm{d}x$' terms are cancelled

$$uv = \int u\mathrm{d}v + \int v\mathrm{d}u \tag{5.110}$$

This is then rearranged to

$$\int u\mathrm{d}v = uv - \int v\mathrm{d}u \tag{5.111}$$

This is the formula for integration by parts and gives the integral of a product of functions, $u\mathrm{d}v$, in terms of the functions themselves and integral of the related product, $v\mathrm{d}u$. This is particularly useful if the derivative of one part of the product is known and the integral of the other is relatively easy.

Another form of this equation labels the two components as the first and second functions. This is then written as

$$\int \text{1st} \times \text{2nd} = \text{1st} \int \text{2nd} - \int \left[\text{derivative of 1st} \times \int \text{2nd} \right] \tag{5.112}$$

Comparing this with the formal equation, (5.111), it is seen that v must be obtained by integrating $\mathrm{d}v$ and is equivalent to the second function. v is therefore chosen as the function which is easier to integrate. u, on the other hand, is the function for which the derivative is known or easily calculable and is equivalent to the first function in equation (5.112).

The main pitfall in using this method of integration is the choice of u and dv. In many cases either designation will give the correct result but one will be much quicker to derive than the other. As is always the case, only experience leads to the quickest or simplest method first time.

This method will be illustrated by proving one of the integrals given in Table 5.1.

Example—integral of ln(x)

To find this integral we need to define $\ln(x)$ as the function for which we know the derivative (this being $1/x$). This is then designated u in equation (5.111). dv must therefore be equal to dx. This must be integrated and we already know the integral of dx, as x. The various terms in equation (5.111) are given as follows.

$$u = \ln(x) \tag{5.113}$$

therefore

$$\frac{du}{dx} = \frac{1}{x} \tag{5.114}$$

and

$$du = \frac{dx}{x} \tag{5.115}$$

By comparison with (5.111)

$$dv = dx \tag{5.116}$$

Integrating we obtain

$$v = x \tag{5.117}$$

We can now substitute all the relevant terms in equation (5.111):

$$\begin{aligned} \int \ln(x)dx &= \ln(x) \cdot x - \int x\frac{dx}{x} \\ &= x \cdot \ln(x) - \int dx \\ &= x \cdot \ln(x) - x \end{aligned} \tag{5.118}$$

This particular integral is important in some statistical equations where we need to estimate the value of $\ln(x!)$. The usual approach is called *Stirling's approximation*. By the normal rules concerning logarithms $\ln(x!)$ is given by

$$\ln(x!) = \ln(1 \cdot 2 \cdot 3 \cdot 4 \cdot 5 \ldots x) \tag{5.119}$$

This can be expressed using normal summation notation

$$\ln(x!) = \sum_{x=1}^{x} \ln(x) \tag{5.120}$$

If x is very large and the intervals between values relatively small then this summation may be replaced by an integration (which is by definition a summation)

$$\sum_{x=1}^{x} \ln(x!) = \int_{1}^{x} \ln(x)\mathrm{d}x \tag{5.121}$$

the result of which is given above.

5.4.6 Numerical Integration

There are numerous cases, particularly in experimental chemistry, when integration of a function is used to supply more information than the raw data itself. In spectroscopic, chromatographic and electrochemical measurements the area under a curve is proportional to the amount of substance which is present or has reacted. Although the precise nature of the mathematical relation is unknown integration may be performed digitally by computer.

Computers may also be used when an integral of interest is impossible to find using normal means. An example is the function

$$y = \int_{x_1}^{x_2} \exp(-x^2)\mathrm{d}x \tag{5.122}$$

This integral occurs frequently in quantum mechanical equations and in statistical thermodynamics. Depending on the limits of integration it may be very difficult or even impossible to solve by analytical means, i.e., those described in this chapter. From the plot of the function, Figure 5.8, we can see that it exists and that there should be an integral related to the area under the curve. The approach to solving difficult problems such as this is to use numerical methods. Many of these are similar to or even based on the geometric method used in Section 5.3 to prove that integration is the same as finding an area under a curve.

Two methods of numerical integration are usually referred to in chemistry texts and these are the *trapezium* and *Simpson's* rules. Whereas we used a number of rectangles to estimate the area under a curve a closer approximation may be obtained by using trapezia

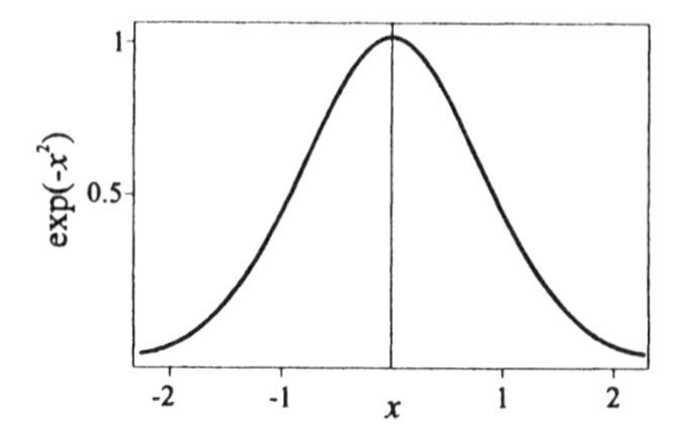

Figure 5.8 The function $y = \exp(-x^2)$

or other shapes. In Figure 5.9(a) we have divided the area under the arbitrary function $y = f(x)$ into two trapezia, each of which has the same width, h. On this diagram there are three points of interest on the ordinate, the lower limit, x_1, the middle point, x_2, and the upper limit of integration, x_3. The corresponding points on the curve are y_1, y_2 and y_3 respectively. Elementary geometry tells us that the area of a trapezium (Figure 5.10) is given by

$$\text{Area} = \frac{1}{2}\,\text{width} \times (\text{sum of lengths of sides}) \tag{5.123}$$

Referring to Figure 5.9(a) it can be seen that the areas of the two trapezia are given by

$$\text{Area 1} = \frac{h}{2}(y_1 + y_2) \tag{5.124}$$

$$\text{Area 2} = \frac{h}{2}(y_2 + y_3) \tag{5.125}$$

The total area is therefore

$$\begin{aligned}\text{Area} &= \frac{h}{2}(y_1 + y_2) + \frac{h}{2}(y_2 + y_3)\\ &= \frac{h}{2}(y_1 + y_2 + y_2 + y_3) \qquad (5.126)\\ &= \frac{h}{2}(y_1 + 2y_2 + y_3)\end{aligned}$$

Figure 5.9(b) shows an extension of this method which uses many more trapezia in order to minimise errors. Noting that the values of y which are not on the limits of the integration appear twice (as y_2 does in the simple equation) then the general equation for the area under the curve is

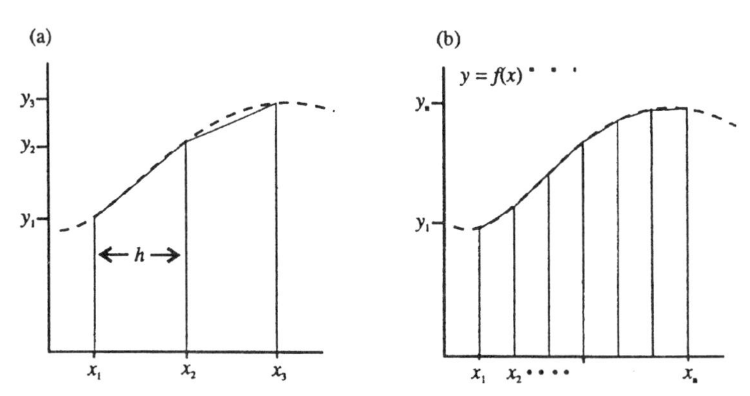

Figure 5.9 Greater accuracy is gained by estimating the area under a curve using trapezia. Using more trapezia leads to better estimates of the true area

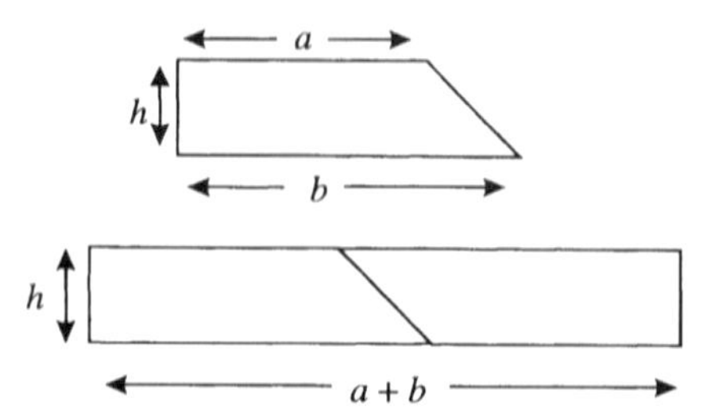

Figure 5.10 The area of a trapezium is found by placing two trapezia end to end to make a rectangle. The area of one is therefore $h(a + b)/2$

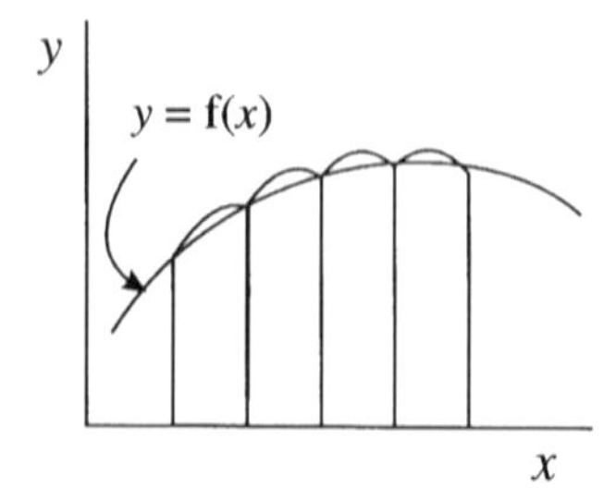

Figure 5.11 Exaggerated illustration of Simpson's rule. This is similar to the trapezium rule but the tops are parabolae rather than straight lines

$$\int_{x_1}^{x_n} y\mathrm{d}x = \frac{h}{2}(y_1 + 2y_2 + 2y_3 + 2y_4 + \ldots + 2y_{n-1} + y_n) \tag{5.127}$$

where

$$h = \frac{x_n - x_1}{n - 1} \tag{5.128}$$

$n - 1$ appears in this equation because if we divide the x axis up into a number of evenly spaced points, $x_1, x_2, x_3, \ldots x_n$, where x_1 and x_n are at the ends then this allows us $n - 1$ trapezia.

Simpson's rule uses a similar calculation but assumes that the tops of the trapezia are curved (Figure 5.11). These curves are parabolae which are mathematically well defined. The theory is more complex than that for the trapezium rule but the result is better. The x axis is divided into an even number of equally spaced sections with x_1 and x_n, as the limits of the integration.

$$\int_{x_1}^{x_n} y\mathrm{d}x = \frac{h}{3}(y_1 + 4y_2 + 2y_3 + 4y_4 + 2y_5 + \ldots + 4y_{n-1} + y_n) \tag{5.129}$$

h is again given by

$$h = \frac{x_n - x_1}{n - 1} \tag{5.128}$$

Both of these equations are easy to use and adaptable for programming by computer, where after the initial programming stage one might be able to increase the number of

Table 5.3 Calculation of a definite integral using trapezium and Simpson's rules

No. of divisions	Trapezium rule	Simpson's rule
2	3.022	2.558
5	2.563	2.276
6	2.335	2.226
8	2.275	2.208
10	2.257	2.205
20	2.210	2.198
30	2.203	2.197
50	2.199	—
100	2.197	—

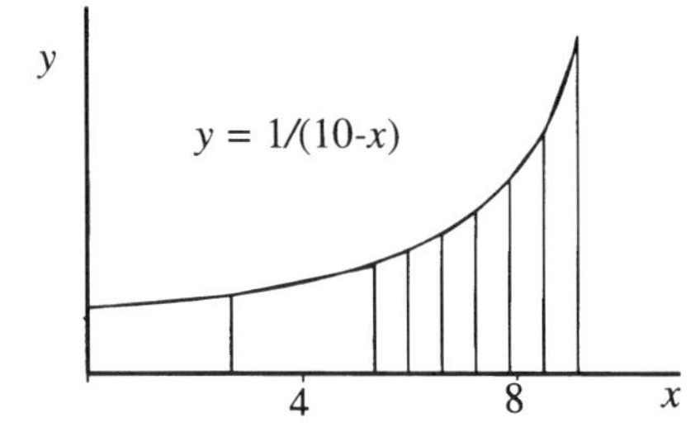

Figure 5.12 Plot of $y = 1/(10 - x)$. An increase in accuracy may be obtained by using an uneven distribution of trapezia, using more where the function curves most

trapezia at will, increasing the accuracy of the calculation with no extra effort to the user. Table 5.3 lists some results of the use of these formulae for the integral

$$\int_1^9 \frac{\mathrm{d}x}{10 - x} \tag{5.130}$$

Solving this equation using the method of substitution ($u = 10 - x$, see Section 5.2.3) gives an analytical result of ln(1/9) which is 2.197, which can be compared with the results in this table.

Comparing these results with the analytical result it is seen that Simpson's rule is always closer than the trapezium rule. Using ten divisions the error using Simpson's rule is less than 0.5%, whilst it is greater than 2% using the trapezium method. Simpson's rule reproduces the analytical result (for this integral at least) using about 36 divisions. Similar accuracy is only achieved using up to 100 divisions for the trapezium rule.

There are, of course, even more accurate methods. These include using an uneven distribution of subdivisions. This uses more when a function is more curved than when it is less curved. Figure 5.12 shows a plot of $1/(10 - x)$ and it can be seen that the greatest error would be expected as the x approaches 9 as the plot curves rapidly in this region. In this area more divisions would increase accuracy whereas they might not have much effect near $x = 1$. Another computational method produces a rectangle defined by the limits of the integration. A random number generator is then used to produce coordinates and the computer tells us whether they are within the boundary of the function (below the curve) or outside it. If the coordinates are truly random and enough are generated then the number within the boundary should be proportional to the area under the curve. To be accurate the number of random numbers must be very large and other computational methods are often used in preference.

5.5 Differentials, Line Integrals and Thermodynamic Concepts

In closing this chapter we will briefly discuss the use of calculus in thermodynamics. In Chapter 4 we presented the equation for an exact differential which is a function of two independent variables.

$$\mathrm{d}A = \left(\frac{\partial A}{\partial x}\right)_y \mathrm{d}x + \left(\frac{\partial A}{\partial y}\right)_x \mathrm{d}y \tag{5.131}$$

If a variable, such as A, which is a function of the two independent variables x and y obeys this equation then the differential $\mathrm{d}A$ is said to be an exact differential. Another important property of exact differentials is that if we integrate them

$$\int_a^b \mathrm{d}A = [A]_a^b \tag{5.132}$$

then the value of the integral depends only on the initial and final values of the function A (in this case a and b).

Thermodynamic properties whose differentials have these properties are called *state functions*. These include internal energy, U, enthalpy H, entropy S and the Gibbs function, G. They are called state functions because they depend only on the state of the system of interest and not the way that it is prepared. For example, the volume of a sample of a gas is another state function. It is a quantity we may measure or even calculate, but it is difficult to define other than that it is a measure of the space the gas takes up, a measure of the state of the gas. Although we may change the volume, the new value is still a measure of the state of the gas at that particular time and it does not matter how we go about changing it.

On the other hand, the functions defining the way we implement these changes are not state functions and the values depend on the way they occur. A common analogy is that state functions are the coordinates of places on a map. These are where they are and that is that. We may devise an infinite number of routes to get from Town A to Town B, all of which have different values. However many routes we choose the coordinates of these two towns remain the same.

The routes are these other thermodynamic functions such as the heat or work involved in bringing about changes. They are sometimes called path functions because they depend on the path, or route taken.

A good example is the work, w, involved in expanding a gas. This is usually defined by $w = -P\mathrm{d}V$ and we would find the work by integrating this equation. This can be done if the pressure is known as a function of volume. Under isothermal conditions P *is given as* nRT/V and the integral is

$$\begin{aligned} w &= -nRT\int_{V_1}^{V_2} \frac{\mathrm{d}V}{V} \\ &= -nRT \cdot \ln\left(\frac{V_2}{V_1}\right) \end{aligned} \tag{5.133}$$

There is, however, more than one way to expand a gas. If this is done under adiabatic conditions (so that no energy is exchanged with the surroundings) then the relationship between P and V is given by $PV^{\gamma} = k$ where k and γ are constants. In this case the integral is

$$\begin{aligned} w &= -k \int_{V_1}^{V_2} \frac{\mathrm{d}V}{V^{\gamma}} \\ &= \frac{k(V_2^{(1-\gamma)} - V_1^{(1-\gamma)})}{1-\gamma} \end{aligned} \tag{5.134}$$

Clearly the result depends on how we bring about the change. This may also be represented graphically. Figure 5.1 shows a hypothetical change in volume. From this it is clearly seen that the different function leads to a different area between the curve and the ordinate.

In pure mathematics these integrals are called *line integrals* and their equivalent in thermodynamics is the *path function.* In both cases the integral can only be solved if we know how the variables involved depend on each other, as illustrated in this example. Extensive use of this is used in calculation on the Carnot cycle which involves alternating isothermal expansion, adiabatic expansion, isothermal compression and adiabatic compression. If we know the limits of each stage of the cycle and the expression which relates P to V then the total work involved may be calculated.

Finally it is the property of integrals of state functions which makes them so important in chemistry. Because the change in the function does not depend on the way it is brought about it is possible to devise any means, theoretical or practical, to find what that change may be. The overall result should be independent of the method or route chosen.

Problems

5.1 Find the following indefinite integrals.

(a) $\int 4\mathrm{d}x$ (P5.1)

(b) $\int x\mathrm{d}x$ (P5.2)

(c) $\int 6x^2\mathrm{d}x$ (P5.3)

(d) $\int (3x^2 + 4x)\mathrm{d}x$ (P5.4)

(e) $\frac{1}{x}\mathrm{d}x$ (P5.5)

5.2 Find the following definite integrals.

(a) $\int x\mathrm{d}x$ between $x = 0$ and $x = 6$ (P5.6)

(b) $\int \cos(x)\mathrm{d}x$ between $x = 0$ and $x = \pi/2$ (P5.7)

(c) $\int e^x \mathrm{d}x$ between $-\infty$ and ∞ (P5.8)

(d) $\int \ln(x)\mathrm{d}x$ between $x = 1$ and $x = e$ (P5.9)

(e) $\int \frac{1}{x^2}\mathrm{d}x$ between $x = a_0$ and $x = a$ (P5.10)

5.3 This problem concerns the following integral

$$\int \frac{\mathrm{d}x}{\sqrt{x+1}} \tag{P5.11}$$

(a) If $u = x + 1$ find the derivative $\mathrm{d}u/\mathrm{d}x$ and thus express $\mathrm{d}x$ in terms of $\mathrm{d}u$.
(b) Substitute u and $\mathrm{d}u$ for x and $\mathrm{d}x$ in equation (P5.11) and solve the resultant integral.
(c) Rewrite the result in terms of x.

5.4 This problem concerns the following integral

$$\int \frac{\mathrm{d}x}{\sqrt{2x+1}} \tag{P5.12}$$

(a) If $u = 2x + 1$ find the derivative $\mathrm{d}u/\mathrm{d}x$ and thus express $\mathrm{d}x$ in terms of $\mathrm{d}u$.
(b) Substitute u and $\mathrm{d}u$ for x and $\mathrm{d}x$ in equation (P5.12) and solve the resultant integral.
(c) Rewrite the result in terms of x.

5.5 This problem concerns the following integral

$$\int \frac{\mathrm{d}x}{\sqrt{3x+2}} \tag{P5.13}$$

(a) If $u = 3x + 2$ find the derivative $\mathrm{d}u/\mathrm{d}x$ and thus express $\mathrm{d}x$ in terms of $\mathrm{d}u$.
(b) Substitute u and $\mathrm{d}u$ for x and $\mathrm{d}x$ in equation (P5.13) and solve the resultant integral.
(c) Rewrite the result in terms of x.

5.6 This problem concerns the following integral

$$\int (2x+3)^2 \mathrm{d}x \tag{P5.14}$$

(a) If $u = 2x + 3$ find the derivative $\mathrm{d}u/\mathrm{d}x$ and thus express $\mathrm{d}x$ in terms of $\mathrm{d}u$.
(b) Substitute u and $\mathrm{d}u$ for x and $\mathrm{d}x$ in equation (P5.14) and solve the resultant integral.
(c) Rewrite the result in terms of x.

5.7 This problem concerns the following general integral

$$\int (bx + c)^n \mathrm{d}x \tag{P5.15}$$

where b and c are constants.
(a) If $u = bx + c$ find the derivative $\mathrm{d}u/\mathrm{d}x$ and thus express $\mathrm{d}x$ in terms of $\mathrm{d}u$.
(b) Substitute u and $\mathrm{d}u$ for x and $\mathrm{d}x$ in equation (P5.15) and solve the resultant integral.
(c) Rewrite the result in terms of x.

5.8 This problem concerns the following integral

$$\int \frac{\mathrm{d}x}{x + 1} \tag{P5.16}$$

(a) If $u = x + 1$ find the derivative $\mathrm{d}u/\mathrm{d}x$ and thus express $\mathrm{d}x$ in terms of $\mathrm{d}u$.
(b) Substitute u and $\mathrm{d}u$ for x and $\mathrm{d}x$ in equation (P5.16) and solve the resultant integral.
(c) Rewrite the result in terms of x.

5.9 This problem concerns the following general integral

$$\int \frac{\mathrm{d}x}{bx + c} \tag{P5.17}$$

where b and c are constants.
(a) Using the same methods as above substitute for x and $\mathrm{d}x$ and solve the resultant integral.
(b) Rewrite the result in terms of x.

5.10 Suggest substitutions and solve the following integrals

(a) $$\int \cos(5x)\mathrm{d}x \tag{P5.18}$$

(b) $$\int \frac{1}{x \cdot \ln(x)} \mathrm{d}x \tag{P5.19}$$

(c) $$\int \exp(-kx)\mathrm{d}x \tag{P5.20}$$

5.11 This problem concerns the following integral

$$\int x \cdot \cos(x)\mathrm{d}x \tag{P5.21}$$

This can be solved by the method of integration by parts. Use the formula given for this method and take $u = x$ and $\cos(x)\mathrm{d}x = \mathrm{d}v$.

(a) Find u and $\mathrm{d}v$.
(b) Solve the integral.

5.12 The following integral may be solved using the method of integration by parts.

$$\int x \cdot \exp(-kx)\mathrm{d}x \tag{P5.22}$$

Solve this using the substitutions $u = x$ and $\exp(-kx)\mathrm{d}x = \mathrm{d}v$.

5.13 Find the indefinite integrals. (Use the method of integration by parts.)

(a) $$\int \cos(5x)\mathrm{d}x \tag{P5.23}$$

(b) $$\int \frac{1}{x \cdot \ln(x)}\mathrm{d}x \tag{P5.24}$$

5.14 Express the following as partial fractions and then solve the integrals.

(a) $$\int \frac{\mathrm{d}x}{x(x+1)} \tag{P5.25}$$

(b) $$\int \frac{1}{x^2 - 1}\mathrm{d}x \tag{P5.26}$$

6 Differential Equations

6.1 Introduction

A differential equation tells us how one variable changes with respect to another. This quantity is often directly measurable by experiment. For example, in a car travelling at speed, we can measure this speed using the car's speedometer and record it at known time intervals. Since speed is the derivative of distance, s, with respect to time, t, we can write directly

$$\text{Speed} = \frac{\mathrm{d}s}{\mathrm{d}t} = \mathrm{f}'(t) \tag{6.1}$$

The actual distance travelled may well be of more interest to the driver and passengers than the speed and so we need to find an expression which relates distance directly to time. This is achieved by solving the differential equation (6.1).

The same procedure is used in chemistry in following the progress of chemical reactions. It may be possible to measure the rate of reaction relatively easily but in order to find the function which describes the relation between concentration and time we must solve the appropriate differential equation. The solution is usually another equation which relates concentration directly to time. This equation may allow us to calculate a number of useful parameters such as the rate constant and activation energy.

Similarly, differential equations are used to solve problems in thermodynamics and quantum mechanics. In thermodynamics we are looking at how the energy of systems behave when the system is changed and in quantum mechanics at how particles (or waves!) behave with respect to each other in space.

In all cases the problem involves two parts. The first is deriving the correct differential equation. The second is solving it. Although students may not have to perform these tasks themselves, proofs given will usually involve these steps, and so some knowledge of the process is useful.

In this chapter we will consider a number of differential equations, their derivation, solution and some related manipulations. As in Chapters 4 and 5 we will concentrate in the areas of kinetics and thermodynamics.

6.2 Definitions

As much notation and jargon goes with differential equations as with normal equations. It is therefore worthwhile listing a few relevant definitions.

The *order* of the equation is the index (n) of the highest derivative. So the equation

$$\frac{dy}{dx} = x^2 \tag{6.2}$$

is a first order equation whilst the equation

$$\frac{d^2y}{dx^2} = x^2 \tag{6.3}$$

is a second order equation. Fortunately most equations found will be first order although theories describing diffusion (look up Fick's laws in any physical chemistry textbook) and quantum mechanics involve second order equations.

The solution to a differential equation is another equation relating the variables involved to each other. Sometimes in chemistry we are content to rearrange a differential equation into a form where the derivatives are directly measurable quantities. This may still allow us to calculate useful parameters.

Many physical problems, however different they may be in practice, can be solved by application of similar differential equations. For example, the flow of heat through a body and the diffusion of ions in a solution can be modelled using identical differential equations.

Many of the methods of analysis using differential equations are over two hundred years old. Therefore many types of solution are already known and can be found in standard mathematics texts. Once a differential equation is correctly derived it may well be the case that the solution may be given directly without going through the process of solving it conventionally. Equations of the same, or similar form have similar solutions and the problem remaining is to tailor the solution to the exact problem under investigation.

These well known solutions may be termed *general solutions*. A general solution involves arbitrary constants. The tailoring process involves assigning values to these constants so that the solution suitably describes the system being considered. Where integration is the method used to solve the equation the constants probably arise as the constants of integration (Chapter 5).

An important rule is that a diffential equation of order n has a general solution containing n arbitrary constants. For example, a second order equation describing the acceleration of a particle towards a point is*

$$\frac{d^2x}{dt^2} = -kx^2 \tag{6.4}$$

This has a general solution

$$x = A\sin(kt) + B\cos(kt) \tag{6.5}$$

*This is the equation describing simple harmonic motion as observed in the oscillations of a weight on a spring (obeying Hookes law) or atoms in a molecule. It also describes the generalised motion of a wave.

This form of solution is known from experience and it is often the case that no further explanation as to the method of solution will be given. Assigning values to A and B relies on knowledge of the experiment or system and the limits of that system. The solution will be considered to be correct if it predicts the observed behaviour of the system correctly.

Having been successful in finding values for constants in general solutions we say that we have obtained a *particular solution.* A differential equation may have several particular solutions, depending on the limits set. This is in the same sense that the integral of x^2 may have several values depending on the value of the constant of integration.

Another form of solution which is not a particular solution is called a *singular solution.* These types of solution, when they occur, are usually of great importance and examples may be found in maths textbooks but need not concern us here.

First order equations are subject to further terminology. They may be *simple*, in which case they may be expressed in the form

$$\mathrm{d}y = \mathrm{f}(x)\mathrm{d}x \tag{6.6}$$

where $\mathrm{f}(x)$ is a function of x. Many thermodynamic equations are written in this form and they may be solved directly using the methods of integration described in Chapter 5. An equation is separable* if it can be written as equation (6.6) or in the more general form

$$\mathrm{g}(y)\mathrm{d}y = \mathrm{f}(x)\mathrm{d}x \tag{6.7}$$

where $\mathrm{g}(y)$ is a function of y. These can also be solved by integration. Equations involving derivatives such as equation (6.2) are considered to be separable because they can be rearranged to give, in this case,

$$\mathrm{d}y = x^2\mathrm{d}x \tag{6.8}$$

Another term which may crop up is that equations are *linear* in the dependent variable. An equation is *linear in y* (being the dependent variable) if it can be written in the form

$$\frac{\mathrm{d}y}{\mathrm{d}x} + Py = Qx \tag{6.9}$$

where P and Q are independent of y and may be constants or functions of x. It is often the case that such equations can be rearranged to become separable equations and can be solved by methods of integration. These types of equation are not often found in elementary courses and we will discuss only one example in this chapter.

6.3 Differential Equations in Kinetics

6.3.1 Solutions Involving Integration: Separable Equations

Every text on chemical kinetics will tell you that it is not possible to determine the rate or mechanism of a chemical reaction from a stoichiometric equation. This information can only come from experimental measurements. This is not a ploy designed to justify endless laboratory sessions but a simple fact. It is therefore important to state the definition of rate of reaction.

*Some textbooks call them *variables separable.*

The rate of reaction is the derivative of concentration with respect to time.

Thus an experiment will usually involve recording some parameter which is a known function of concentration (of either reactant or product) at known time intervals. From a plot of concentration (c) against time (t) the rate may be determined as the slope (dc/dt). Note that at this stage it is easy to make this plot from experimental data, but we do not yet know the mathematical relationship between these two quantities. The purpose of the experiment is to find it.

Information concerning the mechanism can be derived from the relationship between rate and concentration, as described in Chapter 2. However, it may not be easy to measure the slope if insufficient or inaccurate data is obtained and so the problem is usually approached by applying the methods of calculus. That is, we derive and solve differential equations.

This approach is summarised as follows.

(1) Propose a mechanism or *model* for the reaction being studied.
(2) Derive a differential equation (rate equation) consistent with the suggested model.
(3) Solve this equation with known limits of experimental procedure.
(4) This solution will be another equation which relates concentration directly to time. The experimental data are plotted according to this equation.
(5) Whether the model, the differential equation and its solution, are correct is tested by the quality of this fit (is it a straight line?). If it is good then the proposed model may be deemed to be suitable. If it is not the above steps will be repeated for a new model.

This all appears quite straightforward but as the complexity of the chemical system increases then the process becomes more and more complex itself. At the elementary level many solutions have been found and fortunately many real systems fit these models. To demonstrate we will now consider some standard kinetic models.

Simple theory tells us that the rate (the derivative of concentration with respect to time) is related to the concentration of reactants $[A]$, $[B]$ and $[C]$ by an equation of the type

$$\text{Rate} = \pm\frac{dc}{dt} = k[A]^x[B]^y[C]^z \tag{6.10}$$

The values of the powers x, y and z may be found by the method described in Chapter 2. We have used the sign $\pm$ to indicate that the rate may be increasing or decreasing depending on whether we are following the change in concentration of product or reactant. Even when we know the values of x, y and z we still need to find the exact relationship between c and t in order to find the rate constant k. To do this we must solve the differential equation.

Because the indices in equation (6.10) are often integers and for many reactants may be zero it is possible to describe a number of idealised cases. These are zero, first and second order reactions and these will now be described in turn.

Example 1—zero order reaction

We will consider the case where we are following the concentration, c, of a single reactant. Since the order of reaction is the power to which the concentration is raised, the rate equation is

$$-\frac{dc}{dt} = kc^0 \tag{6.11}$$

which is the same as

$$-\frac{dc}{dt} = k \tag{6.12}$$

Note the use of the negative sign because we are following the loss of a reactant. It is convenient to follow the loss of a reactant because we then determine directly the influence of this reactant on the rate. Equation (6.12) is a separable equation and can be rewritten as

$$-dc = kdt \tag{6.13}$$

This may be solved by integration. To do this we write in the integral signs and in order to find a solution relevant to the conditions of our experiment we must decide on the limits or boundary conditions of the integral. In keeping the derivation general, so that we may then apply it to any system, these limits are given as variable quantities. These are usually as follows.

Condition 1 When $t = 0$, $c = a$.
This states that at the start of the experiment t is time zero and the concentration is some known (constant) value a.

Condition 2 When $t = t$, $c = c$.
At some given time, t, the concentration will be equal to c (which we will measure in our experiment).

Now we may write the differential equation (6.13) as an equality involving two integrals

$$-\int_a^c dc = \int_0^t kdt \tag{6.14}$$

The solution of this is (Chapter 5)

$$-[c]_a^c = k[t]_0^t \tag{6.15}$$

Putting the values of the limits in gives

$$-[c - a] = k[t - 0] \tag{6.16}$$

which becomes

$$a - c = kt \tag{6.17}$$

This is the *integrated rate equation* for zero order kinetics.

It is always worth investigating a result like this to see if it is logical. As time increases, the term on the right hand side of equation (6.17) increases. As c decreases (we are observing loss of reactant) then $a - c$ increases (c is always less than a). That is, both sides of the equation increase as they should and the result is therefore acceptable. Note that concentration decreases in a linear fashion. The rate of reaction is constant (as suggested by the rate equation (6.11)) and independent of concentration. Because of this a zero order reaction mechanism is always easy to recognise because the concentration–time plot is linear.

This is in contrast to first and second order reactions where the concentration time plots are curved (Figure 6.1). These curves are similar in shape and it is not always easy to distinguish between experimental plots, especially if data is subject to large random errors. Unless enough data is obtained to allow us to draw a smooth curve through the points such that tangents (to estimate the rate) may be drawn calculus must be applied.

Example 2—first order reaction

The rate equation for a first order reaction is

$$-\frac{\mathrm{d}c}{\mathrm{d}t} = kc \tag{6.18}$$

This is again a separable equation and can be written

$$-\frac{\mathrm{d}c}{c} = k\mathrm{d}t \tag{6.19}$$

The limits of the integration are the same as above so we put the integral signs in and solve.

$$-\int_a^c \frac{\mathrm{d}c}{c} = k\int_0^t \mathrm{d}t \tag{6.20}$$

From Chapter 5 this becomes

$$-[\ln(c)]_a^c = k[t]_0^t \tag{6.21}$$

$$-[\ln(c) - \ln(a)] = kt \tag{6.22}$$

$$-\left[\ln\left(\frac{c}{a}\right)\right] = kt \tag{6.23}$$

Recalling that $-\ln(x) = \ln(1/x)$ (Chapter 2) this may be written as

$$\ln\left(\frac{a}{c}\right) = kt \tag{6.24}$$

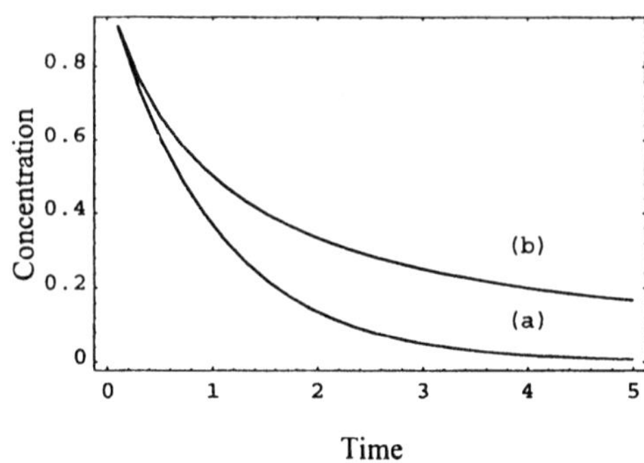

Figure 6.1 Concentration time plots for (a) first order and (b) second order kinetics. Experimentally these may be difficult to distinguish, especially at short times

This is the integrated rate equation for first order kinetics. Again we may check its validity. The right hand side must increase as time increases. Since c decreases with time and is always less than a then the ratio a/c increases and so does the term $\ln(a/c)$. Thus both sides increase together and the equation is valid. Finally a plot of $\ln(a/c)$ versus time should be linear if first order kinetics apply.* It is also useful to note that this equation involves a ratio of concentrations. Recalling the use of units described in Chapter 1 we see that the units of concentration will cancel in this equation. It is, therefore, possible to use completely arbitrary units of concentration in order to simplify mathematical manipulation of actual results. For example, if the concentration is being found by titration, the volume of titre could be used, or alternatively absorbance could be used in spectrophotometric measurements.

Example 3—second order kinetics: one reactant

This is the class of reactions where the rate determining step is of the kind

$$\mathrm{A} + \mathrm{A} \longrightarrow \text{product} \tag{6.25}$$

and presents a different case to those second order reactions which involve two different reacting species. These are discussed below.

For this case the rate equation shows that the rate is proportional to the square of the concentration.

$$-\frac{\mathrm{d}c}{\mathrm{d}t} = kc^2 \tag{6.26}$$

This is again separable and the limits are identical to those given above. Rearranging and writing in the limits of integration gives

$$-\int_a^c \frac{\mathrm{d}c}{c^2} = k\int_0^t \mathrm{d}t \tag{6.27}$$

This can be solved directly as shown in Chapter 5.

$$-\left[-\frac{1}{c}\right]_a^c = k[t]_0^t \tag{6.28}$$

The minus signs inside and outside the square bracket on the left cancel and this gives

$$\frac{1}{c} - \frac{1}{a} = k(t - 0) \tag{6.29}$$

This is the integrated rate equation for second order kinetics. Once again we can check its validity. As time increases (right hand side) then c decreases and $1/c$ increases. $1/a$ is constant and, since a is larger than c, $1/a$ is always less than $1/c$. Therefore the left hand side also increases with time.

*A plot of $\ln(c)$ versus time should also be linear. Why?

Example 4—second order kinetics: two reactants

When two reactants are important in the rate determining step we are forced to take a different approach to solving the differential equation. As it turns out, this new approach is much more general and may be applied to many more problems.

If we consider the chemical reaction

$$\mathrm{A} + \mathrm{B} \longrightarrow \text{products} \tag{6.30}$$

the differential equation for this system must include the concentration of both species, and this is

$$-\frac{\mathrm{d}[\mathrm{A}]}{\mathrm{d}t} = k[\mathrm{A}][\mathrm{B}] \tag{6.31}$$

where as usual the square brackets indicate the concentration of the species inside. To solve this, we must know how the concentration of B varies and how it relates to the concentration of A. Because of the stoichiometry of the reaction we know that every time a molecule of A is used up, a molecule of B is also used up. There must, therefore, be a direct relationship between [A] and [B]. To find this we note the stoichiometry and introduce the parameter x. This is the amount of A which has reacted at a given time, t. Because of the stoichiometry of the reaction, when x of A has been used up, then x of B must have also been used up. If the initial concentration of A is given by a, then the concentration of A at time t must be $a - x$. Similarly the concentration of B at the same time is $b - x$, using b as the initial concentration of B. We can now rewrite the differential equation (6.31) as

$$-\frac{\mathrm{d}[\mathrm{A}]}{\mathrm{d}t} = k(a - x)(b - x) \tag{6.32}$$

Also, replacing the concentration term in the derivative by $a - x$,

$$-\frac{\mathrm{d}(a - x)}{\mathrm{d}t} = k(a - x)(b - x) \tag{6.33}$$

Recalling that a derivative of several terms, as we have on the left hand side of this equation, may be given as the sum or difference of the derivative of each individual term, the left hand side of (6.33) can be written

$$\begin{aligned} -\frac{\mathrm{d}(a - x)}{\mathrm{d}t} &= -\left[\frac{\mathrm{d}a}{\mathrm{d}t} - \frac{\mathrm{d}x}{\mathrm{d}t}\right] \\ &= \frac{\mathrm{d}x}{\mathrm{d}t} - \frac{\mathrm{d}a}{\mathrm{d}t} \end{aligned} \tag{6.34}$$

Because the initial concentration of A, a, is a constant $\mathrm{d}a/\mathrm{d}t$ is equal to zero. Therefore we can write our differential equation as

$$\frac{\mathrm{d}x}{\mathrm{d}t} = k(a - x)(b - x) \tag{6.35}$$

This is the differential equation which must be solved. Note that the derivative on the left is now positive since x increases as the reaction proceeds. It is a separable equation and can be written in the form of two integrals. Before doing this it is useful to consider the limits of integration which we will use.

These are essentially the same as those used above, but we must recast them in terms of x.

Condition 1 When $t = 0$, $x = 0$.
Before the reaction begins, neither reactant has been used up and so x must be zero.

Condition 2 When $t = t$, $x = x$.

This is the general statement concerning concentration. Although it appears trivial to state this, we must do so to avoid confusion.

We may now write the full equation as

$$\int_0^x \frac{\mathrm{d}x}{(a-x)(b-x)} = k\int_0^t \mathrm{d}t \tag{6.36}$$

The solution to the right hand side should by now be trivial but the left hand side is more complex. This may be solved by the method of partial fractions described in Chapter 5. First the integrand is written in terms of the partial fraction*

$$\frac{-1}{a-b}\int_0^x \left(\frac{1}{a-x} - \frac{1}{b-x}\right)\mathrm{d}x \tag{6.37}$$

We now solve each term of the integrand by substituting u for $a - x$ and $b - x$ (again see Chapter 5), i.e., for $\mathrm{d}x/(a - x)$

$$\int \frac{\mathrm{d}x}{a-x} = -\int \frac{\mathrm{d}u}{u} \tag{6.38}$$

The result of this is $-\ln(u)$. Resubstituting $a - x$ for u and putting in the limits of the integration we get

$$\begin{aligned} -[\ln(a-x)]_0^x &= -[\ln(a-x) - \ln(a)] \\ &= -\ln\left(\frac{a-x}{a}\right) \\ &= \ln\left(\frac{a}{a-x}\right) \end{aligned} \tag{6.39}$$

Both terms in equation (6.37) are computed in this way and so the integrated equation is

$$\frac{-1}{a-b}\left[\ln\left(\frac{a}{a-x}\right) - \ln\left(\frac{b}{b-x}\right)\right] = kt \tag{6.40}$$

The two logarithmic terms may be combined and then turned upside down (since we have the minus sign outside the bracket). The final result is then

$$\frac{1}{a-b}\ln\left[\frac{b(a-x)}{a(b-x)}\right] = kt \tag{6.41}$$

*Note that in Chapter 5 we had $1/(b - a)$ which is the same as $-1/(a - b)$.

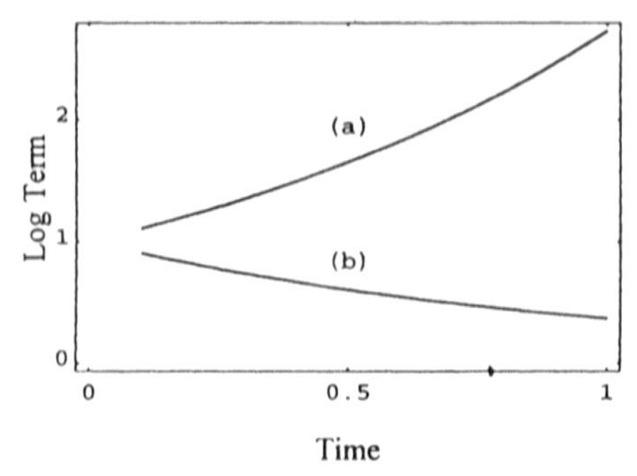

Figure 6.2 Plots of the ln term in equation (6.41) for (a) a greater than b and (b) b greater than a

This integrated rate equation is less easy to validate than the equations given above. However, Figure 6.2 shows theoretical plots of the logarithmic terms versus time for the two cases (a) where a is greater than b and (b) where b is greater than a. In the former the slope is positive and, noting that $1/(a - b)$ is also positive, then this side of equation (6.41) increases as time increases. If b is greater than a then the slope is negative. Note, however, that the term $1/(a - b)$ is also negative and so the left hand side of this equation does increase as t increases. The equation is therefore valid.

Another important point to note is that if a is equal to b then $1/(a - b)$ becomes equal to $1/0$ which is infinite. In this case the result is no longer valid. Fortunately, if this happens to be the case we can use the result obtained above for the second order reaction – single reactant.

This method of using the amount of reactant which has been used is particularly useful and you should derive expressions for the previous cases described above (see the problems at the end of this chapter). It is useful because it may be applied to more and more complex systems and also because many experimental techniques measure this quantity rather than the actual concentration of reactant at a given time.

Example 5—reactions going towards equilibrium

Many reactions do not go to completion in terms of reactant becoming one hundred per cent product but reach some point where reactants and products are in equilibrium. The simplest case is the reaction

$$\mathrm{A} \underset{k_b}{\overset{k_f}{\rightleftharpoons}} \mathrm{B} \tag{6.42}$$

where k_f and k_b are the rate constants for the forward and reverse reactions respectively. Analysis of this scheme is another example where we can use the amount of reactant that has reacted as the variable in our differential equation. This equation states that the rate of change of concentration of A is the sum of the forward and reverse reactions.

$$\frac{\mathrm{d}[\mathrm{A}]}{\mathrm{d}t} = -k_f[\mathrm{A}] + k_b[\mathrm{B}] \tag{6.43}$$

A is lost by the forward reaction (hence the minus sign in the first term), and is produced by the reverse reaction. To solve this equation we need some relation between [A] and

[B]. This is again found using x, the amount of A used up at time t. The stoichiometry tells us that if x moles of A have been used, x moles of B have been formed. In a constant volume system, if the initial concentration of A is a, then the concentration at time t is $a - x$ and the concentration of B is x. In a manner similar to that used for the second order reaction we rewrite equation (6.43) as

$$\frac{d(a-x)}{dt} = -k_f(a-x) + k_b x \tag{6.44}$$

Because $d(a - x)/dt$ is equal to $-dx/dt$ and taking care with minus signs we obtain

$$\frac{dx}{dt} = k_f(a-x) - k_b x \tag{6.45}$$

This equation can be rearranged to give dx/dt as a function of x

$$\frac{dx}{dt} = k_f a - x(k_f + k_b) \tag{6.46}$$

This may be written as

$$\frac{dx}{dt} + (k_f + k_b)x = k_f a \tag{6.47}$$

which is in the form of a linear differential equation (see Section 6.2, Definitions). Although this may be solved the method is more difficult than finding the solution to a simple separable equation, and we are still left at the end with the constants k_f and k_b combined. In order to maximise information gained (values for both constants) it is useful to try and find a relation between the two before we start solving the differential equation. After all, this is quite a simple step and it may lead to further simplifications.

The only useful information we have, albeit implicit information, concerns the equilibrium condition. At equilibrium the concentration of A will not be changing. Therefore the derivative dx/dt will be equal to zero

$$\frac{dx}{dt} = 0 \tag{6.48}$$

Consequently

$$0 = k_f(a - x_e) - k_b x_e \tag{6.49}$$

where x_e is used to signify x at equilibrium.

This may be written as the equality

$$k_f(a - x_e) = k_b x_e \tag{6.50}$$

Although we are introducing another constant, x_e, it is one which may be measured directly. Equation (6.50) can be rearranged to give k_b in terms of k_f and measurable quantities.

$$k_b = \frac{k_f(a - x_e)}{x_e} \tag{6.51}$$

This may be substituted in equation (6.45) and simplified in the following steps.

$$\frac{dx}{dt} = k_f(a - x) - \frac{k_f(a - x_e)}{x_e} x \tag{6.52}$$

Expanding each term

$$\begin{aligned} \frac{dx}{dt} &= k_f a - k_f x - \frac{k_f a x}{x_e} + k_f x \\ &= k_f a - \frac{k_f a x}{x_e} \end{aligned} \tag{6.53}$$

This can be made into one fraction rather than two separate terms by multiplying $k_f a$ by x_e/x_e (which is of course equal to one).

$$\begin{aligned} \frac{dx}{dt} &= \frac{x_e k_f a}{x_e} - \frac{k_f a x}{x_e} \\ &= \frac{k_f a}{x_e}(x_e - x) \end{aligned} \tag{6.54}$$

This is now separable and can be solved directly from our knowledge of integration.

$$\int_0^x \frac{dx}{(x_e - x)} = \frac{k_f a}{x_e} \int_0^t dt \tag{6.55}$$

The right hand side becomes $k_f at/x_e$ and the left hand side can be integrated by the substitution method to give

$$\begin{aligned} \int_0^x \frac{dx}{(x_e - x)} &= -\left[\ln(x_e - x)\right]_0^x \\ &= -\left[\ln(x_e - x) - \ln(x_e - 0)\right] \\ &= -\left[\ln\left(\frac{x_e - x}{x_e}\right)\right] \\ &= \ln\left(\frac{x_e}{x_e - x}\right) \end{aligned} \tag{6.56}$$

The full integrated rate equation for this system is therefore

$$\frac{x_e}{a} \cdot \ln\left(\frac{x_e}{x_e - x}\right) = k_f t \tag{6.57}$$

In this equation we have put all concentration terms on the left and time on the right. A plot of the left hand side against time should yield a straight line with slope equal to k_f. From the relationship given in equation (6.52) we can calculate k_f using this value and, because the equilibrium constant is defined as the ratio of these two constants, we can also calculate this parameter (which is in turn related to other thermodynamic parameters).

6.3.2 Solutions Involving Integration: First Order Linear Equations

As defined in Section 6.2, first order linear equations are differential equations which may be written in the form

$$\frac{dy}{dx} + Py = Qx \tag{6.58}$$

where P and Q are either constant or functions of x. If both P and Q are constants then the equation is separable. For example,

$$\frac{dy}{dx} + 2y = 5 \tag{6.59}$$

may be rearranged to give

$$\frac{dy}{dx} = 5 - 2y \tag{6.60}$$

This is then separable and can be integrated in the form

$$\int \frac{dy}{5 - 2y} = \int dx \tag{6.61}$$

Equation (6.58) is also separable if Q is zero.

$$\frac{dy}{dx} + y = 0 \tag{6.62}$$

Rearrangement of this produces

$$\frac{dy}{dx} = -y \tag{6.63}$$

which is of the same form as the equations described in Section 6.3.1 and can be solved by integration.

If neither of these criteria applies then the equation is not separable and alternative tactics must be applied. This method involves multiplying each term by some factor, called an *integrating factor*. This has the effect of turning each term into something which may be integrated. The integrating factor is different for each equation but may be defined generally for the differential equation (6.58).

$$\text{Integrating factor} = \exp\left(\int P \cdot dx\right) \tag{6.64}$$

Although this appears as something of a magic formula, there is a logical derivation which will be found in most advanced mathematics texts. To see how this works we will describe how it is used, first in a purely mathematical example and then in one relating to a chemical problem.

Example—solution of dy/dx + 4y/x = 2x²

The equation

$$\frac{dy}{dx} + \frac{4}{x}y = 2x^2 \tag{6.65}$$

is a first order linear equation. By comparison with equation (6.58) we see that P is equal to $4/x$ and Q is equal to $2x^2$. To apply the integrating factor method we must first find the integrating factor itself. This factor, IF, is defined by

$$\begin{aligned} \text{IF} &= \exp\left(\int P\mathrm{d}x\right) \\ &= \exp\left(\int \frac{4}{x}\mathrm{d}x\right) \end{aligned} \tag{6.66}$$

We know from Chapter 5 that $4/x$ integrates to $4 \cdot \ln(x)$.* The integrating factor is therefore

$$\text{IF} = \mathrm{e}^{4 \cdot \ln(x)} \tag{6.67}$$

From our knowledge of logarithms and the exponential function this becomes

$$\text{IF} = \mathrm{e}^{\ln(x^4)} = x^4 \tag{6.68}$$

We therefore multiply each term in (6.65) by x^4.

$$x^4\frac{dy}{dx} + x^4\frac{4}{x}y = x^4 2x^2 \tag{6.69}$$

which is the same as

$$x^4\frac{dy}{dx} + 4x^3 y = 2x^6 \tag{6.70}$$

Now note that the second term on the left hand side, $4x^3$, is the derivative of x^4. We may therefore rewrite this equation as

$$x^4\frac{dy}{dx} + y\frac{d(x^4)}{dx} = 2x^6 \tag{6.71}$$

Referring back to Chapter 4 and Table 4.2, the left hand side may be recognisable as the formula for the derivative of a product of functions, i.e., yx^4, differentiated with respect to x. Equation (6.71) therefore becomes

$$\frac{d(yx^4)}{dx} = 2x^6 \tag{6.72}$$

This is now separable and may be integrated.

$$\int d(yx^4) = \int 2x^6 dx \tag{6.73}$$

$d(yx^4)$ integrates to yx^4 and the right hand side integrates to $2x^7/7$. We must not, of course, forget that this has been an indefinite integration and we must include a constant of integration for each integral. Because we have an as yet undefined constant on each

*The constant of integration is taken care of in the theory and is omitted.

side we may combine them into a single constant, C, just on the right (since the sum of two constants is another constant).

$$yx^4 = \frac{2}{7}x^7 + C \tag{6.74}$$

y may now be given as a function of x by rearranging this equation

$$y = \frac{2}{7}x^3 + \frac{C}{x^4} \tag{6.75}$$

In an experimental situation we could now define C by imposing the correct limits on the equation. For example, if $y = 1$ when $x = 1$

$$1 = \frac{2}{7} + \frac{C}{1} \tag{6.76}$$

Therefore C is 5/7, and the complete solution to (6.65) is

$$y = \frac{2}{7}x^3 + \frac{5}{7x^4} \tag{6.77}$$

Example—successive first order reactions

This reaction scheme occurs when the product of a reaction is also reactive and goes on to form a further product. That is,

$$\mathrm{A} \xrightarrow{k_1} \mathrm{B} \xrightarrow{k_2} \mathrm{C} \tag{6.78}$$

where k_1 is the rate constant for the first reaction and k_2 is the rate constant for the second reaction. In this procedure we will indicate concentrations of the species A, B and C by the lower case letters, a, b and c. Concentrations of these at the beginning of the reaction, when $t = 0$, are a_0, b_0 and c_0 respectively. To simplify matters we will make the conditions such that at time zero only A is present, so that both b_0 and c_0 are equal to zero.

Having done this we can write differential equations describing the rate of change of each concentration with respect to time. For reactant A this is the normal first order decay

$$-\frac{\mathrm{d}a}{\mathrm{d}t} = k_1 a \tag{6.79}$$

For B the rate of change, $\mathrm{d}b/\mathrm{d}t$, is the combination of the increase due to decay of A and the loss due to B forming C.

$$\frac{\mathrm{d}b}{\mathrm{d}t} = k_1 a - k_2 b \tag{6.80}$$

For C, the increase in concentration is the rate of loss of B.

$$\frac{\mathrm{d}c}{\mathrm{d}t} = k_2 b \tag{6.81}$$

The solution to (6.79) is the normal first order integrated rate equation

$$\ln\left(\frac{a_0}{a}\right) = k_1 t \tag{6.82}$$

To solve equation (6.80) we can rearrange it into the form of a first order linear equation.

$$\frac{\mathrm{d}b}{dt} + k_2 b = k_1 a \tag{6.83}$$

We now need a as a function of time and to get this we rewrite the integrated rate equation (6.82). We first turn the logarithm upside down, introducing a negative term (Chapter 2),

$$\ln\left(\frac{a}{a_0}\right) = -k_1 t \tag{6.84}$$

make it into an exponential function,

$$\frac{a}{a_0} = \mathrm{e}^{-k_1 t} \tag{6.85}$$

and rearrange to give a as the dependent variable.

$$a = a_0 \mathrm{e}^{-k_1 t} \tag{6.86}$$

Substituting this value for a in equation (6.83)

$$\frac{\mathrm{d}b}{\mathrm{d}t} + k_2 b = a_0 k_1 \mathrm{e}^{-k_1 t} \tag{6.87}$$

This is now in the form of a first order linear differential equation where $P = k_2$ and Q is the whole of the term on the right hand side. The integrating factor is

$$\mathrm{IF} = \mathrm{e}^{\int P \mathrm{d}t} = \mathrm{e}^{\int k_2 \mathrm{d}t} = \mathrm{e}^{k_2 t} \tag{6.88}$$

Each term in (6.83) is multiplied by this factor.

$$\mathrm{e}^{k_2 t}\frac{\mathrm{d}b}{\mathrm{d}t} + \mathrm{e}^{k_2 t} b k_2 = k_1 a_0 \mathrm{e}^{k_2 t}\mathrm{e}^{-k_1 t} \tag{6.89}$$

There are two things to note here. One is that the exponential terms on the right hand side may be combined using the normal rules concerning products of exponential functions (Chapter 1).

$$\mathrm{e}^{k_2 t}\mathrm{e}^{-k_1 t} = \mathrm{e}^{k_2 t - k_1 t} = \mathrm{e}^{(k_2 - k_1)t} \tag{6.90}$$

The other, more important one is that in the second term on the left we have the factor $k_2 \cdot \exp(k_2 t)$ which we know from Chapter 4 to be the derivative of $\exp(k_2 t)$ with respect to t. Therefore, we may write equation (6.90) as

$$\mathrm{e}^{k_2 t}\frac{\mathrm{d}b}{\mathrm{d}t} + b\frac{\mathrm{d}(\mathrm{e}^{k_2 t})}{\mathrm{d}t} = k_1 a_0 \mathrm{e}^{(k_2 - k_1)t} \tag{6.91}$$

The left hand side of this equation should now be recognisable as the formula for the derivative of a product of functions (Chapter 4, Table 4.2), Having noted this, the equation can be simplified

$$\frac{\mathrm{d}(b \cdot \mathrm{e}^{k_2 t})}{\mathrm{d}t} = k_1 a_0 \mathrm{e}^{(k_2 - k_1)t} \tag{6.92}$$

This is now separable and can be rearranged with integral signs in place.

$$\int \mathrm{d}(b \cdot \mathrm{e}^{k_2 t}) = k_1 a_0 \int \mathrm{e}^{(k_2 - k_1)t} \mathrm{d}t \tag{6.93}$$

Note that although we know the limits of the experiment, that is when $t = 0$, $b = 0$, we will initially solve this equation as an equality of indefinite integrals. This will hopefully avoid any confusion arising from the fact that both variables are combined on the left hand side.

The left hand side, where the integrand is solely a differential, is solved directly. The right hand side must be solved by the method of substitution as described in Chapter 5. We substitute $u = (k_2 - k_1)t$, the $\mathrm{d}t$ is given by $\mathrm{d}u/(k_2 - k_1)$ and the integrand becomes e^u. The solution to this is of course e^{u}. The result is then

$$b \cdot \mathrm{e}^{k_2 t} = \frac{k_1 a_0}{k_2 - k_1} \mathrm{e}^{(k_2 - k_1)t} + C \tag{6.94}$$

Although both indefinite integrals have associated constants of integration, we have combined these together into one single constant, C on the right hand side. To find the value of C we must apply the known limits of the experiment. To do this we write the equation with values of $t = 0$ and $b = 0$. When we do this all the exponential terms become equal to e^0, which is unity. The only term in b is on the left and this becomes equal to zero.

$$0 = \frac{k_1 a_0}{k_2 - k_1} + C \tag{6.95}$$

Therefore

$$C = -\frac{k_1 a_0}{k_2 - k_1} \tag{6.96}$$

The complete version of (6.94) is therefore

$$b \cdot \mathrm{e}^{k_2 t} = \frac{k_1 a_0}{k_2 - k_1} \mathrm{e}^{(k_2 - k_1)t} - \frac{k_1 a_0}{k_2 - k_1} \tag{6.97}$$

Finally to give b as a function of time we divide through by $\exp(k_2 t)$. Consistent with rules of division by exponential functions (Chapter 1) we combine them by subtracting the index. Thus the term $\exp[(k_2 - k_1)t]$ when divided by $\exp(k_2)$ becomes $\exp[(k_2 - k_1 - k_2)t]$ which is $\exp(-k_1 t)$. The second term is written in a form consistent with this. Therefore the time dependence of b is given by

$$b = \frac{k_1 a_0}{k_2 - k_1} \mathrm{e}^{-k_1 t} - \frac{k_1 a_0}{k_2 - k_1} \mathrm{e}^{-k_2 t} \tag{6.98}$$

Factorising common terms

$$b = \frac{k_1 a_0}{k_2 - k_1} \cdot [\mathrm{e}^{-k_1 t} - \mathrm{e}^{-k_2 t}] \tag{6.99}$$

The last step in analysing this problem is to find a function which describes the change in concentration of C as the reaction proceeds. Fortunately for those already weary of differential equations the need for further integration is avoided by considering the original experimental limitation. If only A is present the total quantity of material present must be the same. That is, the sum of quantities a, b and c must be equal to a_0. Arithmetically

$$a_0 = a + b + c \tag{6.100}$$

Since we have just derived expressions for a and b and a_0 is presumably known then c can be calculated as

$$c = a_0 - a - b \tag{6.101}$$

Manipulations such as this have been considered in Chapter 1, but the result is

$$c = a_0\left[1 + \frac{1}{k_2 - k_1}(k_2 e^{-k_1 t} - k_1 e^{-k_2 t})\right] \tag{6.102}$$

Figure 6.3 shows plots of each of these concentration terms versus time. It is interesting to note that as a decreases and c increases (as might be expected), the concentration of B rises to a maximum and then decreases. The time taken for the concentration to reach a maximum would be of great interest should B be the most important product. Then we could calculate the time for maximum yield by differentiating the equation (6.99) and setting the derivative equal to zero as shown in Chapter 4.

6.3.3 Simplifications: Avoiding Complex Integration

Even from this one example (consecutive first order reactions, Section 6.3.2) it can be deduced that as systems become more complex than the idealised cases described in Section 6.3.1, the problem of solving the differential equations increases greatly. Real systems are often much more complex: they may involve parallel reactions of different reactants producing the same products, equilibria coupled with irreversible reactions and in the case of free radical reactions there may be several pathways and reactive species all competing at once. In some cases it is possible to control the experimental conditions such that the reaction behaves as if it were mechanistically simpler. For example, a second order reaction, carried out in excess of one of the reactants behaves as if it were first order

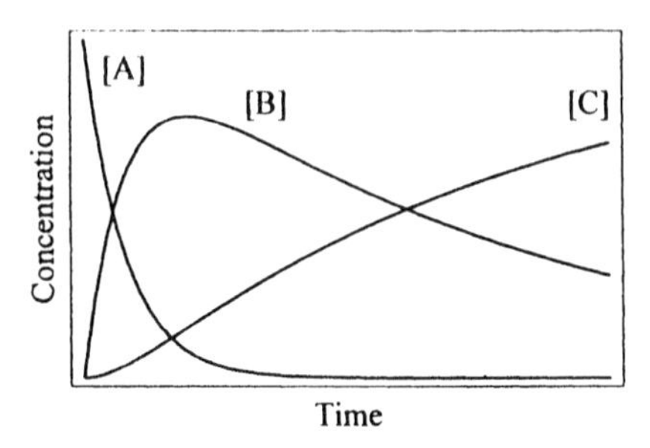

Figure 6.3 Concentration–time plots for the three species, A, B and C involved in the consecutive first order reaction scheme

overall. This is because the concentration of the excess reactant hardly changes as the reaction proceeds.

Such practical constraints may not always be possible or applicable and so it is the mathematical analysis which must be simplified. The most common simplification is called the *stationary state* or *steady state assumption* (SSA).

This assumption arises in the following manner. Figure 6.4 shows plots for the consecutive reaction A to B and B to C as discussed in Section 6.3.2. The difference between this and Figure 6.3 is that k_2, the rate constant for the second reaction, is much greater than k_1, the rate constant for the first reaction. These curves have been generated entirely by the theory described in this previous section. It can be seen that the product B hardly appears to take part in the reaction at all. It would be easy with a single glance to assume that it was a simple first order decay of A with concomitant rise in the product C. On closer inspection it can still be seen that the profile of [B] rises and falls as expected. This, however, occurs very rapidly and by comparison the concentration of B, always much less than that of A or C, is more or less constant. (Note that we have said that this is by comparison!)

We can also justify this chemically. If k_2 is very large compared with k_1 then the second reaction occurs very rapidly. As soon as any B is produced it is immediately converted into C. Thus the concentration of B is never given the chance to rise above a very small value.

Consideration of the mathematics also suggests this result. Considering equation (6.99),

$$b = \frac{k_1 a_0}{k_2 - k_1} \cdot [e^{-k_1 t} - e^{-k_2 t}] \tag{6.99}$$

if k_2 is very large then the term $\exp(-k_2 t)$ approaches zero (Chapter 2). If k_1 is small then the term $\exp(-k_1 t)$ approaches $\exp(0)$ which is equal to unity. Under these conditions the term $k_2 - k_1$ can be approximated as k_2 and thus this equation reduces to

$$b \approx a_0 \frac{k_1}{k_2} \tag{6.103}$$

The ratio k_1/k_2 is very small and so the concentration of B is very small compared to a_0. Notably it is also defined solely by constant terms. Thus the concentration of B is (approximately) constant. If this is the case then the derivative of b with respect to time must be zero

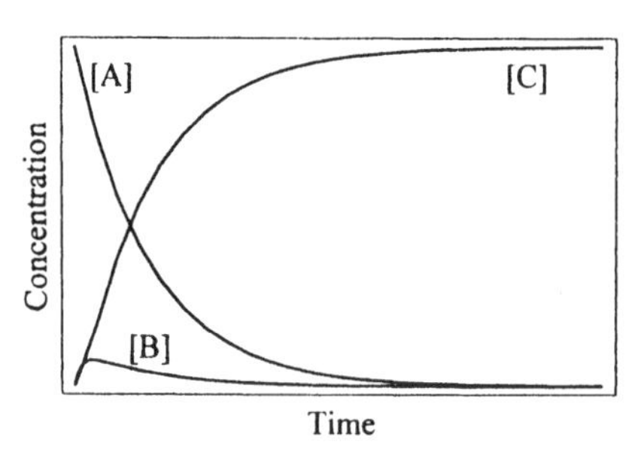

Figure 6.4 Plots like those in Figure 6.3 with $k_1 \ll k_2$. Note that the intermediate, B, hardly appears to take part in the reaction

$$\frac{db}{dt} = 0 \tag{6.104}$$

This is the essence of the stationary state assumption. In words it is

> The concentration of a reactive intermediate is small and constant. Therefore the rate of change of this concentration with time is zero.

Applying equation (6.104) to the analysis in Section 6.3.2 greatly simplifies the whole procedure. Equation (6.80) becomes

$$\frac{db}{dt} = k_1 a - k_2 b = 0 \tag{6.105}$$

Therefore

$$k_1 a = k_2 b \tag{6.106}$$

This rearranges to produce an equation similar to equation (6.103)

$$b = \frac{k_1 a}{k_2} \tag{6.107}$$

This may then be substituted in equation (6.81) in order to derive an expression for the concentration–time dependence of C.

$$\begin{aligned}\frac{dc}{dt} &= k_2 b \\ &= k_2 \cdot \frac{k_1 a}{k_2}\end{aligned} \tag{6.108}$$

The k_2 cancels and we can now substitute the expression for a

$$\frac{dc}{dt} = k_1 a_0 e^{-k_1 t} \tag{6.109}$$

This equation is separable and may be solved by integration, using the method of substitution ($u = -k_1 t$). The solution is

$$c = -a_0 e^{-k_1 t} + D \tag{6.110}$$

where D is the constant of integration. This is found by noting that when t is zero c is also equal to zero (none is present at the start of the reaction), i.e.,

$$0 = -a_0 e^0 + D \tag{6.111}$$

Therefore

$$D = a_0 \tag{6.112}$$

and (6.111) becomes

$$c = -a_0 e^{-k_1 t} + a_0 \tag{6.113}$$

which rearranges to give the final result

$$c = a_0(1 - e^{-k_1 t}) \tag{6.114}$$

There are three things to note about this equation. One is that the second term on the right is the usual term for the concentration of a (equation (6.86)). This equation therefore represents c as the difference between the initial concentration of A, a_0, and the concentration at time t. This must be correct if our original assumptions that there is only A present at the start and the concentration of B is very small (recalling equation (6.100), putting $b = 0$).

The second point is that the same equation may be arrived at by imposing the assumption that k_2 is much greater than k_1 on the full equation for c, equation (6.102). If k_2 is large then the term $k_1\exp(-k_2 t)$ tends to zero and the term $1/(k_1 - k_2)$ is approximately equal to $-1/k_2$. Under these circumstances k_2 cancels and we are left with equation (6.114).

The final point is that the rate of production of C depends only on the magnitude of k_1. Implicit in this is the idea of the rate determining step, a concept often introduced early in kinetics courses. Here we have shown using mathematics that in a two-step mechanism, if one step is much faster than the other, then the rate overall is determined by the rate of the slower.

6.3.4 Overview of Differential Equations in Kinetics

In the few examples given here we have shown how mathematics may be applied to the solution of problems of rates of reaction and reaction mechanisms in chemistry. It will probably be apparent that as the complexity of the system under investigation increases, so does the complexity of the mathematical model. Moreover, not only do different reactions have to have their own differential equations, but we must also account for different experimental techniques.

In any course on chemical kinetics most of the above examples, and many more, will be discussed. The approach to modelling the system is, however, the same. A mechanism is proposed and an appropriate differential equation derived. This may be modified to accommodate the technique being used and in any case the technique will probably provide the limits or boundary condition for the differential equation. The solution may be considered satisfactory if it can be shown to represent the proposed mechanism or perhaps reproduce the correct result for a known and well characterised reaction. Of course, the solution to the equation may be correct but the proposed mechanism be at fault. The whole process must therefore be considered to be an iterative one where the model is changed until a satisfactory result is obtained.

6.4 Differential Equations in Thermodynamics

Compared with the examples given in the previous section many of the differential equations used in thermodynamics are relatively trivial. They are of the form

$$dU = dq + dw \tag{6.115}$$

You will recall from Chapter 5 that the dq and dw terms are path-dependent integrals. That is, they depend on the way the energy is changed. In mathematical terms we must specify the function representing q. The corresponding thermodynamic view is that q and

w are not state functions. U is a state function and the integral of $\mathrm{d}U$ does not depend on the path and we do not have to specify the function. If we, therefore, solve equation (6.116) between the hypothetical limits of thermodynamic states A and B:

$$\int_A^B \mathrm{d}U = \int_A^B \mathrm{d}q + \int_A^B \mathrm{d}w \tag{6.116}$$

The result is usually given as

$$U_B - U_A = q + w \tag{6.117}$$

or commonly

$$\Delta U = q + w \tag{6.118}$$

We cannot say that $\mathrm{d}q$ becomes Δq and $\mathrm{d}w$ becomes Δw because we do not necessarily know the functions which describe q and w. Giving the result as some undefined quantity allows us to continue with our mathematical manipulations. We can always define these non-state functions at a later date.

In the remainder of this section we will present a few examples of how differential equations are used in thermodynamics. Many manipulations are purely algebraic. We usually begin with some abstract definition of a thermodynamic quantity—internal energy, enthalpy, entropy or Gibbs function. The purpose of the derivation is to turn this into something more tangible, something which we can actually measure in the laboratory. Measurable quantities are heat, mechanical and electrical work, temperature, volume and pressure.

The following examples are only a few of very many found in the study of thermodynamics.

6.4.1 Heat Capacities

Heat capacity is a way of relating the somewhat abstract quantity of heat to the readily understood measurement of temperature. It is defined by the differential equation

$$C = \frac{\mathrm{d}q}{\mathrm{d}T} \tag{6.119}$$

This is separable and may be solved as

$$\int \mathrm{d}q = C\int_{T_1}^{T_2} \mathrm{d}T \tag{6.120}$$

Therefore

$$q = C(T_2 - T_1) \tag{6.121}$$

or

$$q = C\Delta T \tag{6.122}$$

As above we need to specify just how the heat was applied. This equation is very general and it is possible to define heat capacity in different ways depending on the experimental conditions. The first way is to rewrite equation (6.115), replacing work, w, by $-P\mathrm{d}V$, the product of the pressure, P, and the change in volume, $\mathrm{d}V$.

$$\mathrm{d}U = \mathrm{d}q - P\mathrm{d}V \tag{6.123}$$

If we now specify that the change occurs at constant volume, in a bomb calorimeter for example, then $\mathrm{d}V$ is equal to zero and this equation reduces to

$$(\mathrm{d}U)_V = (\mathrm{d}q)_V \tag{6.124}$$

This is typical of the methodology of thermodynamic manipulation. A mathematical statement is made which may be quite complicated. It is then simplified by specifying certain conditions—constant volume in this case.

The definition in equation (6.119) may now be written as

$$C_V = \left(\frac{\partial U}{\partial T}\right)_V \tag{6.125}$$

where C_V is the heat capacity at constant volume. Note that we now use partial derivative notation to ensure that the reader knows that we have specified that the heat capacity is measured at constant volume, but is a function of another variable, temperature.

This is also separable and may be solved by integration. Because U is a state function this integrates directly to $U_\mathrm{B} - U_\mathrm{A}$ if A and B are the limits of the experiment or

$$\Delta U = C_V \Delta T \tag{6.126}$$

Heat capacity can also be measured at constant pressure. This is more useful because it is easier to perform experiments at constant pressure than at constant volume (in the open laboratory for instance). We do, however, need to define a new parameter; this is the enthalpy, H, which is the internal energy plus the product of the system's pressure and volume.

$$H = U + PV \tag{6.127}$$

In differential form we write $\mathrm{d}H$ as

$$\mathrm{d}H = \mathrm{d}U + \mathrm{d}(PV) \tag{6.128}$$

On the right hand side we have the differential of a product which from Chapter 4 is given as

$$\mathrm{d}H = \mathrm{d}U + P\mathrm{d}V + V\mathrm{d}P \tag{6.129}$$

$\mathrm{d}U$ has been given above as $\mathrm{d}q - P\mathrm{d}V$ and so the full equation is

$$\mathrm{d}H = \mathrm{d}q - P\mathrm{d}V + P\mathrm{d}V + V\mathrm{d}P \tag{6.130}$$

The terms in $P\mathrm{d}V$ cancel and if we specify that we are operating at constant pressure the $\mathrm{d}P$ term becomes zero. The equation then rather conveniently reduces to

$$(\mathrm{d}H)_P = (\mathrm{d}q)_P \tag{6.131}$$

This may be cast in equation (6.119) as

$$C_P = \left(\frac{\partial H}{\partial T}\right)_P \tag{6.132}$$

This is again separable and integration results in

$$\Delta H = C_P \Delta T \tag{6.133}$$

This in itself implies that the enthalpy of a reaction may be calculated if we know the heat capacity and the change in temperature. The meaning of this is not always clear. The enthalpy of a reaction is known as the difference in enthalpies of products and reactants but how do we define the heat capacity of the reaction? It may be measured by the rise in temperature of the surroundings of the reaction, for which the heat capacity may be known, but the usual equation given is called Kirchhoff's equation. In this we say that from equation (6.132), since H is proportional to C_P, then ΔH must be proportional to ΔC_P, where this is the difference in heat capacities of the products and reactants.

$$\left(\frac{\partial(\Delta H)}{\partial T}\right)_P = \Delta C_P \tag{6.134}$$

This, Kirchhoff's equation, is separable and may be integrated in the normal way. This may, however, be derived in a much more satisfactory way by considering the enthalpy change of a reaction as the difference between enthalpies of products and reactants

$$\Delta H = H_{\text{products}} - H_{\text{reactants}} \tag{6.135}$$

If we now differentiate each term with respect to temperature, specifying that the change occurs at constant pressure, we get

$$\left(\frac{\partial(\Delta H)}{\partial T}\right)_P = \left(\frac{\partial H_{\text{products}}}{\partial t}\right)_P - \left(\frac{\partial H_{\text{reactants}}}{\partial T}\right)_P \tag{6.136}$$

We have already defined the $(\partial H/\partial T)_P$ terms as the heat capacities and so this equation may be written in the form

$$\left(\frac{\partial(\Delta H)}{\partial T}\right)_P = C_P(\text{products}) - C_P(\text{reactants}) \tag{6.137}$$

$$= \Delta C_P$$

The use of Kirchhoff's equation is in calculating the enthalpy of a reaction at a different temperature. For example, data in tables will usually list enthalpies at standard temperature and pressure. The researcher may wish to know whether a reaction would occur more efficiently at a higher or a lower temperature. If heat capacities of components are known, Kirchhoff's equation may be used to calculate the new enthalpy. For example, integrating between temperatures T_1 and T_2, with enthalpies ΔH_1 and ΔH_2 equation (6.137) becomes

$$\int_{\Delta H_1}^{\Delta H_2} \mathrm{d}(\Delta H) = \int_{T_1}^{T_2} \Delta C_P \cdot \mathrm{d}T \tag{6.138}$$

the solution to which is

$$\Delta H_2 - \Delta H_1 = \Delta C_P(T_2 - T_1) \tag{6.139}$$

This is of course only true if C_P is independent of temperature. Sometimes C_P will be given as a function of temperature, for example

$$C_P = a + bT + cT^{-2} \tag{6.140}$$

Substitution of this into (6.139) and integrating each term results in

$$\Delta H_2 - \Delta H_1 = \Delta a(T_2 - T_1) + \Delta b(T_2^2 - T_1^2) + \Delta c\left(\frac{1}{T_2} - \frac{1}{T_1}\right) \tag{6.141}$$

where Δa, Δb and Δc represent the differences in these parameters for products and reactants.

The application of Kirchhoff's law to chemical problems can be likened to the application of Hess's law. In Figure 6.5 we wish to know the enthalpy of reaction, ΔH_2. If we know the enthalpies for each of the conversions A to A′, B to B′, C to C′ and D to D′, as indicated by the arrows, and the enthalpy ΔH_1 for the reaction A plus B to C and D, then the unknown enthalpy may be calculated as

$$\Delta H_2 = \Delta H_1 + (-\Delta H_3) + (-\Delta H_4) + \Delta H_5 + \Delta H_6 \tag{6.142}$$

where the minus signs indicate that in our cycle we are moving backwards along these particular arrows. Now suppose that the only difference between the reaction of interest (2) and the known reaction (1) is a change in temperature. Each of the enthalpy terms for the reactions 3, 4, 5 and 6 may be replaced by $C_P \Delta T$. Equation (6.142) then becomes

$$\Delta H_2 = (-C_{P,\mathrm{A}}\Delta T) + (-C_{P,\mathrm{B}}\Delta T) + \Delta H_1 + C_{P,\mathrm{C}}\Delta T + C_{P,\mathrm{D}}\Delta T \tag{6.143}$$

where $C_{P,\mathrm{A}}$ is the heat capacity of A and likewise for components B, C and D. The ΔT is a common factor term in each of these terms and so we may write

$$\Delta H_2 = \Delta H_1 + \Delta T[(C_{P,\mathrm{C}} + C_{P,\mathrm{D}}) - (C_{P,\mathrm{A}} + C_{P,\mathrm{B}})] \tag{6.144}$$

This may be compared to equation (6.139) where we see that ΔC_P is the difference between the sums of the heat capacities of products and reactants and that ΔT is $T_2 - T_1$.

Finally in this section heat capacities are also related to the entropy, S, of a system. The change in entropy, dS, may be defined as the amount of heat exchanged at a constant temperature.

$$\mathrm{d}S = \frac{\mathrm{d}q}{T} \tag{6.145}$$

In some texts it may also be specified that the change occurs reversibly, but as we are treating this as a purely mathematical manipulation we will not worry too much about the implications. Note, however, that a rearrangement of this equation gives dq as being

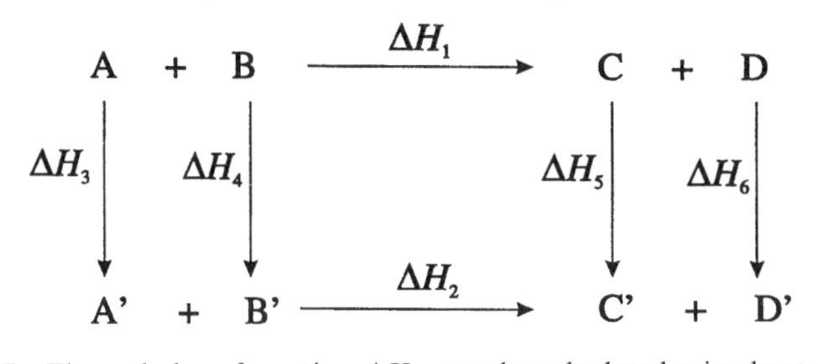

Figure 6.5 The enthalpy of reaction, ΔH_2, may be calculated using heat capacities in a Hess's law cycle. The temperature is changed from A′ and B′, this known reaction is allowed to proceed, and then the temperature of products C and D is changed to C′ and D′. This is essentially the basis of Kirchoff's law

equal to $T\mathrm{d}S$, a term which appears in other thermodynamic equations. Since $\mathrm{d}q$ is also given at constant pressure by $C_P\mathrm{d}T$, equation (6.145) may be written as

$$\mathrm{d}S = C_P \frac{\mathrm{d}T}{T} \tag{6.146}$$

This may be integrated between T_2 and T_1, with the corresponding values of entropy, S_2 and S_1.

$$\int_{S_1}^{S_2} \mathrm{d}S = C_P \int_{T_1}^{T_2} \frac{\mathrm{d}T}{T} \tag{6.147}$$

the result of which is

$$S_2 - S_1 = C_P \cdot \ln\left(\frac{T_2}{T_1}\right) \tag{6.148}$$

assuming that C_P is independent of temperature. If S_1 is a standard entropy at T_1 we now have an equation which is in the same form as those thermodynamic equations discussed in Chapter 2.

6.4.2 The Gibbs Function—Pressure and Temperature Dependence

The Gibbs function (symbol G) is probably the most important thermodynamic parameter encountered. This is because it is used as a criterion for the equilibrium condition (ΔG is zero) and because its sign is used to indicate whether a reaction will proceed in the forwards (ΔG is negative) or reverse (ΔG is positive) direction. It is also directly related to the potential difference of an electrochemical cell, the equilibrium constant, enthalpy and entropy change of all reactions. The Gibbs function itself is defined by the following equation

$$G = U + PV - TS \tag{6.149}$$

This may be written in differential form, remembering that as usual we must take full account of the products.

$$\mathrm{d}G = \mathrm{d}U + \mathrm{d}(PV) - \mathrm{d}(ST) \tag{6.150}$$

Which is the same as

$$\mathrm{d}G = \mathrm{d}U + P\mathrm{d}V + V\mathrm{d}P - S\mathrm{d}T - T\mathrm{d}S \tag{6.151}$$

If we specify that changes will occur at constant temperaure and pressure then the $\mathrm{d}P$ and $\mathrm{d}T$ terms disappear.

$$\mathrm{d}G = \mathrm{d}U + P\mathrm{d}V - T\mathrm{d}S \tag{6.152}$$

We have already defined $\mathrm{d}H$ under the condition of constant pressure as equal to $\mathrm{d}U + P\mathrm{d}V$, equation (6.127), and so (6.152) is

$$\mathrm{d}G = \mathrm{d}H - T\mathrm{d}S \tag{6.153}$$

This may be integrated between states one and two

$$\int_{G_1}^{G_2} \mathrm{d}G = \int_{H_1}^{H_2} \mathrm{d}H - T\int_{S_1}^{S_2} \mathrm{d}S \tag{6.154}$$

Each of these is a state function and integrates directly to produce

$$G_2 - G_1 = (H_2 - H_1) - T(S_2 - S_1) \tag{6.155}$$

or

$$\Delta G = \Delta H - T\Delta S \tag{6.156}$$

Note that both of the terms on the right hand side are forms of heat we have defined above.The change in Gibbs function may therefore be thought of as the difference between the heat absorbed in the conventional sense and that which is used to increase the entropy of the system. If this equation is differentiated with the assumption that ΔH is independent of temperature the derivative is equal to $-\Delta S$. A plot of ΔG versus temperature should have a slope of ΔS and an intercept at zero temperature of ΔH.

Equation (6.152) may be simplified in another manner. The term $\mathrm{d}U$ is usually defined as the sum of the heat exchanged, $\mathrm{d}q$, and the work, $\mathrm{d}w$.

$$\mathrm{d}U = \mathrm{d}q + \mathrm{d}w \tag{6.115}$$

If this work is only due to the expansion of a gas then $\mathrm{d}w$ is equated with $-P\mathrm{d}V$ and $\mathrm{d}q$ is equated with $T\mathrm{d}S$

$$\mathrm{d}U = T\mathrm{d}S - P\mathrm{d}V \tag{6.157}$$

(6.151) then becomes

$$\mathrm{d}G = T\mathrm{d}S - P\mathrm{d}V + P\mathrm{d}V + V\mathrm{d}P - S\mathrm{d}T - T\mathrm{d}S \tag{6.158}$$

The $T\mathrm{d}S$ and the $P\mathrm{d}V$ terms all cancel and we are left with

$$\mathrm{d}G = V\mathrm{d}P - S\mathrm{d}T \tag{6.159}$$

If we now specify constant temperature this reduces even further

$$\mathrm{d}G = V\mathrm{d}P \tag{6.160}$$

This may of course be integrated but is is often written as a partial derivative

$$\left(\frac{\partial G}{\partial P}\right)_T = V \tag{6.161}$$

which implies that the rate of change of free energy with pressure (of a gas) is merely its volume. We may also simplify equation (6.159) by specifying the condition of constant pressure.

$$\mathrm{d}G = -S\mathrm{d}T \tag{6.162}$$

This is, again, usually written as the partial derivative.

$$\left(\frac{\partial G}{\partial T}\right)_P = -S \tag{6.163}$$

Thus a plot of free energy versus temperature should have a slope of $-S$.

These results may also be obtained by considering G, a state function, as a function of temperature and pressure. A state function may be written as an exact differential, which from Chapter 4 is a form which includes the partial derivatives with respect to these other variables,

$$dG = \left(\frac{\partial G}{\partial P}\right)_T dP + \left(\frac{\partial G}{\partial T}\right)_P dT \qquad (6.164)$$

Comparison of this with equation (6.159) leads directly to equations (6.161) and (6.163).

To solve equation (6.161) by integration we note that V is related to P by the ideal gas equation and so the right hand side may be written purely in terms of P. If we integrate between the limits of states indicated by subscripts 1 and 2 we obtain

$$\int_{G_1}^{G_2} dG = nRT \int_{P_1}^{P_2} \frac{dP}{P} \qquad (6.165)$$

Assuming that T is constant this integrates to

$$G_2 - G_1 = nRT \cdot \ln\left(\frac{P_2}{P_1}\right) \qquad (6.166)$$

This is another equation which should be familiar from Chapter 2. We can see that the ln term arises from the integration of the $1/P$ function. If we set G_1 to be a standard value, G^0, at a standard pressure, say 1 atmosphere, then this equation may be written in standard form

$$G = G^0 + nRT \cdot \ln(P) \qquad (6.167)$$

This is a very important equation. Not just because it links the free energy of a gas to a measurable quantity, pressure, but also because it is the starting point of many thermodynamic descriptions of real systems. It is used to calculate the Gibbs function of liquids and solutions and manipulation leads to the Nernst equation used in electrochemistry.

The temperature dependence of the Gibbs function is found by a less obvious method. Taking the following version of equation (6.159) or (6.166) we substitute for $-S$ from differential equation (6.163)

$$-S = \frac{G}{T} - \frac{H}{T} \qquad (6.168)$$

Therefore

$$\left(\frac{\partial G}{\partial T}\right)_P = \frac{G}{T} - \frac{H}{T} \qquad (6.169)$$

This is then normally rearranged

$$\left(\frac{\partial G}{\partial T}\right)_P - \frac{G}{T} = -\frac{H}{T} \qquad (6.170)$$

For reasons which will become clear, each term is now divided by T.

$$\frac{1}{T}\left(\frac{\partial G}{\partial T}\right)_P - \frac{G}{T^2} = -\frac{H}{T^2} \qquad (6.171)$$

It happens that the function G/T is occasionally a more useful parameter than G itself. Now we see what happens if we try to generate a differential equation including the derivative of this combined function. In Chapter 4 we gave (Table 4.2) a number of equations or reduction formulae for this purpose. The equation for the derivative of a ratio of functions, u/v, where both of these are functions of x, is

$$\frac{\mathrm{d}(u/v)}{\mathrm{d}x} = \frac{v\dfrac{\mathrm{d}u}{\mathrm{d}x} - u\dfrac{\mathrm{d}v}{\mathrm{d}x}}{v^2} \tag{6.172}$$

If we consider both G and T to be functions of T (even though T is merely equal to T), then with G equivalent to u and T equivalent to v we obtain

$$\frac{\mathrm{d}(G/T)}{\mathrm{d}T} = \frac{T\dfrac{\mathrm{d}G}{\mathrm{d}T} - G\dfrac{\mathrm{d}T}{\mathrm{d}T}}{T^2} \tag{6.173}$$

and

$$\frac{\mathrm{d}(G/T)}{\mathrm{d}T} = \frac{1}{T}\frac{\mathrm{d}G}{\mathrm{d}T} - \frac{G}{T^2} \tag{6.174}$$

since $\mathrm{d}T/\mathrm{d}T$ is equal to one. This equation may now be compared to equation (6.171), and we see why it was necessary to write this equation in this form. The right hand side of equation (6.174) is the same as the left hand side of equation (6.171). The only difference is that in (6.171) we have a partial derivative because we have specified that the change occurs at constant pressure. By making this comparison we may therefore equate the left hand side of (6.174) with the right hand side of (6.171).

$$\frac{\mathrm{d}(G/T)}{\mathrm{d}T} = -\frac{H}{T^2} \tag{6.175}$$

Usually, because we are more interested in changes in the Gibbs function and enthalpy this is written as

$$\frac{\mathrm{d}(\Delta G/T)}{\mathrm{d}T} = -\frac{\Delta H}{T^2} \tag{6.176}$$

This is the Gibbs–Helmholtz equation which relates the free energy change to temperature and the enthalpy change. Below are two examples of how this equation is used.

Example—the Clausius–Clapeyron equation

This relates the vapour pressure of a liquid to the temperature. We can obtain an equation for the standard change in Gibbs function upon vaporisation for one mole of a vapour which is

$$\Delta G^0_{\mathrm{vap}} = -RT \cdot \ln(P) \tag{6.177}$$

or

$$\frac{\Delta G^0_{\mathrm{vap}}}{T} = -R \cdot \ln(P) \tag{6.178}$$

Differentiating both sides with respect to T produces

$$\left(\frac{\mathrm{d}(\Delta G^0_{\mathrm{vap}}/T)}{\mathrm{d}T}\right) = -R\cdot\frac{\mathrm{d}\ln(P)}{\mathrm{d}T} \tag{6.179}$$

This may then be substituted into the Gibbs–Helmholtz equation as

$$\frac{\mathrm{d}\ln(P)}{\mathrm{d}T} = \frac{1}{R}\frac{\Delta H^0_{\mathrm{vap}}}{T^2} \tag{6.180}$$

This is separable and may be solved by integration. This is usually done as an indefinite integral to produce

$$\ln(P) = -\frac{\Delta H^0_{\mathrm{vap}}}{RT} + C \tag{6.181}$$

where C is the constant of integration and we have assumed that ΔH is independent of temperature. Plots are made of $\ln(P)$ versus $1/T$, and since R is known the slope is a measure of the standard enthalpy of vaporisation.

Example—the elevation of boiling point of solvent

It is a common observation that when an involatile solute is added to a solvent the boiling point of the solvent is raised. We can show this to be the case mathematically by using the Gibbs–Helmholtz equation. The Gibbs function of vaporisation of the solvent is related to the mole fraction, x_{solv}, of the solvent by the equation

$$\Delta G^0_{\mathrm{vap}} = RT\cdot\ln(x_{\mathrm{solv}}) \tag{6.182}$$

Therefore by rearrangement

$$\frac{\Delta G^0_{\mathrm{vap}}}{T} = R\cdot\ln(x_{\mathrm{solv}}) \tag{6.183}$$

Dividing by R and differentiating each side with respect to T we get

$$\frac{\mathrm{d}(\Delta G^0_{\mathrm{vap}}/T)}{\mathrm{d}T} = \frac{\mathrm{d}\ln(x_{\mathrm{solv}})}{\mathrm{d}T} \tag{6.184}$$

This is then related to ΔH_{vap} by the Gibbs–Helmholtz equation.

$$\frac{\mathrm{d}\ln(x_{\mathrm{solv}})}{\mathrm{d}T} = -\frac{\Delta H^0_{\mathrm{vap}}}{RT^2} \tag{6.185}$$

This is solved by integration between the normal boiling point, T_{bp} and some variable temperature T. We take note that the solvent will only boil at T_{bp} when it is pure, which is when its mole fraction x_{solv} is equal to one. The limits of integration for the left hand side are therefore 1 and x_{solv}.

$$\int_1^{x_{\mathrm{solv}}} \mathrm{d}\ln(x_{\mathrm{solv}}) = -\frac{\Delta H^0_{\mathrm{vap}}}{R}\int_{T_{\mathrm{bp}}}^{T}\frac{1}{T^2}\mathrm{d}T \tag{6.186}$$

The result of this is

$$\ln(x_{\text{solv}}) = \frac{\Delta H^0_{\text{vap}}}{R}\left(\frac{1}{T} - \frac{1}{T_{\text{bp}}}\right) \tag{6.187}$$

It is normal to alter this equation to a form using the concentration of the solute rather than the solvent. This necessitates making a number of approximations. However, even in this form it can be seen that the result is satisfactory. As the concentration of solute increases, the mole fraction of solvent decreases, therefore $\ln(x_{\text{solv}})$ decreases. Because of this one expects the right hand side of this equation to decrease also. This can be seen to be the case if the observed boiling point, T, increases. Therefore $1/T$ decreases and the right hand side decreases in magnitude accordingly.

Similar uses of the Gibbs–Helmholtz equation are made in finding the temperature dependence of other free energy functions such as equilibrium constant, the lowering of freezing point and the solubility of solutes in solution.

6.5 Concluding Remarks

In this chapter we have shown two uses of differential equations in chemistry. One is the derivation and solution of differential equations in modelling kinetic systems. The other is the manipulation of such equations, largely algebraically, to derive equations which represent measurable quantities rather than abstract energy terms.

In both cases we are able to apply these methods because chemistry is a discipline concerning change and the properties of derivatives and differentials suitably concur with the way chemical systems change.

Problems

6.1 Consider the following rate determining step:

$$\text{A} \longrightarrow \text{B} \tag{P6.1}$$

If the reaction is characterised by the rate constant k_1 write a differential equation describing

(a) the loss of A
(b) the production of B

6.2 Consider the following rate determining step:

$$2\text{A} \longrightarrow \text{B} \tag{P6.2}$$

If the reaction is characterised by the rate constant k_1 write a differential equation describing

(a) the loss of A
(b) the production of B

6.3 Consider the following rate determining step:

$$\text{A} + \text{B} \longrightarrow \text{C} \tag{P6.3}$$

If the reaction is characterised by the rate constant k_1 write a differential equation describing

(a) the loss of A
(b) the loss of B
(c) the production of C

6.4 Consider the following equilibrium:

$$A \rightleftharpoons B \qquad (P6.4)$$

If the forward reaction is characterised by the rate constant k_1 and the reverse by k_{-1} write a differential equation describing

(a) the change in concentration of A
(b) the change in concentration of B

6.5 A reaction between two species, A and B, is found to follow the following rate law:

$$\text{rate} = [A]^x[B]^y \qquad (P6.5)$$

where $x = 1$ and $y = 0$.

(a) What is the order of reaction with respect to A?
(b) What is the order of reaction with respect to B?
(c) What is the overall order of reaction?
(d) Write equation (P6.5) as a differential equation in terms of the loss of A and solve it between the limits of $[A] = [A]_0$ at time $t = 0$ and $[A] = [A]$ at time t.

6.6 Many reactions may be characterised by the following reaction scheme,

$$A \rightleftharpoons B \longrightarrow C \qquad (P6.6)$$

The forward reaction of the equilibrium is characterised by the rate constant k_1, the reverse by k_{-1}, and the reaction B to C is characterised by the rate constant k_2.

(a) Write a differential equation for the rate of production of C in terms of the concentration of intermediate B.
(b) Write a differential equation for the *increase* in concentration of B ($d[B]/dt =$) in terms of the concentration of A ([A]).
(c) Write a differential equation for the *loss* of B ($d[B]/dt =$) in terms of the concentration of B itself.
(d) The overall rate of change in [B] is the difference between the rates of formation and destruction of B. Write this as a differential equation in terms of [A] and [B] and the relevant rate constants.
(e) The stationary state assumption has it that the rate of change of [B] is zero. Use this statement to write an equation for the concentration of B.
(f) Use the last result to write a rate equation for the production of C in terms of [A] and the three rate constants.
(g) What can you say about the extreme cases when $k_2 \gg k_{-1}$ and $k_2 \ll k_{-1}$?

6.7 One reaction similar to that in the above problem describes the rate of an electrochemical reaction.

$$A \underset{k_D}{\overset{k_D}{\rightleftharpoons}} A^* \xrightarrow{k_{EC}} C \qquad (P6.7)$$

The pre-equilibrium describes the diffusion (rate constant k_D) to and from the electrode surface and the following reaction is the transfer of electrons to or from the electrode to the surface reactant A*. The current observed is equivalent to the rate of reaction characterised by k_{EC}.

(a) Following the steps used in Problem 6.6 write a differential equation for the rate of production of C.

(b) An important facet of electrochemical studies is that the magnitude of k_{EC} can be controlled by the potential of the electrode. It is therefore easy to change the experimental conditions from $k_{EC} \ll k_D$ to $k_{EC} \gg k_D$ merely by switching the potential of the electrode. Use these two extremes to write approximate equations and state which is the limiting process in each case.

6.8 A parallel reaction scheme has reactant A forming two products B and C in competing reactions

$$A \longrightarrow B \tag{P6.8}$$

$$A \longrightarrow C \tag{P6.9}$$

The former is characterised by the rate constant k_B, the latter by k_C.

(a) Write and solve a differential equation for the rate of loss of A. The boundary conditions are $[A] = [A]_0$ at $t = 0$ and $[A] = [A]$ at $t = t$.

(b) Write the solution in terms of [A] as a function of time, t.

(c) Write and solve a differential equation for the rate of production of B (make substitutions for [B] where necessary).

7 Statistics for Theoretical Chemistry

7.1 Introduction

In this chapter we will look at some of the ways the ideas of statistics are used in the development of theoretical chemistry. We use statistics in this way when we try to rationalise the properties of atoms and molecules, defined by quantum theory, with the bulk thermodynamic properties which we may measure in the laboratory. To take an example which we will expand on later, suppose a sample of only five molecules is analysed and found to have an average of one quantum unit of energy per atom. If there is a total of five quanta in the sample, how do we know how that energy is shared amongst the molecules? Assuming that the molecules may have integer amounts of energy (including zero) we can calculate that there are 126 ways in which this energy may be distributed.

This example is somewhat abstract in that we would never consider five molecules as a bulk sample, but the mathematical methods used in arriving at the result may be extended to cover real quantities. We can already see, however, that even for a tiny sample of five molecules and five quanta of energy, there is a disproportionate number of ways in which they may be arranged. For samples of the order of 6×10^{23} molecules one might expect an astonishing number of ways in which they may be organised. Although the mathematics suggests that this should be the case it also predicts that certain of these ways become dominant, and in modelling the distribution of energy amongst molecules we may only have to consider these dominant arrangements or configurations.

Initially we will introduce the mathematics of *permutations*. Permutations are the different orders in which things may be arranged. We will then show how these ideas may be related to theoretical aspects of chemistry, in particular, the Boltzmann equation.

7.2 Permutations—Calculating Arrangements

Before considering arrangements of molecules and quanta of energy we first look at the mathematics of permutations by considering somewhat more tangible objects—coloured building bricks.

We will consider a number of coloured building bricks: red, yellow, blue and green (denoted R, Y, B and G respectively) and how they may be arranged in a single row from left to right. As they are different colours we note that they are distinguishable objects, all different.

We begin with one red brick and it is apparent that we can only arrange it in one way. To do this we will assume that it does not matter whether bricks are upside down or back to front, it is only their position in line that matters. If we now consider two bricks then we may arrange them in two possible ways:

R Y
or
Y R

To put this on a more mathematical basis we could say that we have two choices (red or yellow) for the first brick and once this has been decided we have only one choice for the second. In equation form we write this as a product,

$$\text{Number of ways, } W = 2 \times 1 = 2 \tag{7.1}$$

We now consider three different coloured bricks, red, yellow and blue. Intuitively we could decide that the possible arrangements are

R	Y	B
R	B	Y
Y	R	B
Y	B	R
B	R	Y
B	Y	R

Thus there are six possible ways in which these three, distinguishable, objects may be arranged.

To continue our mathematical rationalisation we have a choice of three for the first position. For each of these three first choices the remaining two may be arranged (as above) in two possible ways. There are three multiplied by two ways in which to arrange this system,

$$\text{Number of ways, } W = 3 \times 2 = 6 \tag{7.2}$$

Although this seems correct we must remember that the two ways to arrange two objects is really comprised of two for the first choice and one for the second, by the same argument as above these two ways are arrived by

$$\text{Number of ways, } W = 2 \times 1 = 2 \tag{7.3}$$

Thus equation (7.2) is really

$$\text{Number of ways, } W = 3 \times 2 \times 1 = 6 \tag{7.4}$$

Recalling factorial notation, $3 \times 2 \times 1$ is written $3!$, the number of ways of arranging three distinguishable objects in line is given by

$$W = 3! \tag{7.5}$$

Suppose we now add the green brick to our row. We could merely put this in front of the arrangements for the three listed above. That would give six arrangements again. However, there is no reason why any of the colours could not go first and for each of the four first choices there should be six possible arrangements for the following three, i.e., there should be a total of four multiplied by six ways of lining up four different objects,

$$W = 4 \times 6 = 24 \tag{7.6}$$

We can repeat the logic of building up equations (7.1) and (7.4) to redefine equation (7.6). There are four choices for the first in line, three for the second, two for the third and one for the fourth. The number of ways is given by

$$W = 4 \times 3 \times 2 \times 1 = 4! = 24 \tag{7.7}$$

In building up to considering only four distinguishable objects we can already see that the number of ways of arranging them in line, the number of permutations, increases disproportionately compared with the number of objects. If the formula works then there should be 5! or 120 ways of arranging five coloured bricks, 6! or 720 ways of arranging 6 bricks, and so on.

We will now consider what happens if we have a number of indistinguishable objects. If we have two, three or more red bricks then it does not matter which order we put them in. To the observer one arrangement is identical to the next. The expected 2!, 3!, 4! ways of arranging them has been reduced to one. How, then, does this affect the mathematics we have just described?

Say we have one red brick and two yellow bricks. Once again using our intuition we may find the possible arrangements to be

R	Y	Y
Y	R	Y
Y	Y	R

There are three possible arrangements. If we had some inside information such that we could tell the difference between the two yellow bricks we would have said that there should be 3! permutations. We would also say that there are 2! permutations for the two yellow ones alone. To the outsider these 2! permutations are identical and their number has been reduced to one. This can be achieved mathematically if we rewrite equation (7.5) recalling that $3! = 3 \times 2!$. To reduce this by a factor of 2! we divide by the number of permutations which would have been afforded by the two yellow bricks had they been different, i.e., 2!

$$\text{Number of ways, } W = \frac{3 \cdot 2!}{2!} = 3 \tag{7.8}$$

Therefore the total number of different ways of arranging the three objects has been reduced by a factor equal to the number of ways of arranging the identical objects.

$$W = \frac{3!}{2!} = 3 \tag{7.9}$$

If we have four bricks, three red and one yellow the possible arrangements are:

R	R	R	Y
R	R	Y	R
R	Y	R	R
Y	R	R	R

Again we would have expected there to be 4! ways of arranging these four bricks. However, the 3! ways of arranging the red bricks has effectively been reduced to one and the total permutations reduced accordingly, i.e.,

$$W = \frac{4!}{3!} = \frac{4 \times 3 \times 2 \times 1}{3 \times 2 \times 1} = 4 \tag{7.10}$$

A slightly more complicated possibility is if we have two red and two yellow bricks. The initial argument is the same, we would expect 4! ways of arranging these bricks, but each set of identical bricks will reduce the number by a factor equal to the number that they would be expected to contribute had they not been identical. The arrangements are

R	R	Y	Y
R	Y	R	Y
R	Y	Y	R
Y	Y	R	R
Y	R	Y	R
Y	R	R	Y

We have reduced the choice of the first colour from 4 by 2! and the 3! possible ways of arranging the following three to 3!/2!. For this example the number of permutations, W, available is

$$\begin{aligned} W &= \frac{4!}{2! \times 2!} = \frac{4 \times 3 \times 2 \times 1}{2 \times 1 \times 2 \times 1} \\ &= 6 \end{aligned} \tag{7.11}$$

We can generalise these formulae to give the equation applicable to any set of objects which are to be arranged in a particular order. If there are N different objects we would expect there to be $N!$ ways of arranging them. If a number, n_a, of these are identical to each other the total number of ways is reduced by a factor equal to the number of ways it would have been possible to arrange these if they had been different, i.e., $n_a!$. If there are more subsets of identical objects then the total is also reduced by a factor governed by the number in each subset, e.g., $n_a!$, $n_b!$ etc. The master equation is

$$\text{Number of ways, } W = \frac{N!}{n_a!n_b!n_c!\ldots} \tag{7.12}$$

7.3 Configurations and Microstates

In order to define these terms we will consider an example from everyday life rather than from chemistry. This will be the tossing of a coin. We are familiar with the use of a coin to decide who should begin a game. This is because the caller has an even chance of calling correctly or incorrectly, therefore the opponent also has an even chance of starting the game.

It may be found that in three tosses of a coin we observe three heads. This sample might suggest that the coin is not unbiased. If this result is obtained how can we really say that the use of the coin is fair? In reality we have to conclude that the sample is not big enough to truly represent the population and that if we were to consider extending the experiment to 300 tosses we would be much more likely to find an equal number of heads and tails.

That this is so can be shown by considering the possible outcomes for an increasing number of tosses of the coin. For example, if we have two throws of the coin the possible outcomes, labelled *I*, *II* and *III*, are (using H for heads and T for tails)

I	*II*	*III*
HH	HT	TT
	TH	

Each of these four possible outcomes is called a *microstate*. If we are concerned only with the total number of heads and tails produced then we can define a number of *configurations*. These correspond to two heads (Configuration *I*), one head and one tail (Configuration *II*), and two tails (Configuration *III*). In the configuration *II* of one head and one tail the order in which they are achieved is irrelevant. Therefore of the three possible configurations there are two ways of achieving this 'one of each' outcome compared with one way of obtaining two heads or one way of achieving two tails.

If we now consider three tosses of the coin the possible outcomes fall into four configurations:

I	*II*	*III*	*IV*
HHH	HHT	TTH	TTT
	HTH	THT	
	THH	HTT	

We have eight possible outcomes or microstates. The configurations are *I* three heads, *II* two heads and one tail, *III* two tails and one head and *IV* three tails. The numbers of microstates per configuration are one, three, three and one respectively. Although it is possible to estimate the possible outcomes for this experiment quite easily we can also arrive at the number of microstates per configuration using the mathematics of permutations described in Section 7.2, above.

We define W as the number of microstates per configuration, n_H as the number of heads and n_T as the number of tails. For N tosses of the coin we must define the possible configurations and for each we can calculate W by

$$\text{Number of microstates per configuration, } W = \frac{N!}{n_{\mathrm{H}}!n_{\mathrm{T}}!} \tag{7.13}$$

Therefore we might predict the numbers of microstates per configuration for four throws of the coin as follows. The possible outcomes would be:

I Four heads
II Three heads and one tail
III Two heads and two tails
IV Three tails and one head
V Four tails

Configuration *I* has four identical objects which can therefore only be achieved in one way (strictly 4!/4!0!, recalling that by definition 0! is equal to one).

Configuration *II* can be achieved in 4!/3!1! ways, i.e., 4 ways in total.
Configuration *III* can be achieved in 4!/(2! × 2!) ways, i.e., 6 ways.
Configuration *IV* as for *II* can be achieved in four ways.
Configuration *V* as for *I* can be achieved in one way only.

Adding all of these up we predict that the number of microstates is given by

$$\text{Total number of microstates} = 1 + 4 + 6 + 4 + 1 = 16 \tag{7.14}$$

These are:

I	*II*	*III*	*IV*	*V*
HHHH	HHHT	HHTT	TTTT	TTTT
	HHTH	HTHT	TTHT	
	HTHH	HTTH	THTT	
	THHH	TTHH	HTTT	
		THHT		
		THTH		

From these three experiments we may make a number of important conclusions. We can already see that the configuration corresponding to an equal number of heads and tails comprises the greatest number of microstates and is consequently more likely than the others.

We may also note that for N tosses of the coin there are $N + 1$ configurations and the numbers of microstates per configuration corresponds to the coefficients of the binomial expansion met in Chapter 1. These can be predicted using Pascal's triangle (Chapter 1 also) and so it should be possible to predict the numbers of microstates for any number of throws. For example, for ten throws there should be eleven possible configurations and the numbers of microstates for each from Pascal's triangle are

I	*II*	*III*	*IV*	*V*	*VI*	*VII*	*VIII*	*IX*	*X*	*XI*
1	10	45	120	210	252	210	120	45	10	1

It is also possible to illustrate how the number of microstates per configuration varies as the number of throws increases by constructing a frequency plot.

This is shown in Figure 7.1 for experiments making two, four, ten and 100 throws. The

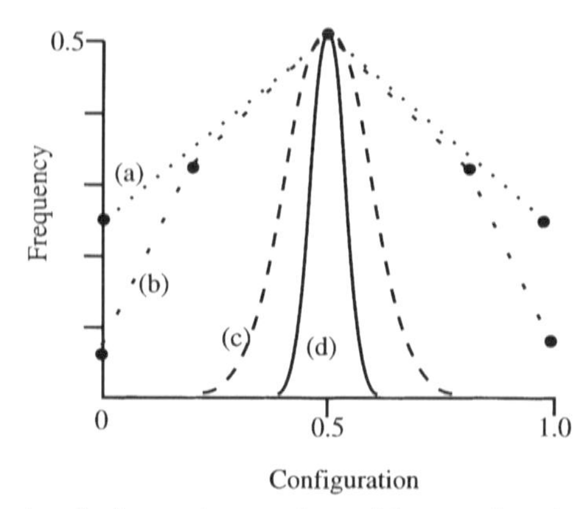

Figure 7.1 Frequency plots for increasing numbers of throws of a coin, (a) two, (b) four, (c) ten and (d) 100 throws. Note that as the number increases then the frequency of the 0.5 configuration dominates the distribution

vertical scale of this plot has been normalised so that the peaks coincide. To do this we divide each frequency by the largest frequency for that experiment, e.g., in the example of four throws of the coin each number has been divided by six, the number of microstates in configuration *III*.

This is then plotted against a number corresponding to the index number of the configuration. Once again we must put all of the plots on the same scale. To do this we convert the index number into a ratio of heads to the total number of throws. Thus index *I* for two throws is 1, *II* is 0.5 and *III* is zero. For the ten throw experiment Configuration *VI* is equal to 0.5 as this configuration has 5 heads and 5 tails.

As the number of throws increases it appears that the plot narrows and the configuration corresponding to equal numbers of heads and tails is coming to dominate the distribution. The likelihood of obtaining all heads or all tails is more and more remote. Thus the original assumption that we are as likely to obtain a head as we are to obtain a tail appears to be verified by extensive experiments.

It is interesting enough that this dominance of one particular configuration occurs with such a simple system and with a relatively small number of samples. When we introduce these ideas to chemical systems, where the numbers increase vastly (to the order of 10^{23}) the effect would be expected to be even more marked.

7.4 Molecular Assemblies

We will now apply these ideas to assemblies of molecules. To do this we will initially define an idealised system. This is a quantised system of a limited number of particles sharing a limited number of energy units. The molecules may be considered to reside in energy levels defined as 0, 1, 2, 3, etc. if they have the corresponding number of energy units (quanta) assigned to them. For example, with no energy they reside in level 0, with one quantum unit they reside in level 1 and so on. Consequently we also consider the levels to be evenly spaced with an energy difference of one unit. Figure 7.2 illustrates the system under consideration. A final, but important, stipulation is that the molecules are identical but may be distinguished by their occupation of different energy levels. This may also be interpreted as being able to distinguish them by their position in space.

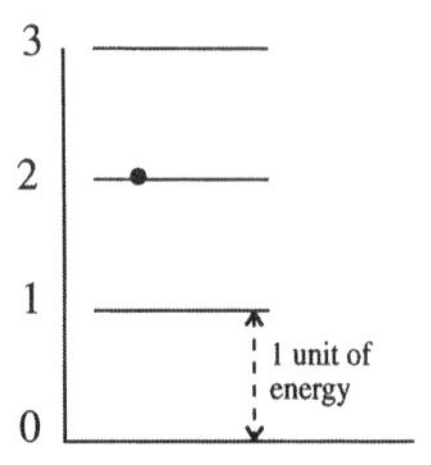

Figure 7.2 Schematic representation of a discrete energy level system. Each level is separated from the next by one unit of energy. Particles may only possess integer numbers of units of energy. In this example the system has two units, both allocated to the particle residing in level 2

The problem is to define the number of possible arrangements in which the total energy may be distributed between the limited number of particles.

Example—three molecules, three quanta

An assembly of three particles is found to have a total of three units or quanta of energy, that is, each particle has an average of one quantum of energy. Individual particles may, however, have more than one unit of energy as long as the total for the three particles is equal to three. When a molecule has more than one unit of energy it occupies the corresponding energy level. Thus these molecules, although essentially identical, can be differentiated by their occupation of these different energy levels.

Figure 7.3 shows the possible configurations for distribution of molecules in these energy levels so that the sum is always three quanta. Configuration *I* allocates all three quanta to one molecule, which therefore occupies energy level 3. This molecule is now different to the other two which are identical in occupying energy level 0. How many ways may this configuration be achieved?

This problem is analogous to the problem of arranging one red and two yellow bricks in a row. We can apply the mathematics of permutations and calculate W from

$$W = \frac{N!}{n_a!n_b!\ldots} \tag{7.15}$$

We have three molecules and therefore N is equal to three and the subsets are n_a equal to 1 and n_b is equal to two. Since 1! is equal to one W is calculated as

$$W = \frac{3!}{2!} = \frac{3 \times 2 \times 1}{2 \times 1} = 3 \tag{7.16}$$

This result is, as we expect, the same as that found for the problem arranging three bricks of which two are identical.

In future calculations such as that in equation (7.16) we will make use of the simplification

$$\frac{3!}{2!} = \frac{3 \times 2 \times 1}{2!} = \frac{3 \times 2!}{2!} \tag{7.17}$$

Although apparently trivial in this case it makes for great simplifications when larger numbers are used.

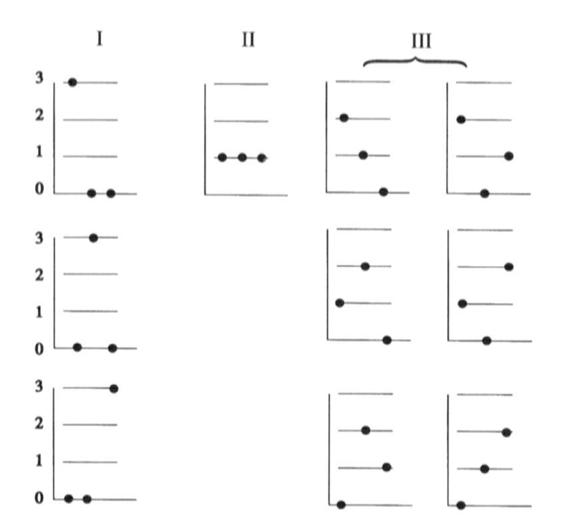

Figure 7.3 The configurations and microstates allowed for the three particle–three quanta system

Configuration *II* in Figure 7.3 is the next simplest arrangement with each molecule possessing one quantum of energy, each occupying energy level 1. We now have three identical particles and there should only be one way of arranging them. To confirm this we use equation (7.15) again. This time n_a is also three and so

$$W = \frac{3!}{3!} \tag{7.18}$$

Configuration *III* is more interesting with one molecule in level 2, one in level 1 and one in level 0. We now have three distinguishable objects and we expect W to be given by

$$W = 3! = 3 \times 2 \times 1 = 6 \tag{7.19}$$

These arrangements are the microstates for this system. There are a total of ten microstates, three of which correspond to Configuration *I*, one of which corresponds to Configuration *II* and six of which correspond to Configuration *III*. This configuration is much more likely to occur than either of the others and we see that even in this system one configuration dominates.

Example—four molecules, four quanta

In this example we will consider a slightly larger system but keep the conditions essentially the same. Each molecule may occupy any of levels 0, 1, 2, 3 or 4, depending on how many quanta it possesses. The only other stipulation is that the total amount of energy adds up to four.

Figure 7.4 shows the configurations possible for this arrangement. The number of microstates for Configurations *I*, *II*, *III*, *IV* and *V* are:

I This is the same as above with all the energy assigned to one molecule. This occupies level 4 and the rest occupy level 0. The number of microstates is therefore given by

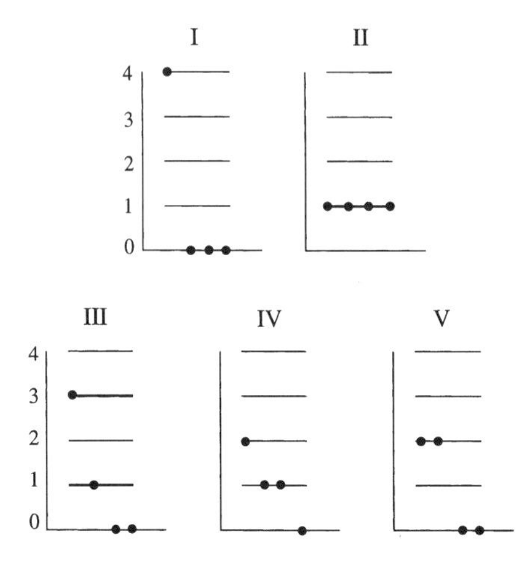

Figure 7.4 The configurations and microstates allowed for the four particle–four quanta system

$$W = \frac{4!}{3!} = \frac{4 \times 3!}{3!} = 4 \tag{7.20}$$

II Analogous to II above, all of the molecules have one quantum of energy and occupy level 1. We therefore have just one microstate.

$$W = \frac{4!}{4!} = 1 \tag{7.21}$$

III One molecule possesses three quanta, one possesses one quantum and the remaining two have zero energy. There is only one subset with more than one identical molecule and therefore n_a is equal to 2. W is therefore calculated by

$$W = \frac{4!}{2!} = \frac{4 \times 3 \times 2!}{2!} = 12 \tag{7.22}$$

since the 2! terms cancel.

IV One molecule possesses two quanta, two possess one quantum and the remaining one has none. Once again there is only one subset with more than one identical molecule and we expect 12 microstates.

$$W = \frac{4!}{2!} = 12 \tag{7.23}$$

V In this configuration two molecules possess two quanta each, using up all four available units. The remaining two therefore occupy the zero level. This corresponds to the example of lining up two red and two yellow bricks, which led to six microstates. n_a and n_b are both equal to 2 and so W is

$$W = \frac{4!}{2!2!} = \frac{4 \times 3 \times 2!}{2 \times 1 \times 2!} = 6 \tag{7.24}$$

In total therefore there are 35 microstates. In increasing the number of particles by one we have increased the number of microstates by more than a factor of three. In this case, however, there is no singly dominant configuration. It is possible to achieve both *III* and *IV* in twelve ways.

Example—eight particles, four quanta

We will now double the number of particles but keep the number of quanta available the same, thus the average energy per molecule falls to one half. This system has been chosen because it allows us to use the same general configurations as the previous example. The only difference is that we add four molecules to the zero energy level, which of course adds nothing to the total energy of the system. The configurations (Figure 7.5) are labelled *Ia*, *IIa*, *IIIa*, and so on. The numbers of microstates may be calculated in the usual manner.

Ia We have one molecule in the upper, four, level and the remaining seven in the lowest, zero, level. N is now equal to eight and therefore W is

$$W = \frac{8!}{7!} = \frac{8 \times 7!}{7!} = 8 \tag{7.25}$$

IIa We can only accommodate four of the eight in level 1. Therefore, four remain in level 0 and consequently there are two subsets containing four molecules, i.e., $n_a = n_b = 4$.

$$W = \frac{8!}{4!4!} = \frac{8 \times 7 \times 6 \times 5 \times 4!}{4 \times 3 \times 2 \times 1 \times 4!} = 70 \tag{7.26}$$

IIIa One molecule possesses three quanta, one has one and the other six occupy the zero level. Therefore n_a is equal to 6 and

$$W = \frac{8!}{6!} = \frac{8 \times 7 \times 6!}{6!} = 56 \tag{7.27}$$

IVa One molecule possesses two quanta, two possess one and five have none. There are two subsets, $n_a = 2$ and $n_b = 5$, and

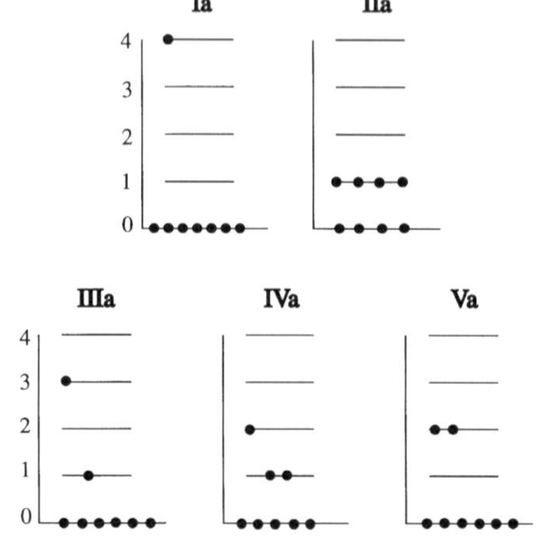

Figure 7.5 The configurations and microstates allowed for the eight particle–four quanta system

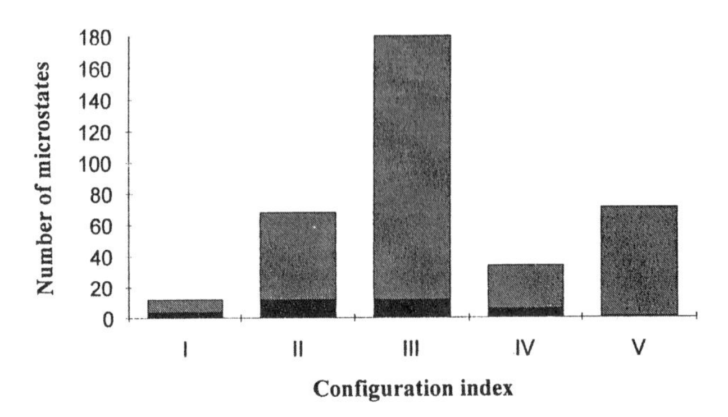

Figure 7.6 Histogram comparing the numbers of microstates for four and eight particles sharing four quanta of energy

$$W = \frac{8!}{2!5!} = \frac{8 \times 7 \times 6 \times 5!}{2 \times 1 \times 5!} = 168 \tag{7.28}$$

Va Two molecules have two quanta each and occupy level 2, the six remaining occupy level 0. Two subsets exist such that n_a is equal to 2 and n_b is equal to 6.

$$W = \frac{8!}{2!6!} = \frac{8 \times 7 \times 6!}{2 \times 1 \times 6!} = 28 \tag{7.29}$$

There are now a total of 330 microstates. This is approximately a ninefold increase, as compared with the doubling of the number of particles. Furthermore, configuration *IVa* can now be seen to dominate the distribution, comprising more than half the total number of microstates for the system. This is illustrated in Figure 7.6, a histogram plotting the number of microstates per configuration. When the number of particles is increased to eight the total increase in microstates is observed as is the dominance of the configuration labelled *IV*/*IVa*.*

7.5 The Importance of $W = N!/n_a!n_b!$

Two important results have been achieved in the previous sections. These are:

(1) As the number of experiments, units or particles considered increases the number of microstates associated with the system increases in a disproportionate manner.
(2) As this increase occurs a very few, or even just one configuration comes to dominate the system.

In Section 7.4 we considered a very limited system of up to eight particles. It may be calculated that for a system containing 1000 particles and 1000 units of energy 10^{600} possible microstates occur. This number is itself of unimaginable magnitude but if we extend the calculation to include real numbers of molecules, 10^{23}, the result is so large as

*An entropic aside. Note that in all cases the dominant configuration is the one which entails occupation of most energy levels and the greatest spread of particles, i.e., the most disordered arrangement. W is related to entropy and the most likely state for the system to be in is the one with greatest entropy as predicted by the second law of thermodynamics.

to be meaningless. It is perhaps fortunate then that the second result also arises. If we cannot identify every possible microstate for every possible configuration we may at least be able to home in on the most likely of these.

As engaging as they are, these conclusions lead in turn to more questions. What, for instance, is the identity of this dominant configuration? How does the population of this configuration compare with the others and can we put this information to good use? What happens to this configuration if we change the system, by adding energy or particles?

In beginning to explore these questions the first step may be to take a closer look at the equation we have made so much use of.

$$W = \frac{N!}{n_a!n_b!n_c!\ldots} \tag{7.12}$$

In our analyses N was the total number of particles in the system. n_a, n_b, n_c etc. represent, statistically, the numbers of identical particles in each of the possible subsets that make up the complete system. In the examples given these are also the populations of the individual energy levels. It, therefore, appears that the number of microstates for a particular configuration is dependent on the populations of the energy levels. This now becomes very interesting. Kinetic theory, rate theory and theories put forward to explain spectroscopic phenomena all rely on calculations of populations of energy levels. The knowledge of what these populations are and how they might change is therefore very useful.

Although the specific identity of the predominant configuration may be difficult to determine from purely statistical considerations we can find the population of the energy level or levels associated with this configuration. This is accomplished by manipulation of equation (7.12) in order to find an expression for n, the population of the energy level(s) in the most likely, or dominant configuration.

We now recall Figure 7.1 showing the distribution of configurations for the tossing of a coin. We noted that as the number of throws of the coin increased the frequency curve became narrower and narrower. This effect also occurred as we increased the number of particles in our hypothetical molecular system in Section 7.4. The histograms drawn to represent these distributions could also be drawn as frequency curves. Figure 7.7 shows a hypothetical curve for a system consisting of real numbers of molecules. The curve is very

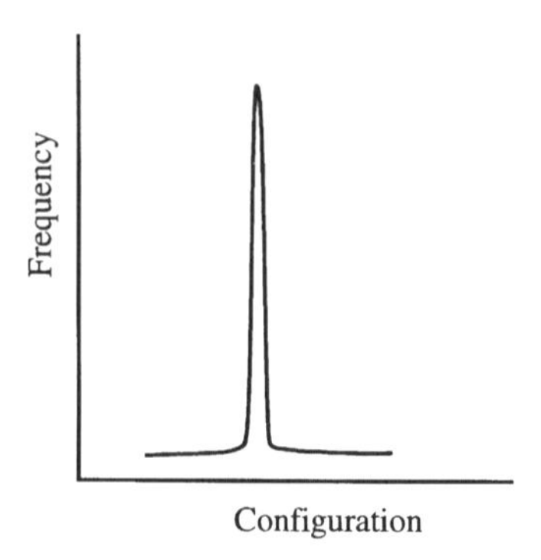

Figure 7.7 Hypothetical distribution curve for real numbers of particles in an assembly. The curve is expected to be very narrow showing the dominance of only one, or a very few configurations

narrow, suggesting that for all practical purposes 100% of the particles reside in this dominant configuration.

The most likely configuration is that associated with the peak in this distribution curve. In order to find the position of the peak, or maximum, in the curve we also recall Chapter 4, concerning differential calculus. At a maximum in any mathematical function the derivative of the function is equal to zero. The only function we have to describe this curve is

$$W = \frac{N!}{\Pi n_x} \tag{7.30}$$

where n_x is now a variable quantity, a population of an energy level. Capital pi, Π, is analogous to the summation sign sigma, Σ, but represents the product of the following numbers or variables. Differentiation of W with respect to n_x should therefore lead to some expression for n_x at the maximum in the curve. Most physical chemistry texts present this analysis which involves the use of Stirling's approximation and a method of finding a derivative called 'Lagrange's method of undetermined multipliers'.

However, in order to keep matters as simple as possible we will present an alternative method, which nevertheless uses some of the assumptions of the method of calculus.

7.6 The Boltzmann Distribution

We begin our derivation by describing a limited energy level system, similar to those used in Section 7.3 (Figure 7.8). We will focus our attention on three energy levels, l, m and n, and the only stipulation we will place on these is that their energies, E_l, E_m and E_n are in the order

$$E_l < E_m < E_n 0 \tag{7.31}$$

In making this stipulation we will note that the labelling of these levels is purely arbitrary and that the levels are not necessarily adjacent to each other, but only the inequality (7.31) is important. The populations of the levels are n_l, n_m and n_n respectively. All of these levels are close to or involved in the dominant configuration for the system, but other levels $(a, b, \ldots, z)$ may also exist.

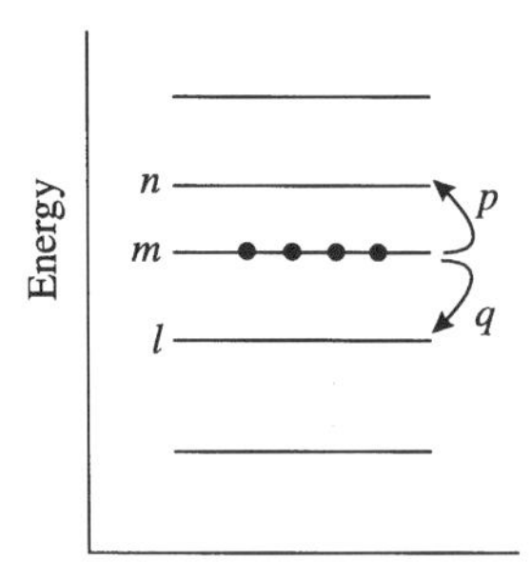

Figure 7.8 Generalised energy level system. The levels are not necessarily evenly spaced but particles may only move with energy corresponding to differences between the levels. Because the system is isolated any change caused by promotion of p particles from m to n must be compensated by demotion of q particles from m to l

The difference in energy between levels n and m is

$$E_n - E_m \tag{7.32}$$

and this amount of energy must be supplied if a particle is transferred from level m to level n.

Therefore if p, an integer number of particles, are promoted from level m to level n the amount of energy involved is

$$p(E_n - E_m) \tag{7.33}$$

Another important assumption is that this system is isolated, so that no particles or energy may be transferred to or from the surroundings. This means that the energy requirement for the promotion of these p particles must be compensated by some loss of energy. This can occur if some particles are also demoted from level m to level l. The difference in energy between these levels is

$$E_m - E_l \tag{7.34}$$

and we will transfer another integer number of particles, q. Therefore, the amount of energy involved in this transfer is

$$q(E_m - E_l) \tag{7.35}$$

To balance the energy we may now write an equality

$$p(E_n - E_m) = q(E_m - E_l) \tag{7.36}$$

In bringing about this rearrangement the population of the level m has decreased by $(p + q)$ particles to $(n_m - p - q)$ with concomitant increases in n_n to $(n + p)$ and n_l to $(n_l + q)$. The final assumption we shall make is that the integers p and q are small compared with the overall populations n_l, n_m and n_n.

We may calculate the number of microstates for the original system using our master equation

$$W = \frac{N!}{n_a! \ldots n_l! n_m! n_n! \ldots n_z!} \tag{7.37}$$

After the transfer of $p + q$ particles from level m we can calculate W to be

$$W = \frac{N!}{n_a! \ldots (n_l + q)!(n_m - p - q)!(n_n + p)! \ldots n_z!} \tag{7.38}$$

Since these levels are at the peak of the distribution curve then the relation

$$\frac{\mathrm{d}W}{\mathrm{d}n_x} = 0 \tag{7.39}$$

applies. W does not change and we may equate the right hand side of equation (7.37) with that of (7.38).

$$\frac{N!}{n_a! \ldots n_l! n_m! n_n! \ldots n_z!} = \frac{N!}{n_a! \ldots (n_l + q)!(n_m - p - q)!(n_n + p)! \ldots n_z!} \tag{7.40}$$

This has many common terms which may be cancelled, leaving us with

$$\frac{1}{n_l!n_m!n_n!} = \frac{1}{(n_l + q)!(n_m - p - q)!(n_n + p)!} \tag{7.41}$$

This may be rearranged, putting all similar terms together

$$\frac{(n_l + q)!}{n_l!} \cdot \frac{(n_n + p)!}{n_n!} = \frac{n_m!}{(n_m - p - q)!} \tag{7.42}$$

We will now recall the relation for all factorial numbers

$$n! = n(n - 1)(n - 2)! \tag{7.43}$$

and that we may extend this series either all the way down to one or stop at some convenient value of $(n - x)!$. As q is a positive integer $(n + q)!$ can be replaced with $(n_l + q)(n_l + q - 1)\ldots(n_l + 1)n_l!$. Likewise $(n_n + p)$ can be replaced with $(n_n + p)(n_n + p - 1)\ldots(n_n + 1)n_n!$. On the right hand side of equation (7.42) we note that n_m is greater than $(n_m - p - q)$ and that $n_m!$ may be replaced by the product $n_m(n_m - 1)(n_m - 2)\ldots(n_m - p - q + 1)(n_m - p - q)!$.

Thus equation (7.41) becomes

$$\frac{(n_l + q)(n_l + q - 1)\ldots(n_l + 1)n_l!}{n_l!} \cdot \frac{(n_n + p)(n_n + p - 1)\ldots(n_n + 1)n_n!}{n_n!}$$
$$= \frac{n_m(n_m - 1)(n_m - 2)\ldots(n_m - p - q + 1)(n_m - p - q)!}{(n_m - p - q)!} \tag{7.44}$$

We can now cancel all of the common terms, a step which has the effect of eliminating each of the factorial terms to produce

$$(n_l + q)\ldots(n_l + 1)\cdot(n + p)\ldots(n_n + 1) = n_m(n_m - 1)\ldots(n_m - p - q + 1) \tag{7.45}$$

We next invoke the assumption that the n terms are all much larger than the integers p and q. If this is the case we can write

$$n_l + q \approx n_l \tag{7.46}$$

so that all the terms $n_l + q, n_l + q - 1$ down to $n_l + 1$ can be approximated to n_l. Likewise

$$n_m + p \approx n_m \tag{7.47}$$

and

$$n_m - p - q \approx n_m \tag{7.48}$$

We now have on the left hand side q terms in n_l (effectively $n_l \times n_l q$ times) and p terms in n_n (effectively $n_n \times n_n p$ times). On the right hand side these manipulations have led to $p + q$ terms in n_m (effectively $n_m \times n_m(p + q)$ times)). Thus equation (7.45) becomes

$$n_l^q \cdot n_n^p = n_m^{(p+q)} \tag{7.49}$$

remembering that $n_m^{(p+q)}$ is the same as $n_m^p n_m^q$ we can rearrange this equation to yield

$$\frac{n_l^q}{n_m^q} = \frac{n_m^p}{n_n^p} \tag{7.50}$$

which is the same as

$$\left(\frac{n_l}{n_m}\right)^q = \left(\frac{n_m}{n_n}\right)^p \tag{7.51}$$

This is a remarkably simple expression considering all we have stipulated about these three energy levels is their order in terms of increasing energy.

We can introduce terms for this energy by taking logarithms of each side and recalling that $\log(x^n)$ is equal to $n \cdot \log(x)$. Using logs to base e,

$$q \cdot \ln\left(\frac{n_l}{n_m}\right) = p \cdot \ln\left(\frac{n_m}{n_n}\right) \tag{7.52}$$

or

$$\frac{q}{p} \cdot \ln\left(\frac{n_l}{n_m}\right) = \ln\left(\frac{n_m}{n_n}\right) \tag{7.53}$$

To find a term for the ratio q/p we go all the way back to equation (7.36) which rearranges to give

$$\frac{q}{p} = \frac{E_n - E_m}{E_m - E_l} \tag{7.54}$$

This can be substituted in (7.53) and rearranged to put all related terms together on each side of the equation,

$$\frac{1}{E_m - E_l} \ln\left(\frac{n_l}{n_m}\right) = \frac{1}{E_n - E_m} \ln\left(\frac{n_m}{n_n}\right) \tag{7.55}$$

This is a very important result. To emphasise the importance we reiterate one of the assumptions made in its derivation. This is that the energy levels are not necessarily adjacent and it is only their order of increasing energy which is important ($E_l < E_m < E_n$). Although we began this derivation by considering relationships between three energy levels what equation (7.55) gives us is a relationship between two sets of energy levels, the $l - m$ on the left and $m - n$ on the right. On the left we have a mathematical expression linking the energy difference between levels l and m to the ratio of their populations (n_l/n_m). This is equal to the same expression for the levels m and n. We could justifiably write an expression for the levels a and b, where E_a was less than E_b:

$$\frac{1}{E_b - E_a} \ln\left(\frac{n_a}{n_b}\right) = \frac{1}{E_n - E_m} \ln\left(\frac{n_m}{n_n}\right) \tag{7.56}$$

That is, for any two quantum states of a particular system this function always has the same value. This constant value is given its own symbol, β. For two quantum states, i and j, in any isolated system we now write

$$\frac{1}{E_i - E_j} \ln\left(\frac{n_j}{n_i}\right) = \beta \tag{7.57}$$

This rearranges to

$$\ln\left(\frac{n_j}{n_i}\right) = \beta(E_i - E_j) \tag{7.58}$$

Here we have not stipulated that E_i is greater or less than E_j but if we say that i is the ground state, with E_i equivalent to zero then this equation becomes

$$\ln\left(\frac{n_j}{n_i}\right) = -\beta E_j \qquad (7.59)$$

When this is written in exponential form

$$\frac{n_j}{n_i} = e^{-\beta E_j} \qquad (7.60)$$

it is the Boltzmann equation.

All that remains to be done is to prove that β is equal to $1/kT$. This is usually given in physical chemistry and thermodynamics textbooks and we will not deal with it further in this chapter. It can, however, be seen that β must have units of 1/energy, since in the exponential we also have an energy term. The Boltzmann constant, k, has units $\mathrm{J\,K^{-1}}$, when multiplied by temperature (T/K) the product has units of energy (J) and this is therefore satisfactory.

We can, finally, see that this equation is correct in form because if E_j is large then $\exp(-\beta E_j)$ must be small (Chapter 2) and the ratio n_j/n_i will be small, i.e., most particles will reside in the ground state.

In closing we will repeat what was stated in Chapter 2 concerning the Boltzmann equation. This relationship between populations of energy levels or states occurs again and again throughout chemistry. It is therefore important to know it and to understand its origins. In this chapter we have shown that it can be arrived at by consideration of the mathematics of permutations and a simple quantum energy level model of matter. Apart from the examples given in Chapter 2, this equation and the equation $N!/n_a!n_b!\ldots$ are also used in development of the ideas of statistical thermodynamics. This branch of chemistry seeks to link the bulk properties of matter, studied in 'normal' chemical thermodynamics, and the quantum mechanical properties of matter studied in the development of atomic and molecular theory.

Problems

7.1 On the roll of a die, how many ways are there of scoring

(a) six
(b) an even number
(c) an odd number?

7.2 For two consecutive rolls of a die list the possible scores and the number of ways it is possible to achieve them.

7.3 Repeat the exercise to find the possible configurations and microstates for a system of

(a) Five particles and five quanta of energy
(b) Ten particles and five quanta of energy

Plot a histogram to illustrate the results.

8 Complex Numbers, Vectors, Determinants and Matrices

8.1 Introduction

In this chapter we introduce four topics which differ substantially from the material in Chapters 1 to 7. Complex numbers, vectors, determinants and matrices are concepts which are used increasingly as a chemistry course advances to its own more complex and abstract ideas. The principles and algebra are not particularly difficult but they each have their own form of algebra which must also be learnt. This chapter is therefore intended to raise the awareness of these subjects and of some areas where they are used.

8.2 Complex Numbers

Throughout this book we have dealt only with real numbers and assumed that an equation such as

$$x^2 + 1 = 0 \tag{8.1}$$

has no solution. If we ignore this assumption and define the solution as

$$x = \mathrm{i} \tag{8.2}$$

such that

$$\mathrm{i}^2 = -1 \tag{8.3}$$

and

$$\mathrm{i} = \sqrt{-1} \tag{8.4}$$

then we introduce a new form of mathematics, that of the *complex numbers*. The utility of complex mathematics is that we can now solve some of the problems which we are unable to do using only real numbers, for example quadratic equations for which there are no real roots (Chapter 1).

A complex number is one which includes a *real* part and an *imaginary* part. The latter is the part which includes i, the square root of minus one (− 1). Most texts define a complex number z by the equation

$$z = x + \mathrm{i}y \tag{8.5}$$

x and y are both real numbers but x is referred to as the real part of the complex number and y is referred to as the imaginary part.

Each complex number has its complex conjugate, z^*, defined by

$$z^* = x - \mathrm{i}y \tag{8.6}$$

A number z where y is zero is 'real' whereas one where x is zero ($z = \mathrm{i}y$) is referred to as *pure imaginary*.

8.2.1 Simple Algebra of Complex Numbers

Complex numbers are subject to many of the normal rules of algebra, particularly that we can only add like with like. Thus the sum of two complex numbers z_1 and z_2 where

$$z_1 = 2 + \mathrm{i} \tag{8.7}$$

and

$$z_2 = 1 + 3\mathrm{i} \tag{8.8}$$

is

$$\begin{aligned} z_1 + z_2 &= (2 + 1) + (\mathrm{i} + 3\mathrm{i}) \\ &= 3 + 4\mathrm{i} \end{aligned} \tag{8.9}$$

That is, we must add the real and imaginary parts quite separately. In general we would write

$$z_1 + z_2 = (x_1 + x_2) + \mathrm{i}(y_1 + y_2) \tag{8.10}$$

where

$$z_1 = x_1 + \mathrm{i}y_1 \tag{8.11}$$

and

$$z_2 = x_2 + \mathrm{i}y_2 \tag{8.12}$$

When multiplying two complex numbers we can take account of the definition of i being the square root of minus one. Thus the product z_1z_2 where these are defined by equations (8.7) and (8.8) is

$$\begin{aligned} z_1z_2 &= (2 + \mathrm{i})(1 + 3\mathrm{i}) \\ &= 2 + 6\mathrm{i} + \mathrm{i} + 3\mathrm{i}^2 \\ &= 2 + 7\mathrm{i} + 3(-1) \\ &= 2 + 7\mathrm{i} - 3 \\ &= -1 + 7\mathrm{i} \end{aligned} \tag{8.13}$$

Example—the product of a complex number and its complex conjugate

An important result is obtained by multiplying z by its complex conjugate. Thus if z is given by $2 + 3i$ and its complex conjugate is $2 - 3i$ the product is

$$\begin{aligned} zz^* &= (2 + 3i)(2 - 3i) \\ &= 4 - 6i + 6i - 9i^2 \\ &= 4 - 9(-1) \\ &= 4 + 9 \\ &= 13 \end{aligned} \tag{8.14}$$

Thus the product is a real number. In general

$$\begin{aligned} zz^* &= (x + iy)(x - iy) \\ &= x^2 - iyx + iyx - i^2y^2 \\ &= x^2 - y^2(-1) \\ &= x^2 + y^2 \end{aligned} \tag{8.15}$$

This result is found in quantum mechanics where the probability of finding a particle in an area of space is proportional to the product of a wavefunction ψ and its complex conjugate ψ^*. Note that this probability, the product $\psi\psi^*$ is therefore a real number. Furthermore since x^2 and y^2 in equation (8.15) are positive then the probability proportional to $\psi\psi^*$ is both real and positive.

Division of complex numbers may also be performed although the method is less obvious. The division of $z_1 = 1 + 2i$ by $z_2 = 3 + 5i$ may be written

$$\frac{z_1}{z_2} = \frac{1 + 2i}{3 + 5i} \tag{8.16}$$

but is not readily solved in this form. Referring to Chapter 1 and the manipulation of equations we can multiply this quotient by one without changing it. The best choice of one is

$$1 = \frac{3 - 5i}{3 - 5i} \tag{8.17}$$

that is, the complex conjugate of the denominator divided by itself. Using this we can solve the quotient (8.16).

$$\begin{aligned} z &= \frac{1 + 2i}{3 + 5i} \cdot \frac{3 - 5i}{3 - 5i} \\ &= \frac{3 - 5i + 6i - 10i^2}{9 - 15i + 15i - 25i^2} \\ &= \frac{3 + i - 10(-1)}{9 - 25(-1)} \end{aligned} \tag{8.18}$$

$$= \frac{3 + \mathrm{i} + 10}{9 + 25}$$

$$= \frac{13 + \mathrm{i}}{34} = \frac{13}{34} + \frac{\mathrm{i}}{34}$$

All we have done is turn the denominator into a real number by multiplying by its complex conjugate.

8.2.2 More about the Complex Conjugate, z*

We have seen that the product of a complex number and its complex conjugate is a real number. The sums and differences are as follows.

$$\begin{aligned} z + z^* &= (x + \mathrm{i}y) + (x - \mathrm{i}y) \\ &= 2x \end{aligned} \tag{8.19}$$

or

$$\frac{1}{2}(z + z^*) = x \text{ (real part)} \tag{8.20}$$

The difference is

$$\begin{aligned} z - z^* &= (x + \mathrm{i}y) - (x - \mathrm{i}y) \\ &= x - x + \mathrm{i}y - (-\mathrm{i}y) \\ &= 2\mathrm{i}y \end{aligned} \tag{8.21}$$

or

$$\frac{1}{2}(z - z^*) = \mathrm{i}y \tag{8.22}$$

In equation (8.15) we saw that

$$zz^* = x^2 + y^2 \tag{8.15}$$

This is sometimes written as

$$|z|^2 = x^2 + y^2 \tag{8.23}$$

or

$$|z| = \sqrt{x^2 + y^2} \tag{8.24}$$

where $|z|$ is the modulus of the complex number and is always positive and is always real.

8.2.3 Graphical Representation of Complex Numbers—the Argand Diagram

One way to represent complex numbers is as points on a graph. A complex number $z = x + \mathrm{i}y$, can be plotted as the coordinates (x, y), Figure 8.1. The x axis is called the real axis and the y axis is called the imaginary axis. Graphs illustrating complex numbers are

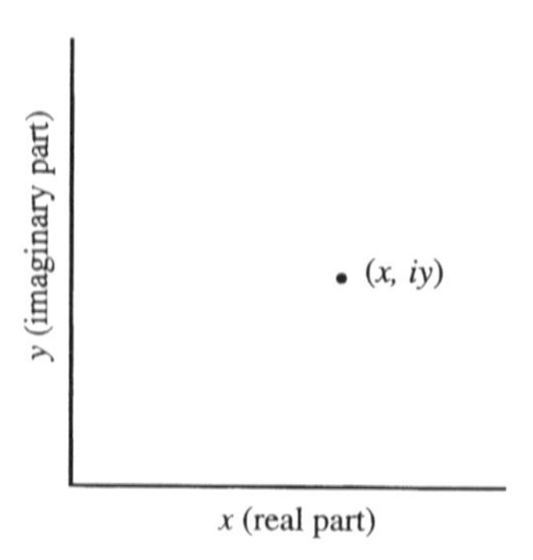

Figure 8.1 The Argand diagram. A complex number, having two parts, is easily represented on a two-dimensional plot—the real part being the x coordinate and the imaginary part being the y coordinate

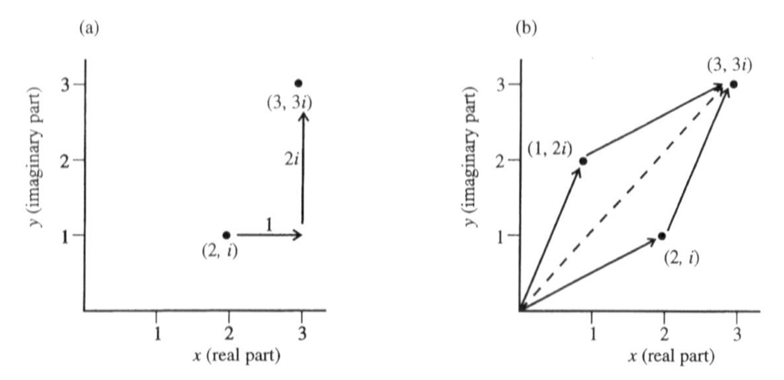

Figure 8.2 Addition of two complex numbers follows normal algebraic rules in that only the real parts are added together and only imaginary parts are added together (a). This is similar to the addition of two vectors as described in Section 8.3 (b)

called *Argand diagrams* after Jean-Robert Argand. Using these diagrams we can show algebraic manipulations such as addition and multiplication. Figure 8.2(a) shows the result of addition of two complex numbers, $z_1 = 1 + 2\text{i}$ and $z_2 = 2 + \text{i}$. If we draw lines from the origin to each of the points (Figure 8.2(b)) we see that addition is akin to the addition of two vectors (Section 8.3) in that the two lines to the two numbers z_1 and z_2 form a parallelogram, the diagonal of which joins the origin to the sum. The multiplication (Figure 8.3) is more abstract in that it appears to include a rotation of the point in space.

This use of lines joining the origin to the point representing the complex number leads to a related representation which uses *polar coordinates* in the Argand diagram. The position of a point can be represented (Figure 8.4) by stating its distance from the origin (symbol r) and the angle the line joining the point with the origin makes with the x axis (symbol θ). Referring to Figure 8.4 we can relate these polar coordinates (r, θ) to the normal *Cartesian* coordinates (x, y) using the trigonometric relations given in Chapter 2.

The x coordinate is

$$x = r\cos(\theta) \tag{8.25}$$

and the y coordinate is

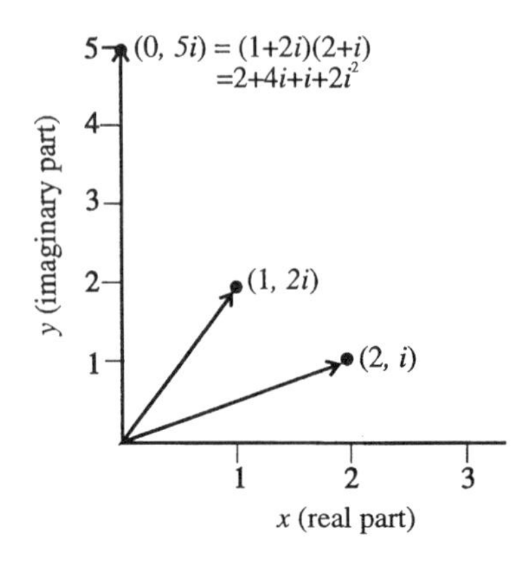

Figure 8.3 Multiplication of two complex numbers 2 + i and 1 + 2i, on the Argand diagram results in a rotation as well as a change in magnitude

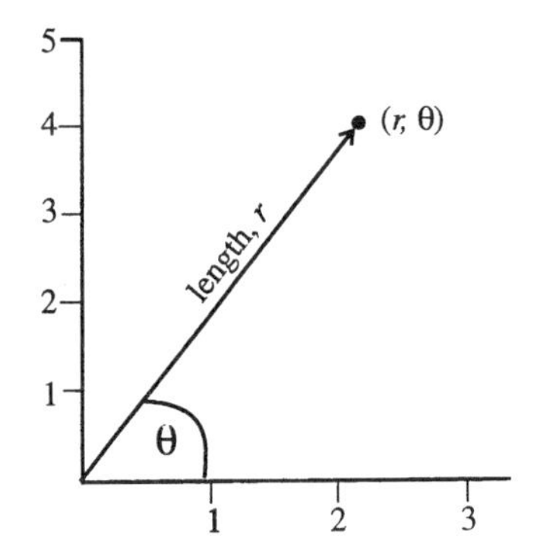

Figure 8.4 The complex number as polar coordinates (r, θ). r is the length of the line joining the point to the origin and θ is the angle this line makes with the x axis. Note that this is related to the (x, y) coordinates by trigonometric rules (Chapter 2)

$$y = r\sin(\theta) \tag{8.26}$$

The definition of the complex number is

$$z = x + \mathrm{i}y \tag{8.5}$$

and substituting for x and y we get

$$\begin{aligned} z &= r\cos(\theta) + \mathrm{i}r\sin(\theta) \\ &= r(\cos(\theta) + \mathrm{i}\sin(\theta)) \end{aligned} \tag{8.27}$$

Similarly the complex conjugate z^* is

$$z^* = r(\cos(\theta) - \mathrm{i}\sin(\theta)) \tag{8.28}$$

We can therefore derive the equivalent of the modulus using this notation.

$$\begin{aligned} zz^* &= r(\cos(\theta) + \mathrm{i}\sin(\theta)) \times r(\cos(\theta) - \mathrm{i}\sin(\theta)) \\ &= r^2(\cos^2(\theta) - \mathrm{i}\sin(\theta)\cos(\theta) + \mathrm{i}\sin(\theta)\cos(\theta) - \mathrm{i}^2\sin^2(\theta)) \\ &= r^2(\cos^2(\theta) + \sin^2(\theta)) \end{aligned} \tag{8.29}$$

Table 2.4 (Chapter 2) gives the result

$$\sin^2(\theta) + \cos^2(\theta) = 1 \quad (8.30)$$

and therefore we can state that

$$zz^* = r^2 \quad (8.31)$$

and since we have defined the modulus, $|z| = \sqrt{(zz^*)}$, we can write

$$|z| = r \quad (8.32)$$

where r is referred to as the modulus of z (mod z) and the angle θ is called the *argument* of z or arg(z). We can now develop more detail of the multiplication of two complex numbers using this notation. Taking a general approach, for z_1 and z_2 defined by

$$z_1 = r_1(\cos(\theta_1) + \mathrm{i}\sin(\theta_1)) \quad (8.33)$$

$$z_2 = r_2(\cos(\theta_2) + \mathrm{i}\sin(\theta_2)) \quad (8.34)$$

We have

$$z_1 z_2 = r_1 r_2(\cos(\theta_1) + \mathrm{i}\sin(\theta_1))(\cos(\theta_2) + \mathrm{i}\sin(\theta_2)) \quad (8.35)$$

Concentrating on the bracketed terms and expanding gives

$$\cos(\theta_1)\cos(\theta_2) + \mathrm{i}\cos(\theta_1)\sin(\theta_2) + \mathrm{i}\sin(\theta_1)\cos(\theta_2) + \mathrm{i}^2\sin(\theta_1)\sin(\theta_2) \quad (8.36)$$

This appears fairly complicated but can be greatly simplified. First there is the term in i^2 on the right which becomes -1 times the product of sines. We can also make the term clearer by factorising the terms which have i as the coefficient. This produces

$$\cos(\theta_1)\cos(\theta_2) + \mathrm{i}(\cos(\theta_1)\sin(\theta_2) + \sin(\theta_1)\cos(\theta_2)) - \sin(\theta_1)\sin(\theta_2) \quad (8.37)$$

We now refer, again, to Table 2.4 which gives the results

$$\sin(x + y) = \sin(x)\cos(y) + \cos(x)\sin(y) \quad (8.38)$$

$$\cos(x + y) = \sin(x)\cos(y) - \cos(x)\sin(y) \quad (8.39)$$

Comparison with equation (8.37) shows that the term in i is therefore equal to $\sin(\theta_1 + \theta_2)$ and that the remaining terms are equal to $\cos(\theta_1 + \theta_2)$. This can now be returned to equation (8.35) to give the result

$$z_1 z_2 = r_1 r_2(\cos(\theta_1 + \theta_2) + \mathrm{i}\sin(\theta_1 + \theta_2)) \quad (8.40)$$

From this result we can see that the modulus of the product, $|z_1 z_2|$, is equal to the product of the individual moduli, $|z_1||z_2|$. We also see that the new angle between the line and the x axis is the sum of the angles for each complex number. This may also be written as

$$\arg(z_1 z_2) = \arg(z_1) + \arg(z_2) \quad (8.41)$$

If both values of r are unity then the multiplication represents a rotation of the point, in an anticlockwise fashion, around a circle of radius one.

8.2.4 The Link between the Exponential and Trigonometric Functions

In Chapter 2 we defined the exponential function e^x using a power series,

$$e^x = 1 + x + \frac{x^2}{2!} + \frac{x^3}{3!} + \frac{x^4}{4!} + \ldots \tag{8.42}$$

If x is a complex number z and this is purely imaginary (having no real part) such that $z = \mathrm{i}\theta$ then the expansion of e^z follows all of the normal rules.

$$e^z = e^{\mathrm{i}\theta} = 1 + \mathrm{i}\theta + \frac{(\mathrm{i}\theta)^2}{2!} + \frac{(\mathrm{i}\theta)^3}{3!} + \frac{(\mathrm{i}\theta)^4}{4!} + \frac{(\mathrm{i}\theta)^5}{5!} + \ldots \tag{8.43}$$

Expanding the terms in brackets,

$$(\mathrm{i}\theta)^2 = \mathrm{i}^2\theta^2 = -\theta^2 \tag{8.44}$$

$$(\mathrm{i}\theta)^3 = \mathrm{i}^3\theta^3 = \mathrm{i}(\mathrm{i}^2)\theta^3 = -\mathrm{i}\theta^3 \tag{8.45}$$

$$(\mathrm{i}\theta)^4 = \mathrm{i}^4\theta^4 = \mathrm{i}^2(\mathrm{i}^2)\theta^4 = \theta^4 \tag{8.46}$$

$$(\mathrm{i}\theta)^5 = \mathrm{i}^5\theta^5 = \mathrm{i}(\mathrm{i}^2)(\mathrm{i}^2)\theta^5 = \mathrm{i}\theta^5 \tag{8.47}$$

$$(\mathrm{i}\theta)^6 = \mathrm{i}^6\theta^6 = (\mathrm{i}^2)(\mathrm{i}^2)(\mathrm{i}^2)\theta^6 = -\theta^6 \tag{8.48}$$

i disappears from even powers of i but remains with the odd powers. Equation 8.43 becomes

$$e^{\mathrm{i}\theta} = 1 + \mathrm{i}\theta - \frac{\theta^2}{2!} - \mathrm{i}\frac{\theta^3}{3!} + \frac{\theta^4}{4!} + \mathrm{i}\frac{\theta^5}{5!} - \frac{\theta^6}{6!} + \ldots \tag{8.49}$$

This can be separated into two distinct parts, those with coefficient i and those without. We then factorise those with i.

$$e^{\mathrm{i}\theta} = \left(1 - \frac{\theta^2}{2!} + \frac{\theta^4}{4!} - \frac{\theta^6}{6!}\ldots\right) + \mathrm{i}\left(\theta - \frac{\theta^3}{3!} + \frac{\theta^5}{5!} - \frac{\theta^7}{5!}\ldots\right) \tag{8.50}$$

Comparison of the two bracketed terms with the power series definitions of the sine and cosine

$$\cos(\theta/\text{radians}) = 1 - \frac{\theta^2}{2!} + \frac{\theta^4}{4!} - \frac{\theta^6}{6!}\ldots \tag{8.51}$$

$$\sin(\theta/\text{radians}) = \theta - \frac{\theta^3}{3!} + \frac{\theta^5}{5!} - \frac{\theta^7}{5!}\ldots \tag{8.52}$$

allows us to write equation (8.50) in terms of the sine and cosine.

$$e^{\mathrm{i}\theta} = \cos(\theta) + \mathrm{i}\sin(\theta) \tag{8.53}$$

which is related to the definition of the complex number in polar coordinates

$$z = r(\cos(\theta) + \mathrm{i}\sin(\theta)) \tag{8.27}$$

Therefore

$$z = re^{\mathrm{i}\theta} \tag{8.54}$$

This is a very important result in that it provides a link between the exponential function, the sine and the cosine. Using the same derivation we can show that the complex conjugate is given by

$$e^{-i\theta} = \cos(\theta) - i\sin(\theta) \quad (8.55)$$

which corresponds to the conjugate of $\cos(\theta) + i\sin(\theta)$. Adding equations (8.53) and (8.55) produces

$$\begin{aligned} e^{i\theta} + e^{-i\theta} &= \cos(\theta) + i\sin(\theta) + \cos(\theta) - i\sin(\theta) \\ &= 2\cos(\theta) \end{aligned} \quad (8.56)$$

and this may be rearranged to define $\cos(\theta)$ in terms of the complex exponential function.

$$\cos(\theta) = \frac{1}{2}(e^{i\theta} + e^{-i\theta}) \quad (8.57)$$

Subtracting (8.55) from (8.53) we get

$$\begin{aligned} e^{i\theta} - e^{-i\theta} &= \cos(\theta) + i\sin(\theta) - \cos(\theta) + \sin(\theta) \\ &= 2i\sin(\theta) \end{aligned} \quad (8.58)$$

which rearranges in a similar manner,

$$\sin(\theta) = \frac{1}{2i}(e^{i\theta} - e^{i\theta}) \quad (8.59)$$

8.2.5 Uses of Complex Numbers in Chemistry

All of this is very neat but what use is it to the chemist? The last result, that the sine and cosine are linked directly to the complex exponential is particularly useful in that this can then be related to functions which are periodic. Periodicity occurs wherever there are repeating units, for example in regular crystals and in the elements of symmetry which are used to describe some molecules and their orbitals (orbital symmetry is also useful in organic chemistry!). We have also mentioned that multiplication of complex numbers is related to rotation and the quantum mechanics of rotation is an important problem in microwave spectroscopy. When the Schrödinger equation is cast in terms of *spherical polar* coordinates (the three-dimensional polar coordinates analogous to Cartesian (x, y, z) representation) the solution involves the complex exponential function. The relations (8.57) and (8.59) can be used to simplify formulae that are perhaps difficult to integrate. In studying the structure of the hydrogen atom we find that the probability of finding a particle in space is proportional to the product of a wavefunction and its complex conjugate. Finally the results of impedance measurements on electrochemical cells are plotted in the complex plane (the Argand diagram).

8.3 Vectors

A *vector* is a quantity which has a direction in addition to a magnitude. To completely specify a vector we must state its direction in addition to its magnitude whereas a *scalar* quantity may be described completely by its magnitude. Examples of scalars are length, mass, time, speed, energy and density. Examples of vectors are displacement, velocity, force and momentum. Note the difference between scalars such as length and speed compared to apparently similar quantities such as displacement and velocity. Also note that many derived quantities of vectors are also vectors. Acceleration, a vector, is the rate of change of velocity (a vector) with time (a scalar). Force is the product of mass (a scalar) and acceleration (a vector).

Linear momentum and angular momentum are vectors which are very important in quantum mechanics. As with other quantum mechanical quantities these vectors are restricted not only in being allowed discrete values of magnitude but in being allowed discrete values of direction. The combination of vector quantities is therefore instrumental in ideas about the shapes of orbitals, molecules and the movement of molecules in space.

8.3.1 Notation and Representation of Vectors

In saying that we must specify both the direction and the magnitude of a vector we have an immediate problem of representing the vector in a compact manner which is open to algebraic manipulation.

The easiest method is to draw a line whose length is proportional to the vector's magnitude and whose direction on paper is that of the vector. In text we usually write the vector as a bold-faced variable letter, e.g., $\boldsymbol{a}, \boldsymbol{b}, \boldsymbol{c}$ etc. If the vector joins two points in space (M and N whose coordinates are known) then it can be represented by $\overrightarrow{\mathrm{MN}}$ in which case the direction of an arrow over the letters tells us where the vector begins and ends, Figure 8.5. The magnitude of the vector is given by the modulus of $\boldsymbol{a}$, $|a|$. A vector in the same direction as $\boldsymbol{a}$ but with a length of one is called the *unit vector*, and given the symbol, $\hat{\boldsymbol{a}}$. The vector $\boldsymbol{a}$ is therefore specified by the product

$$\boldsymbol{a} = |a|\hat{\boldsymbol{a}} \tag{8.60}$$

Unfortunately this does not tell us anything about the direction or the magnitude. It is possible, however, to use such notation to illustrate some vector algebra which includes addition, subtraction and multiplication. The first example is the addition of two vectors.

Example—addition of two vectors

Add the two velocities, $\boldsymbol{v}_1$ and $\boldsymbol{v}_2$ where $\boldsymbol{v}_1$ has a magnitude of $5\,\mathrm{m\,s^{-1}}$ at 0° and $\boldsymbol{v}_2$ has a magnitude of $10\,\mathrm{m\,s^{-1}}$ and a direction of 53°.

We can represent the addition using a normal graph with $\boldsymbol{v}_1$ being an arrow parallel to the x axis with a length of 5 cm. $\boldsymbol{v}_2$ is a line at 53° to the horizontal with a length of 10 cm.

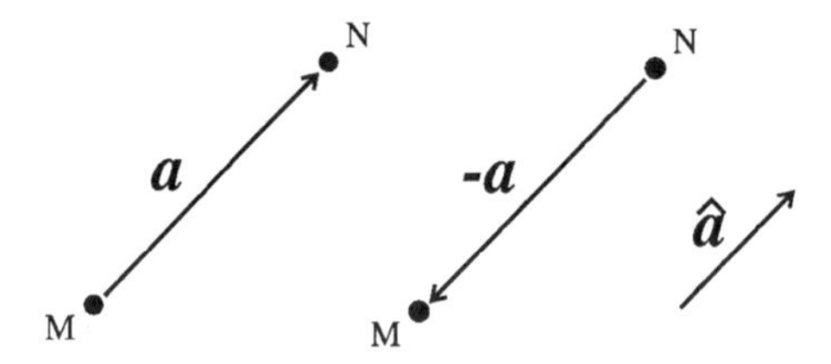

Figure 8.5 The vector $\boldsymbol{a}$ represents a change from point M to point N. $-\boldsymbol{a}$ has the same magnitude but opposite direction—going from N to M. The unit vector $\hat{\boldsymbol{a}}$ has the same direction as $\boldsymbol{a}$ but magnitude of one

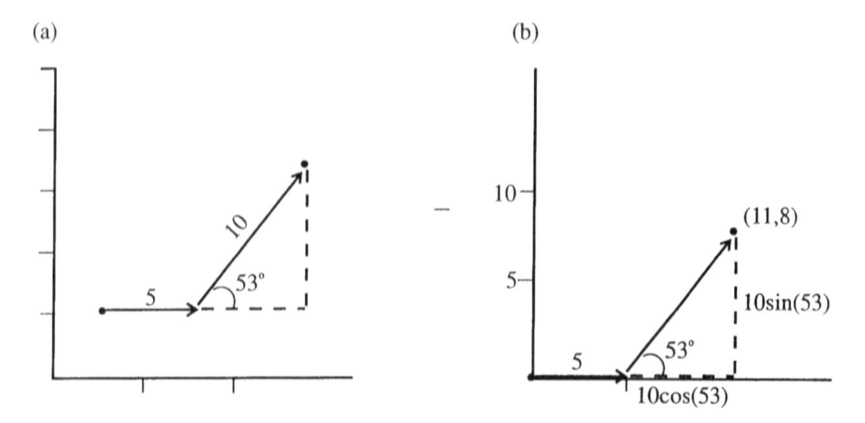

Figure 8.6 Addition of two vectors, (a) in general terms, (b) starting at the origin and using (x, y) coordinates to calculate the new vector

This arrow may be drawn with the aid of a protractor and a ruler. The sum of the two vectors is found by drawing them head to tail. The resultant is a vector joining the start (the tail) of the first to the end (the head) of the second. This is shown in Figure 8.6(a). The length and angle of the resultant may be measured using the tools with which they were drawn. It would, however, be much more elegant to use the methods of trigonometry to determine the result.

If we begin at the origin then $\boldsymbol{v}_1$ is a line from $(0, 0)$ to $(5, 0)$. $\boldsymbol{v}_2$ may also start at the origin but we do not know the coordinates of the end. We can, however, calculate them using the definitions of sine and cosine from Chapter 2.

$$x = 10 \cdot \cos(53^\circ) \tag{8.61}$$

$$y = 10 \cdot \sin(53^\circ) \tag{8.62}$$

$\text{Cos}(\theta)$ is 0.6 and $\sin(\theta)$ is 0.8 and therefore $\boldsymbol{v}_2$ goes from $(0, 0)$ to $(6, 8)$, Figure 8.6(b). We can check that this is correct using Pythagoras' theorem. For a right-angled triangle the length of the hypotenuse (the long side, denoted h in the equation below) is equal to the sum of the squares of the two short sides (x and y).

$$h^2 = x^2 + y^2 \tag{8.63}$$

This gives h^2 as 100 and hence proves $h = 10$.

We can now add the vectors by adding the shifts along the x axis and the shifts up the y axis. This can be seen by inspection of Figure 8.6,

$$\begin{aligned} \text{new } x \text{ position} &= 6 + 5 \\ &= 11 \end{aligned} \tag{8.64}$$

$$\begin{aligned} \text{new } y \text{ position} &= 8 + 0 \\ &= 8 \end{aligned} \tag{8.65}$$

Because we are adding the two numbers it does not matter if we put $\boldsymbol{v}_1$ first or second since $6 + 5$ is the same as $5 + 6$, and $8 + 0$ is the same as $0 + 8$. Whichever way, we arrive at the new coordinates, $(11, 8)$.

The length of this new vector, from $(0, 0)$ to $(11, 8)$ is found using Pythagoras' theorem.

$$\begin{aligned} \text{length} &= \sqrt{x^2 + y^2} \\ &= \sqrt{11^2 + 8^2} \\ &= \sqrt{121 + 64} = \sqrt{185} \\ &= 13.6 \end{aligned} \tag{8.66}$$

The angle can be found from the tangent. Tan(θ) is (8/11) and therefore θ is 36°. The resultant vector is 13.6 m s^{-1} in a direction 36°.

Using the same method of combining coordinates we can also perform subtraction of one vector from another. If we have a vector $\boldsymbol{a}$ then the vector $-\boldsymbol{a}$ has the same magnitude but opposite direction. It is possible to multiply any vector by a number (a scalar quantity) and the resultant vector has the same direction but a magnitude equal to the product of the original and the scalar quantity.

$$\begin{aligned} \boldsymbol{a} &= |a|\hat{\boldsymbol{a}} \\ \therefore x \times \boldsymbol{a} &= x|a|\hat{\boldsymbol{a}} \end{aligned} \tag{8.67}$$

In reversing the direction of the vector we have merely multiplied by the scalar quantity $x = -1$. As in normal algebra where

$$a = b - c \tag{8.68}$$

is the same as

$$a = b + (-c) \tag{8.69}$$

we can write the subtraction of one vector from another as

$$\boldsymbol{a} = \boldsymbol{b} - \boldsymbol{c} \tag{8.70}$$

and this is the same as

$$\boldsymbol{a} = \boldsymbol{b} + (-\boldsymbol{c}) \tag{8.71}$$

Example—subtraction of one vector from another

Perform the vector subtraction $\boldsymbol{v}_1 - \boldsymbol{v}_2$ where $\boldsymbol{v}_1$ and $\boldsymbol{v}_2$ have the same values as in the previous example.

$\boldsymbol{v}_1$ has the same value as before but $\boldsymbol{v}_2$ is drawn in the reverse direction. $-\boldsymbol{v}_2$ may be drawn from the origin to the point $(-6, -8)$. As indicated in Figure 8.7(a) drawing the vector in the opposite direction amounts to drawing a line from the origin to coordinates which have the same values multiplied by minus one. We can add this vector to the first:

$$\begin{aligned} \text{new position} &= (5 - 6, 0 - 8) \\ &= (-1, -8) \end{aligned} \tag{8.72}$$

The result is shown in Figure 8.7. The length of this is found using Pythagoras' theorem again.

$$h^2 = (-1)^2 + (-8)^2 \tag{8.73}$$

Therefore

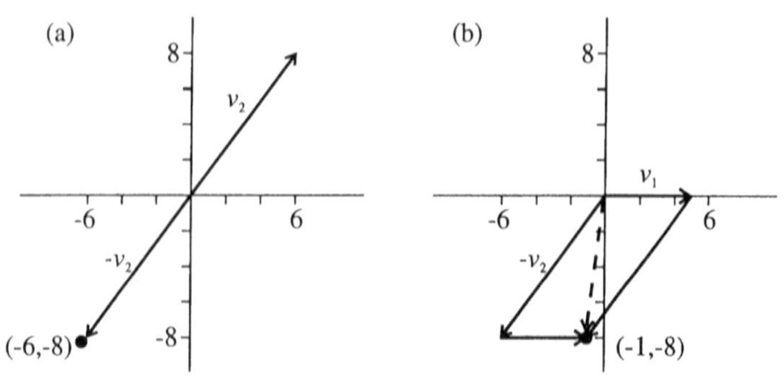

Figure 8.7 Subtraction of vectors, $\boldsymbol{v}_1 - \boldsymbol{v}_2$, (b), is the same as addition of a vector $\boldsymbol{v}_1 + (-\boldsymbol{v}_2)$ (a)

$$
\begin{aligned}
h &= \sqrt{1 + 64} \\
&= \sqrt{65} \\
&= 8.06
\end{aligned}
\tag{8.74}
$$

The angle may be found using the tangent once again recalling that $\tan(\theta) = y/x$.

$$\tan(\theta) = \frac{-8}{-1} = 8 \tag{8.75}$$

or

$$\theta = \tan^{-1}(8) = 82.9^\circ \tag{8.76}$$

In following conventions of direction for rotation we note that the end point is in the third quadrant of the graph and in the conventional sense (anticlockwise) the angle is $(180 + 82.9)^\circ$ or 262.9°.

This method may be used to add any number of vectors even though it is hardly convenient and using trigonometric and geometric functions is still unwieldy. On the other hand drawing lines on a graph is likely to be inaccurate and inappropriate when you move into three dimensions. If we specify a vector as a pair of coordinates then we have all the information to specify its length and its direction; it can be manipulated and extended into three or even more dimensions.

Example—addition of vectors using coordinate notation

Add the two vectors $\boldsymbol{v}_1$ and $\boldsymbol{v}_2$ where $\boldsymbol{v}_1 = (3, 2)$ and $\boldsymbol{v}_2 = (4, 5)$.

The total move in the x direction is the sum of the x coordinates.

$$x_{\text{total}} = 3 + 4 = 7 \tag{8.77}$$

and the total move in the y direction is the sum of the y coordinates.

$$y_{\text{total}} = 2 + 5 = 7 \tag{8.78}$$

The resultant vector is therefore represented by $(7, 7)$, Figure 8.8. The length of this vector is found, as for the previous example, by the application of Pythagoras' theorem.

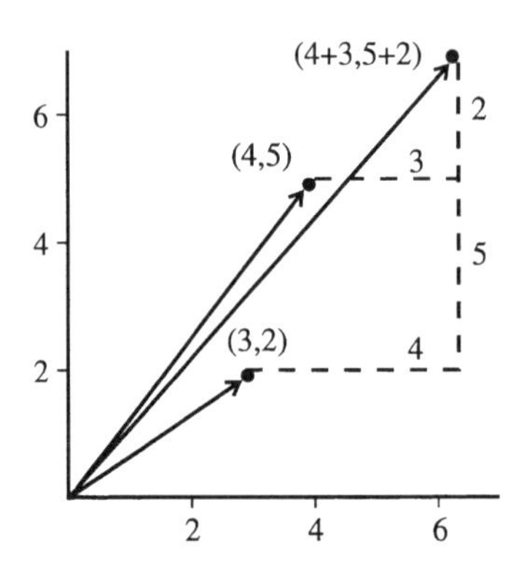

Figure 8.8 Vector addition using coordinates to represent the vectors

$$h^2 = 7^2 + 7^2$$
$$\therefore h = \sqrt{98} = 9.9 \tag{8.79}$$

The angle is calculated from the tangent, $\tan(\theta) = x/y = 1$ and so $\theta = 45°$.

Example—subtraction of vectors

The subtraction $\boldsymbol{v}_1 - \boldsymbol{v}_2$ follows the same rules.

The x value is

$$x_{\text{total}} = 3 - 4 = -1 \tag{8.80}$$

The y value is

$$y_{\text{total}} = 2 - 5 = -3 \tag{8.81}$$

and the resultant vector is $(-1, -3)$. The magnitude of this may be calculated to be $\sqrt{1+9}$ which is 3.16 and the angle is $\tan^{-1}(3)$ which is 71.6°. Remembering that we are in the third quadrant this is actually 108.4° in the clockwise fashion.

Using this notation we can show, for example, that $2\boldsymbol{v}_1$ (multiplication by a scalar quantity) is the same as $\boldsymbol{v}_1 + \boldsymbol{v}_1$ and is a vector with twice the magnitude of v_1 but the same direction.

Example—$2v_1 = v_1 + v_1$

If $\boldsymbol{v}_1$ is written as $(3, 2)$ then its length is $\sqrt{9+4}$ which is 3.6 and the angle of rotation is $\tan^{-1}(2/3)$. Multiplication by two gives the coordinates $(6, 4)$. The length of this vector is $\sqrt{36+16}$ which is 7.2. The angle is $\tan^{-1}(4/6)$ which is equal to $\tan^{-1}(2/3)$, i.e., the same direction.

This rationale is easily extended to three dimensions by specifying three coordinates (x, y, z). The length of the vector is by convention the modulus and this is given by Pythagoras' theorem in three dimensions. For a vector $\boldsymbol{a}$

$$|\boldsymbol{a}|^2 = x^2 + y^2 + z^2 \tag{8.82}$$

This time the direction must be specified using two angles, one in the xy plane and one

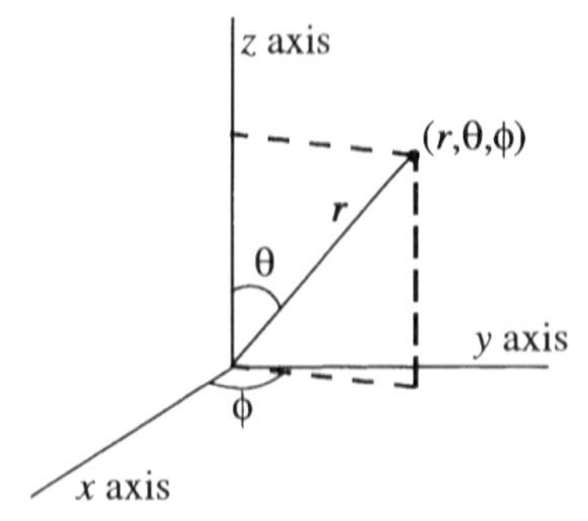

Figure 8.9 A vector in three-dimensional space requires three coordinates. In spherical space these are the length and the angle from the x axis in the xy plane and the angle from the z axis

specifying a rotation through the z axis, Figure 8.9. It is the case, however, that most of the mathematical manipulations will be done using just the notation, graphs being used for illustration only. At this point we will consider some formal notation.

8.3.2 Formal Notation

In representing a two-dimensional vector by its Cartesian coordinates we have shown that a vector may be represented and discussed in terms of two other vectors parallel to the x and y axes respectively. The x coordinate represents a vector from the origin to a point on the x axis though it could be any movement parallel to the x axis. The y coordinate represents any movement parallel to the y axis. In three dimensions use a z coordinate to represent movement parallel to the z axis. Any three-dimensional movement (vector) can then be broken down as a sum of these three steps.

Take three vectors $\boldsymbol{x}, \boldsymbol{y}$ and $\boldsymbol{z}$, which are represented by the following sets of coordinates.

$$\boldsymbol{x} = (3, 0, 0) \tag{8.83}$$

$$\boldsymbol{y} = (0, 2, 0) \tag{8.84}$$

$$\boldsymbol{z} = (0, 0, 1) \tag{8.85}$$

The sum of these vectors, $\boldsymbol{a} = \boldsymbol{x} + \boldsymbol{y} + \boldsymbol{z}$ is given by

$$\begin{aligned} \boldsymbol{a} &= \boldsymbol{x} + \boldsymbol{y} + \boldsymbol{z} = (3 + 0 + 0, 0 + 2 + 0, 0 + 0 + 1) \\ &= (3, 2, 1) \end{aligned} \tag{8.86}$$

$\boldsymbol{a}$ is therefore a three-dimensional vector. In breaking down a vector like this we say we have *resolved* the vector into its components.

We also recall that any vector may be represented as the product of a scalar value and a unit vector. The former is the modulus, a measure of the magnitude of the vector, and the latter has magnitude one and defines the direction of the vector.

$$\boldsymbol{a} = |a|\hat{\boldsymbol{a}} \tag{8.60}$$

Each of the component vectors of $\boldsymbol{a}$ may also be represented by the product of the modulus in that direction and the unit vector. Convention labels the three unit vectors $\hat{\boldsymbol{i}}, \hat{\boldsymbol{j}}$ and $\hat{\boldsymbol{k}}$ in the x, y and z directions respectively. The magnitude or modulus in each direction is labelled (for the vector $\boldsymbol{a}$) a_x, a_y or a_z. Therefore any vector may be written as

$$\boldsymbol{a} = a_x\hat{\boldsymbol{i}} + a_y\hat{\boldsymbol{j}} + a_z\hat{\boldsymbol{k}} \tag{8.87}$$

Using this notation we can not only add and subtract vectors as before but we can easily perform vector multiplication as shown in the next section. We will, nevertheless, give one further example of vector addition.

Example—vector addition and subtraction

Two vectors $\boldsymbol{a}$ and $\boldsymbol{b}$ are defined

$$\boldsymbol{a} = a_x\hat{\boldsymbol{i}} + a_y\hat{\boldsymbol{j}} + a_z\hat{\boldsymbol{k}} \tag{8.88}$$

$$\boldsymbol{b} = b_x\hat{\boldsymbol{i}} + b_y\hat{\boldsymbol{j}} + b_z\hat{\boldsymbol{k}} \tag{8.89}$$

Find the sum and the difference.

The sum is given by

$$\begin{aligned} \boldsymbol{a} + \boldsymbol{b} &= a_x\hat{\boldsymbol{i}} + a_y\hat{\boldsymbol{j}} + a_z\hat{\boldsymbol{k}} + b_x\hat{\boldsymbol{i}} + b_y\hat{\boldsymbol{j}} + b_z\hat{\boldsymbol{k}} \\ &= (a_x + b_x)\hat{\boldsymbol{i}} + (a_y + b_y)\hat{\boldsymbol{j}} + (a_z + b_z)\hat{\boldsymbol{k}} \end{aligned} \tag{8.90}$$

The difference is given by

$$\begin{aligned} \boldsymbol{a} - \boldsymbol{b} &= a_x\hat{\boldsymbol{i}} + a_y\hat{\boldsymbol{j}} + a_z\hat{\boldsymbol{k}} - (b_x\hat{\boldsymbol{i}} + b_y\hat{\boldsymbol{j}} + b_z\hat{\boldsymbol{k}}) \\ &= a_x\hat{\boldsymbol{i}} + a_y\hat{\boldsymbol{j}} + a_z\hat{\boldsymbol{k}} - b_x\hat{\boldsymbol{i}} - b_y\hat{\boldsymbol{j}} - b_z\hat{\boldsymbol{k}} \\ &= (a_x - b_x)\hat{\boldsymbol{i}} + (a_y - b_y)\hat{\boldsymbol{j}} + (a_z - b_z)\hat{\boldsymbol{k}} \end{aligned} \tag{8.91}$$

8.3.3 The Scalar Product

There are two methods by which we may find the product of two vectors. The first of these we call the *scalar* or *dot product*. This is defined as follows. For two vectors $\boldsymbol{a}$ and $\boldsymbol{b}$ which have an angle θ between them (Figure 8.10) the scalar product is

$$\boldsymbol{a} \cdot \boldsymbol{b} = |a||b|\cos(\theta) \tag{8.92}$$

Note that this result is a scalar quantity. By normal rules of geometry the product $|b|\cos(\theta)$ is the projection of the vector $\boldsymbol{b}$ on the vector $\boldsymbol{a}$ and the scalar product is the product of this quantity with the magnitude of $\boldsymbol{a}$. Note also that

$$\boldsymbol{a} \cdot \boldsymbol{b} = \boldsymbol{b} \cdot \boldsymbol{a} \tag{8.93}$$

and if each is multiplied by a scalar quantity

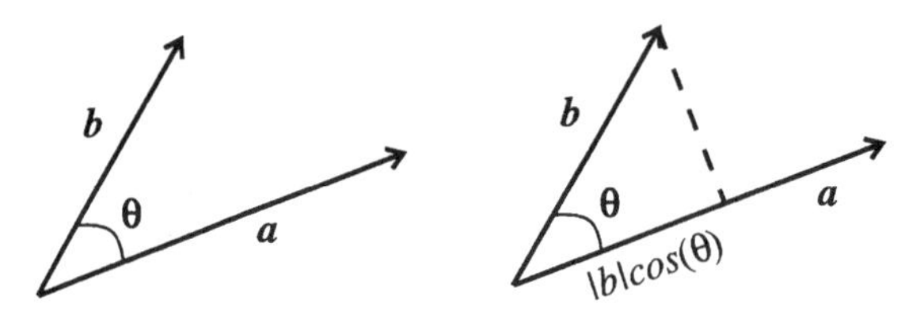

Figure 8.10 The scalar product of two vectors $\boldsymbol{a}$ and $\boldsymbol{b}$ on the left uses the projection of $\boldsymbol{b}$ on $\boldsymbol{a}$ multiplied by the magnitude of $\boldsymbol{a}$. The result is a scalar quantity

$$xa \cdot yb = xy(a \cdot b) \tag{8.94}$$

An example of a scalar product is the definition of work, which is the product of the force acting on a body ($\boldsymbol{F}$) and the body's displacement ($\boldsymbol{s}$). The definition usually given is the product of force and distance as numbers. They are both, strictly speaking, vectors and the product is the scalar product. If the force acts along the direction of displacement then there is no angle between the vectors and cos(0) is one. Therefore the product is merely the product of their magnitudes. When molecules interact with electromagnetic radiation, that is in absorbing light, then the magnitude of the interaction is the scalar product of the electric component of the radiation field ($\boldsymbol{E}$) and the dipole moment ($\boldsymbol{\mu}$) of the molecule. This is related to the wavefunction of the molecule and the excited molecule and is important in theoretical models of spectroscopy. When a molecule with a permanent electric dipole moment ($\boldsymbol{\mu}$ again) is placed in a capacitor and an electric field ($\boldsymbol{E}$) then the energy of the dipole in the field is also the scalar product $-\boldsymbol{\mu}-\boldsymbol{E}=-\mu E\cos(\theta)$. Thus when a material is placed between the plates of a capacitor the molecules tend to line up so as to minimise their energy. To do this they must maximise $\cos(\theta)$ (remembering that there is a minus sign in this equation), which is when θ is zero. This is when the molecules (and their dipole moments) line up with the electric field. Normal thermal motion tends to oppose this lining up and the variation of the polarisation with temperature is used as a method of determining molecular dipole moments.

Example—scalar product of two vectors

Vector $\boldsymbol{a}$ has a magnitude of 5 units and acts at 30° to vector $\boldsymbol{b}$ which has a magnitude of 4.5 units.

Using the definition of the scalar product the result is

$$\begin{aligned} \boldsymbol{a}\cdot\boldsymbol{b} &= 5 \times 4.5 \times \cos(30^\circ) \\ &= 22.5 \times 0.866 \\ &= 19.5 \end{aligned} \tag{8.95}$$

Using this definition we can also produce some important results used in vector multiplication.

Example—scalar product of orthogonal vectors

Orthogonal vectors are those which are perpendicular or at right angles to each other. Good examples are the unit vectors $\hat{\boldsymbol{\imath}}, \hat{\boldsymbol{\jmath}}$ and $\hat{\boldsymbol{k}}$. The scalar product of $\hat{\boldsymbol{\imath}}$ and $\hat{\boldsymbol{\jmath}}$ is given by

$$\begin{aligned} \hat{\boldsymbol{\imath}}\cdot\hat{\boldsymbol{\jmath}} &= 1 \times 1 \times \cos(90^\circ) \\ &= 1 \times 1 \times 0 \\ &= 0 \end{aligned} \tag{8.96}$$

This result is the same for any orthogonal vectors since cos(90°) is always zero. Likewise if we know that the scalar product is zero then the vectors in question must be orthogonal unless one already has zero magnitude (this is called a *null* vector).

Example—the square of a vector

If we find the scalar product of a unit vector with itself, $\hat{\boldsymbol{i}}\cdot\hat{\boldsymbol{i}}$, then the result is

$$\begin{aligned}\hat{\boldsymbol{i}}\cdot\hat{\boldsymbol{i}} &= 1 \times 1 \times \cos(0)\\ &= 1 \times 1 \times 1\\ &= 1\end{aligned} \tag{8.97}$$

These results become useful if we try to define the scalar product in terms of the Cartesian unit vectors $\hat{\boldsymbol{i}}, \hat{\boldsymbol{j}}$ and $\hat{\boldsymbol{k}}$. Again we use the definitions

$$\boldsymbol{a} = a_x\hat{\boldsymbol{i}} + a_y\hat{\boldsymbol{j}} + a_z\hat{\boldsymbol{k}} \tag{8.88}$$

$$\boldsymbol{b} = b_x\hat{\boldsymbol{i}} + b_y\hat{\boldsymbol{j}} + b_z\hat{\boldsymbol{k}} \tag{8.89}$$

Then $\boldsymbol{a}\cdot\boldsymbol{b}$ is found by expanding the product,

$$\boldsymbol{a}\cdot\boldsymbol{b} = (a_x\hat{\boldsymbol{i}} + a_y\hat{\boldsymbol{j}} + a_z\hat{\boldsymbol{k}})\cdot(b_x\hat{\boldsymbol{i}} + b_y\hat{\boldsymbol{j}} + b_z\hat{\boldsymbol{k}}) \tag{8.98}$$

This is done in the usual way.

$$\begin{aligned}\boldsymbol{a}\cdot\boldsymbol{b} = {} & a_xb_x\hat{\boldsymbol{i}}\cdot\hat{\boldsymbol{i}} + a_xb_y\hat{\boldsymbol{i}}\cdot\hat{\boldsymbol{j}} + a_xb_z\hat{\boldsymbol{i}}\cdot\hat{\boldsymbol{k}}\\ & + a_yb_x\hat{\boldsymbol{j}}\cdot\hat{\boldsymbol{i}} + a_yb_y\hat{\boldsymbol{j}}\cdot\hat{\boldsymbol{j}} + a_yb_z\hat{\boldsymbol{j}}\cdot\hat{\boldsymbol{k}}\\ & + a_zb_x\hat{\boldsymbol{k}}\cdot\hat{\boldsymbol{i}} + a_zb_y\hat{\boldsymbol{k}}\cdot\hat{\boldsymbol{j}} + a_zb_z\hat{\boldsymbol{k}}\cdot\hat{\boldsymbol{k}}\end{aligned} \tag{8.99}$$

This appears a little unwieldy until we apply the results of equations (8.96) and (8.97). When we do this all the terms with orthogonal vectors ($\hat{\boldsymbol{i}}\cdot\hat{\boldsymbol{j}}, \hat{\boldsymbol{i}}\cdot\hat{\boldsymbol{k}}$ and $\hat{\boldsymbol{j}}\cdot\hat{\boldsymbol{k}}$) become zero and all the like terms ($\hat{\boldsymbol{i}}\cdot\hat{\boldsymbol{i}}$, $\hat{\boldsymbol{j}}\cdot\hat{\boldsymbol{j}}$ and $\hat{\boldsymbol{k}}\cdot\hat{\boldsymbol{k}}$) become one. Equation (8.99) therefore reduces to

$$\boldsymbol{a}\cdot\boldsymbol{b} = a_xb_x + a_yb_y + a_zb_z \tag{8.100}$$

Using this result we can find the scalar product knowing only the components of the two vectors. We can use this to find the product and then by rearranging equation (8.90) find the angle between the vectors.

Example—scalar product of two vectors

Find the scalar product of the two vectors $\boldsymbol{a}$ and $\boldsymbol{b}$.

$$\boldsymbol{a} = 2\hat{\boldsymbol{i}} + 3\hat{\boldsymbol{j}} + 4\hat{\boldsymbol{k}} \tag{8.101}$$

$$\boldsymbol{b} = \hat{\boldsymbol{i}} + 5\hat{\boldsymbol{j}} + \hat{\boldsymbol{k}} \tag{8.102}$$

From equations (8.88) and (8.89) the values of the coefficients of $\hat{\boldsymbol{i}}$, $\hat{\boldsymbol{j}}$ and $\hat{\boldsymbol{k}}$ are

$$\begin{aligned} a_x &= 2 \quad b_x = 1\\ a_y &= 3 \quad b_y = 5\\ a_z &= 4 \quad b_z = 1\end{aligned} \tag{8.103}$$

From equation (8.100) $\boldsymbol{a}\cdot\boldsymbol{b}$ is

$$\boldsymbol{a}\cdot\boldsymbol{b} = (2 \times 1) + (3 \times 5) + (4 \times 1)$$

$$= 2 + 15 + 4 \quad (8.104)$$
$$= 21$$

It is possible to use this formula in conjunction with equation (8.92) to find the angle between two vectors. We need to calculate the magnitude of each vector using Pythagoras' theorem. $|a|$ and $|b|$ are given by

$$|a| = \sqrt{2^2 + 3^2 + 4^2} = \sqrt{29} \quad (8.105(a))$$

$$|b| = \sqrt{1^2 + 5^2 + 1^2} = \sqrt{27} \quad (8.105(b))$$

The cosine of the angle is

$$\cos(\theta) = \frac{\boldsymbol{a} \cdot \boldsymbol{b}}{|a||b|} \quad (8.106)$$

θ is therefore

$$\theta = \cos^{-1}\left(\frac{21}{\sqrt{29}\sqrt{27}}\right)$$
$$= \cos^{-1}(0.777) \quad (8.107)$$
$$= 38.9^{\circ}$$

8.3.4 The Vector Product

The *vector* or *cross product* is defined as

$$\boldsymbol{a} \wedge \boldsymbol{b} = |a||b|\sin(\theta)\hat{\boldsymbol{n}} \quad (8.108)$$

where $\hat{\boldsymbol{n}}$ is a unit vector. The $\wedge$ sign signifies the cross product though this is often replaced by the multiplication sign, $\times$. If we arrange the two vectors as if we were adding them together, they then form a parallelogram in the same plane, Figure 8.11. The area of this parallelogram is given by

$$\text{area} = |a||b|\sin(\theta) \quad (8.109)$$

where $|a|$ and $|b|$ are the lengths of the sides. The vector product therefore represents a vector whose magnitude is equivalent to the area of the parallelogram formed by the two

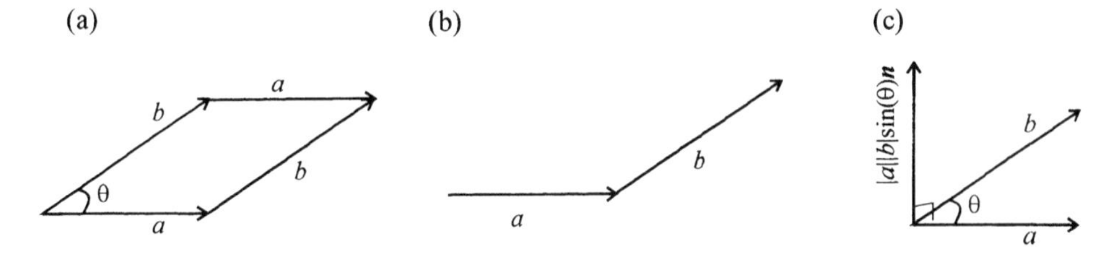

Figure 8.11 The magnitude of the vector product $\boldsymbol{a} \wedge \boldsymbol{b}$ is the area of the parallelogram formed by the vectors (a). The product $\boldsymbol{a} \wedge \boldsymbol{b}$ means $\boldsymbol{a}$ followed by $\boldsymbol{b}$ as shown in (b)—suggesting an anticlockwise motion. The direction is perpendicular to the plane of the parallelogram in (a). The anticlockwise sense of (b) means $\boldsymbol{n}$ points up from the plane of the parallelogram (c)

vectors. All that remains is the direction which is specified by the unit vector $\hat{\boldsymbol{n}}$. Convention says that $\hat{\boldsymbol{n}}$ is normal (at right angles) to the plane of the parallelogram. Figure 8.11 indicates the direction $\hat{\boldsymbol{n}}$ takes. We can place the vectors head to tail in two ways. $\boldsymbol{a}$ followed by $\boldsymbol{b}$ has an anticlockwise sense and $\boldsymbol{b}$ followed by $\boldsymbol{a}$ has a clockwise sense. Convention states that $\boldsymbol{a} \wedge \boldsymbol{b}$ has vector $\boldsymbol{a}$ first and as in Figure 8.11 the arrangement has an anticlockwise sense and the resultant vector points up. If the vectors form a clockwise movement then the vector points down and the product should be written $\boldsymbol{b} \wedge \boldsymbol{a}$. We know that a vector multiplied by -1 has the opposite direction but the same magnitude and so

$$\boldsymbol{b} \wedge \boldsymbol{a} = -(\boldsymbol{a} \wedge \boldsymbol{b}) \tag{8.110}$$

Example—vector product of two identical vectors

The product $\hat{\boldsymbol{\imath}} \wedge \hat{\boldsymbol{\imath}}$ can be calculated knowing that the angle between them, θ, is zero. The sine of zero is also zero and therefore

$$\begin{aligned} \hat{\boldsymbol{\imath}} \wedge \hat{\boldsymbol{\imath}} &= |1||1|\sin(0) \\ &= 0 \end{aligned} \tag{8.111}$$

By the same reasoning, that the sine of the angle between them is zero, any two vectors which are parallel have a vector product which is zero.

Example—vector product of orthogonal vectors

Find the vector product, $\hat{\boldsymbol{\imath}} \wedge \hat{\boldsymbol{\jmath}}$. The angle θ is 90° and the sine of this is 1. Therefore

$$\begin{aligned} \hat{\boldsymbol{\imath}} \wedge \hat{\boldsymbol{\jmath}} &= |1||1|\sin(90)\hat{\boldsymbol{n}} \\ &= \hat{\boldsymbol{n}} \end{aligned} \tag{8.112}$$

In this case the unit vector formed from the anticlockwise rotation (along the x axis and along the y axis) is parallel to the z axis and is the unit vector $\hat{\boldsymbol{k}}$.

$$\hat{\boldsymbol{\imath}} \wedge \hat{\boldsymbol{\jmath}} = \hat{\boldsymbol{k}} \tag{8.113}$$

Likewise we can show that

$$\hat{\boldsymbol{\jmath}} \wedge \hat{\boldsymbol{k}} = \hat{\boldsymbol{\imath}} \tag{8.114}$$

and

$$\hat{\boldsymbol{k}} \wedge \hat{\boldsymbol{\imath}} = \hat{\boldsymbol{\jmath}} \tag{8.115}$$

Also from equation (8.106) we can show that

$$\hat{\boldsymbol{\jmath}} \wedge \hat{\boldsymbol{\imath}} = -\hat{\boldsymbol{k}} \tag{8.116}$$

$$\hat{\boldsymbol{k}} \wedge \hat{\boldsymbol{\jmath}} = -\hat{\boldsymbol{\imath}} \tag{8.117}$$

$$\hat{\boldsymbol{\imath}} \wedge \hat{\boldsymbol{k}} = -\hat{\boldsymbol{\jmath}} \tag{8.118}$$

As for the scalar product we can describe the vector product in terms of the unit vectors. Again we will use the vectors $\boldsymbol{a}$ and $\boldsymbol{b}$.

$$\boldsymbol{a} \wedge \boldsymbol{b} = (a_x\hat{\boldsymbol{\imath}} + a_y\hat{\boldsymbol{\jmath}} + a_z\hat{\boldsymbol{k}}) \wedge (b_x\hat{\boldsymbol{\imath}} + b_y\hat{\boldsymbol{\jmath}} + b_z\hat{\boldsymbol{k}}) \tag{8.119}$$

The products involving scalars will act as they do in normal algebra but the vectors will expand as vector products. Expansion therefore produces

$$\begin{aligned} \boldsymbol{a} \wedge \boldsymbol{b} &= a_xb_x\hat{\boldsymbol{\imath}} \wedge \hat{\boldsymbol{\imath}} + a_xb_y\hat{\boldsymbol{\imath}} \wedge \hat{\boldsymbol{\jmath}} + a_xb_z\hat{\boldsymbol{\imath}} \wedge \hat{\boldsymbol{k}} \\ &+ a_yb_x\hat{\boldsymbol{\jmath}} \wedge \hat{\boldsymbol{\imath}} + a_yb_y\hat{\boldsymbol{\jmath}} \wedge \hat{\boldsymbol{\jmath}} + a_yb_z\hat{\boldsymbol{\jmath}} \wedge \hat{\boldsymbol{k}} \\ &+ a_zb_x\hat{\boldsymbol{k}} \wedge \hat{\boldsymbol{\imath}} + a_zb_y\hat{\boldsymbol{k}} \wedge \hat{\boldsymbol{\jmath}} + a_zb_z\hat{\boldsymbol{k}} \wedge \hat{\boldsymbol{k}} \end{aligned} \tag{8.120}$$

Taking account of the squares which reduce to zero and the results in equations (8.113) to (8.118) we get

$$\boldsymbol{a} \wedge \boldsymbol{b} = a_xb_y\hat{\boldsymbol{k}} + a_xb_z(-\hat{\boldsymbol{\jmath}}) + a_yb_x(-\hat{\boldsymbol{k}}) + a_yb_z\hat{\boldsymbol{\imath}} + a_zb_x\hat{\boldsymbol{\jmath}} + a_zb_y(-\hat{\boldsymbol{\imath}}) \tag{8.121}$$

This is the same as

$$\begin{aligned} \boldsymbol{a} \wedge \boldsymbol{b} &= a_xb_y\hat{\boldsymbol{k}} - a_xb_z(\hat{\boldsymbol{\jmath}}) - a_yb_x(\hat{\boldsymbol{k}}) + a_yb_z\hat{\boldsymbol{\imath}} + a_zb_x\hat{\boldsymbol{\jmath}} - a_zb_y(\hat{\boldsymbol{\imath}}) \\ &= (a_yb_z - a_zb_y)\hat{\boldsymbol{\imath}} + (a_zb_x - a_xb_z)\hat{\boldsymbol{\jmath}} + (a_xb_y - a_yb_x)\hat{\boldsymbol{k}} \end{aligned} \tag{8.122}$$

Example—vector multiplication of two vectors in the xy plane

Find the product of the two vectors $\boldsymbol{a}$ and $\boldsymbol{b}$.

$$\boldsymbol{a} = 2\hat{\boldsymbol{\imath}} + 2\hat{\boldsymbol{\jmath}} \tag{8.123}$$

$$\boldsymbol{b} = 3\hat{\boldsymbol{\imath}} + 4\hat{\boldsymbol{\jmath}} \tag{8.124}$$

In both cases the coefficient of the unit vector $\hat{\boldsymbol{k}}$ is zero, $a_z = b_z = 0$. The other coefficients are

$$\begin{aligned} a_x &= 2 \quad a_y = 2 \\ b_x &= 3 \quad b_y = 4 \end{aligned} \tag{8.125}$$

In this case equation (8.122) becomes

$$\begin{aligned} \boldsymbol{a} \wedge \boldsymbol{b} &= (0 - 0)\hat{\boldsymbol{\imath}} + (0 - 0)\hat{\boldsymbol{\jmath}} + (2 \times 4 - 2 \times 3)\hat{\boldsymbol{k}} \\ &= (8 - 6)\hat{\boldsymbol{k}} \\ &= 2\hat{\boldsymbol{k}} \end{aligned} \tag{8.126}$$

which is a vector of magnitude 2 parallel to the z axis.

8.3.5 Vectors in Chemistry

This presents only the beginning of vector mathematics. In advanced courses you may encounter topics such as vector calculus. We have already mentioned that force and derived quantities are vectors and much of the study of the properties of molecules and subatomic particles is of the effect of forces on charged entities. We also meet the concept of a field—magnetic fields and electric fields being the regions of influence of a magnet and an electric charge respectively. Fields are themselves vector quantities and as they have an effect on whatever is within their area of influence then understanding these

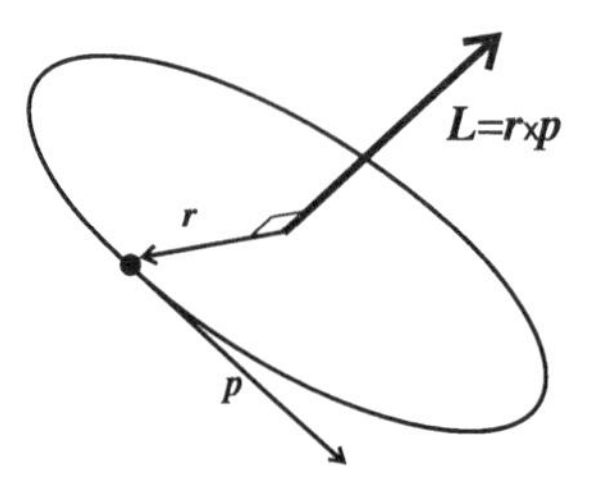

Figure 8.12 Angular momentum is the vector product of the linear momentum ***p*** and the positional vector ***r***. The resultant is, as with other vector products, perpendicular to the plane of circular motion. Depending on the direction of spin the vector points up or down

phenomena necessitates some appreciation of methods of combining vectors.

The angular momentum of a spinning and/or orbiting particle is also represented by a vector. This vector is drawn as an arrow along the axis of spin or rotation and points up for anticlockwise spin and down for clockwise (corresponding to the resultant for vector or cross product). More formally it can be shown that angular momentum is the vector or cross product of the position vector ***r*** and the linear momentum ***p*** of the object moving under the influence of a central force (Figure 8.12). Much is made of the component of the angular momentum vector in the z direction, knowledge of the other two components being restricted by the uncertainty principle. The z component can, as is typical in quantum theory, only take certain values and this has the result that in quantum mechanics not only is magnitude quantised but so is direction. Finally an object that is moving in an orbit and simultaneously spinning (as the Earth is around the Sun) has both orbital and spin angular momentum. Its total angular momentum is, almost obviously, the sum of these.

Even having a full algebraic method of treating vector mathematics chemists are still likely to find the 'arrow' approach used widely. The awareness of vectors and their representation as arrows is in many ways more important than detailed knowledge of their algebraic manipulation.

8.4 Determinants

8.4.1 Definitions

A determinant is a square array of numbers which is written as

$$\begin{vmatrix} 1 & 2 \\ 3 & 4 \end{vmatrix} \tag{8.127}$$

or more generally

$$\begin{vmatrix} a_1 & b_1 \\ a_2 & b_2 \end{vmatrix} \tag{8.128}$$

The value of this determinant is

$$\begin{vmatrix} a_1 & b_1 \\ a_2 & b_2 \end{vmatrix} = a_1b_2 - a_2b_1 \tag{8.129}$$

Thus determinant (8.127) has a value

$$\begin{vmatrix} 1 & 2 \\ 3 & 4 \end{vmatrix} = (1 \times 4) - (3 \times 2) = -2 \tag{8.130}$$

This 2×2 determinant, having two rows and two columns, is called a *second order* determinant. A third order determinant would look like

$$\begin{vmatrix} a_1 & b_1 & c_1 \\ a_2 & b_2 & c_2 \\ a_3 & b_3 & c_3 \end{vmatrix} \tag{8.131}$$

and have a value defined by

$$\begin{vmatrix} a_1 & b_1 & c_1 \\ a_2 & b_2 & c_2 \\ a_3 & b_3 & c_3 \end{vmatrix} = a_1 \begin{vmatrix} b_2 & c_2 \\ b_3 & c_3 \end{vmatrix} - b_1 \begin{vmatrix} a_2 & c_2 \\ a_3 & c_3 \end{vmatrix} + c_1 \begin{vmatrix} a_2 & b_2 \\ a_3 & b_3 \end{vmatrix} \tag{8.132}$$

The second order determinants may be expanded to give the final result.

$$\begin{vmatrix} a_1 & b_1 & c_1 \\ a_2 & b_2 & c_2 \\ a_3 & b_3 & c_3 \end{vmatrix} = a_1 b_2 c_3 - a_1 b_3 c_2 - b_1 a_2 c_3 + b_1 a_3 c_2 + c_1 a_2 b_3 - c_1 a_3 b_2 \tag{8.133}$$

Determinants of higher order are solved in a similar manner, breaking them down into smaller and smaller determinants until the 2×2 determinant is reached.

8.4.2 Determinants for the Solution of Simultaneous Equations

The determinant is one way of representing and finding the solution to simultaneous equations. If we have two linear equations

$$a_1 + b_1 y = c_1 \tag{8.134}$$

$$a_2 x + b_2 y = c_2 \tag{8.135}$$

as shown in Chapter 1 it is possible to solve these by elimination of x and y as appropriate. For example, if we multiply (8.134) by a_2

$$a_2 a_1 x + a_2 b_1 y = a_2 c_1 \tag{8.136}$$

and (8.135) by a_1

$$a_1 a_2 x + a_1 b_2 y = a_1 c_2 \tag{8.137}$$

and then subtracting (8.136) from (8.137) we get

$$a_1 b_2 y - a_2 b_1 y = a_1 c_2 - a_2 - a_2 c_1 \tag{8.138}$$

which rearranges to give y

$$y = \frac{a_1 c_2 - a_2 c_1}{a_1 b_2 - a_2 b_1} \tag{8.139}$$

x can be found by a similar elimination process.

$$x = \frac{c_1 b_2 - c_2 b_1}{a_1 b_2 - a_2 b_1} \tag{8.140}$$

If we now compare the results for x and y with equation (8.129), we find that the denominators and numerators of these fractions are the values of determinants. Both have the same denominator,

$$a_1 b_2 - a_2 b_1 = \begin{vmatrix} a_1 & b_1 \\ a_2 & b_2 \end{vmatrix} \tag{8.141}$$

From the result for y

$$a_1 c_2 - a_2 c_1 = \begin{vmatrix} a_1 & c_1 \\ a_2 & c_2 \end{vmatrix} \tag{8.142}$$

and from the result for x

$$c_1 b_2 - c_2 b_1 = \begin{vmatrix} c_1 & b_1 \\ c_2 & b_2 \end{vmatrix} \tag{8.143}$$

Thus we can write results for x and y.

$$x = \frac{\begin{vmatrix} c_1 & b_1 \\ c_2 & b_2 \end{vmatrix}}{\begin{vmatrix} a_1 & b_1 \\ a_2 & b_2 \end{vmatrix}} \tag{8.144}$$

$$y = \frac{\begin{vmatrix} a_1 & c_1 \\ a_2 & c_2 \end{vmatrix}}{\begin{vmatrix} a_1 & b_1 \\ a_2 & b_2 \end{vmatrix}} \tag{8.145}$$

Example—solution of simultaneous equations

In Chapter 1 we met the equations

$$3x - y = -1 \tag{8.146}$$

$$2y - y = -2 \tag{8.147}$$

We can define the following constants

$$\begin{aligned} a_1 &= 3 \quad a_2 = 2 \\ b_1 &= -1 \quad b_2 = -1 \\ c_1 &= -1 \quad c_2 = -2 \end{aligned} \tag{8.148}$$

and the solution is therefore

$$x = \frac{\begin{vmatrix} -1 & -1 \\ -2 & -1 \end{vmatrix}}{\begin{vmatrix} 3 & -1 \\ 2 & -1 \end{vmatrix}} = \frac{1-2}{-3+2} = \frac{-1}{-1} = 1 \tag{8.149}$$

$$y = \frac{\begin{vmatrix} 3 & -1 \\ 2 & -2 \end{vmatrix}}{\begin{vmatrix} 3 & -1 \\ 2 & -1 \end{vmatrix}} = \frac{-6+2}{-3+2} = \frac{-4}{-1} = 4 \tag{8.150}$$

This solution, $x = 1$, $y = 4$ agrees with that given in Chapter 1.

This is a fairly trivial example but determinants may be used to solve greater numbers of simultaneous equations. For these a more general notation is introduced. For three simultaneous equations we label the variables x_1, x_2, and x_3 and the coefficients according to their position in the determinant.

$$a_{11}x_1 + a_{12}x_2 + a_{13}x_3 = b_1 \tag{8.151}$$

$$a_{21}x_1 + a_{22}x_2 + a_{23}x_3 = b_2 \tag{8.152}$$

$$a_{31}x_1 + a_{32}x_2 + a_{33}x_3 = b_3 \tag{8.153}$$

By the same process of elimination we can arrive at solutions for the equations in terms of determinants. The denominator is

$$\begin{vmatrix} a_{11} & a_{12} & a_{13} \\ a_{21} & a_{22} & a_{23} \\ a_{31} & a_{32} & a_{33} \end{vmatrix} \tag{8.154}$$

where we can now see that the subscript refers to the row and then the column number of the determinant. The solution in terms of x_1, x_2 and x_3 is

$$x_1 = \frac{\begin{vmatrix} b_1 & a_{12} & a_{13} \\ b_2 & a_{22} & a_{23} \\ b_3 & a_{32} & a_{33} \end{vmatrix}}{\begin{vmatrix} a_{11} & a_{12} & a_{13} \\ a_{21} & a_{22} & a_{23} \\ a_{31} & a_{32} & a_{33} \end{vmatrix}} \tag{8.155}$$

$$x_2 = \frac{\begin{vmatrix} a_{11} & b_1 & a_{13} \\ a_{21} & b_2 & a_{23} \\ a_{31} & b_3 & a_{33} \end{vmatrix}}{\begin{vmatrix} a_{11} & a_{12} & a_{13} \\ a_{21} & a_{22} & a_{23} \\ a_{31} & a_{32} & a_{33} \end{vmatrix}} \tag{8.156}$$

$$x_3 = \frac{\begin{vmatrix} a_{11} & a_{12} & b_1 \\ a_{21} & a_{22} & b_2 \\ a_{31} & a_{32} & b_3 \end{vmatrix}}{\begin{vmatrix} a_{11} & a_{12} & a_{13} \\ a_{21} & a_{22} & a_{23} \\ a_{31} & a_{32} & a_{33} \end{vmatrix}} \tag{8.157}$$

Although this seems a lot to remember there is a pattern—for x_1 the first column is replaced by b values, for x_2 the second column is replaced and for x_3 the third column is replaced. This pattern is repeated for higher order determinants, for greater numbers of simultaneous equations.

And so we can produce determinants for more and more simultaneous equations, the solution of which is made systematic but not necessarily any easier or quicker. There are many rules describing the properties of determinants which can make their manipulation and use easier. In the following sections, however, we will merely describe some of the uses made of determinants in chemistry.

8.4.3 Representation of the Vector Product

In discussing vectors and in particular the vector product we derived equation (8.122)

$$\boldsymbol{a} \wedge \boldsymbol{b} = (a_y b_z - a_z b_y)\hat{\boldsymbol{\imath}} + (a_z b_x - a_x b_z)\hat{\boldsymbol{\jmath}} + (a_x b_y - a_y b_x)\hat{\boldsymbol{k}} \tag{8.122}$$

If we compare this with equation (8.133) describing the expansion of a third order determinant

$$\begin{vmatrix} a_1 & b_1 & c_1 \\ a_2 & b_2 & c_2 \\ a_3 & b_3 & c_3 \end{vmatrix} = a_1b_2c_3 - a_1b_3c_2 - b_1a_2c_3 + b_1a_3c_2 + c_1a_2b_3 - c_1a_3b_2 \tag{8.133}$$

$$= a_1(b_2c_3 - b_3c_2) + b_1(a_3c_2 - a_2c_3) + c_1(a_2b_3 - a_3b_2)$$

we can write the vector product as a determinant.

$$\boldsymbol{a} \wedge \boldsymbol{b} = (a_y b_z - a_z b_y)\hat{\boldsymbol{\imath}} + (a_z b_x - a_x b_z)\hat{\boldsymbol{\jmath}} + (a_x b_y - a_y b_x)\hat{\boldsymbol{k}}$$

$$= \begin{vmatrix} \hat{\boldsymbol{\imath}} & \hat{\boldsymbol{\jmath}} & \hat{\boldsymbol{k}} \\ a_x & a_y & a_z \\ b_x & b_y & b_z \end{vmatrix} \tag{8.158}$$

Once again this does not make the calculation any simpler but it is certainly easier to remember.

8.4.4 Slater Determinants

A Slater determinant is a convenient method of representing atomic and molecular wavefunctions which allows for the number of permutations of electrons (or more properly the labels we give them) and allows for the fact that electrons are indistinguishable from each other. They also adhere to certain rules of quantum mechanics. The properties of determinants used are as follows.

The first is that if any two rows or columns in a determinant are identical then the determinant has zero value. For example, the second order determinant (8.159),

$$\begin{vmatrix} 1 & 2 \\ 1 & 2 \end{vmatrix} = 2 - 2 = 0 \tag{8.159}$$

and the third order determinant (8.160) are both equal to zero.

$$\begin{aligned} \begin{vmatrix} 1 & 2 & 3 \\ 1 & 2 & 3 \\ 3 & 2 & 1 \end{vmatrix} &= 1(2-6) - 2(1-9) + 3(2-6) \\ &= -4 + 16 - 12 \\ &= 0 \end{aligned} \tag{8.160}$$

The second is that if two rows or two columns are interchanged then the sign of the determinant is changed, positive to negative or *vice versa*.

$$\begin{vmatrix} 1 & 2 \\ 3 & 4 \end{vmatrix} = 4 - 6 = -2 \tag{8.161}$$

and

$$\begin{vmatrix} 3 & 4 \\ 1 & 2 \end{vmatrix} = 6 - 4 = 2 \tag{8.162}$$

In a Slater determinant the elements (the numbers or functions constituting the array) are labels for electrons in, for example, an atom. Helium has two electrons in a 1s orbital. They are labelled 1s for the orbital they occupy, α or β for the spin of the electron and 1 or 2 for the electron number. A label is written as $1s(1)\alpha(1)$, which means electron (1) in a 1s orbital with spin α. The wavefunction (ψ) for helium may be written as

$$\psi = \frac{1s(1)1s(2)}{\sqrt{2}}[\alpha(1)\beta(2) - \alpha(2)\beta(1)] \tag{8.163}$$

This looks a bit formidable but remember that these are merely labels. If we expand the product we have the difference between the possible arrangements in terms of the spins the electrons may have in this orbital.

$$\psi = \frac{1}{\sqrt{2}}[1s(1)\alpha(1)1s(2)\beta(2) - 1s(1)\beta(1)1s(2)\alpha(2)] \tag{8.164}$$

The term in square brackets looks like, because it is, the expansion of a determinant and the wavefunction for helium may be written as such.

$$\psi = \frac{1}{\sqrt{2}}\begin{vmatrix} 1s(1)\alpha(1) & 1s(1)\beta(1) \\ 1s(2)\alpha(2) & 1s(2)\beta(2) \end{vmatrix} \tag{8.165}$$

We seem to be straying to abstract territory, but keep reminding yourself that these are still labels, nothing more. This effectively states the permutation of electron arrangements, that one must have spin α and one spin β, but we do not know which is which. Furthermore if we were to change the labels so that both did have spin α, we would have

$$\psi = \frac{1}{\sqrt{2}}\begin{vmatrix} 1s(1)\alpha(1) & 1s(1)\alpha(1) \\ 1s(2)\alpha(2) & 1s(2)\alpha(2) \end{vmatrix} \tag{8.166}$$

Then as in equation (8.159) the determinant has zero value and the arrangement is therefore not allowed, i.e., the wavefunction has zero value.

If we swap the labels over, 1 for 2,

$$\begin{vmatrix} 1s(2)\alpha(2) & 1s(2)\alpha(2) \\ 1s(1)\alpha(1) & 1s(1)\alpha(1) \end{vmatrix} \tag{8.167}$$

then as in equations (8.161) and (8.162) the sign of the determinant changes to $-\psi$. This is in accordance with the *Pauli* principle which states that if the labels on any two electrons are exchanged then the sign of the wavefunctions changes. A special case is the *Pauli exclusion* principle which states that no two electrons can have the same set of four quantum numbers. A consequence of this is that if there are two electrons in one orbital (which takes care of three quantum numbers) their spins must be anti-parallel, i.e., one α and one β.

8.4.5 Secular Equations and Secular Determinants

As progress is made from the description of the hydrogen atom to the hydrogen molecule, to helium and then larger molecules we are forced to make more and more approximations. The number of variables (numbers of particles, number of interactions) increases and the equations become more and more difficult if not impossible to solve. In the time honoured tradition, then, broad assumptions are made, equations simplified and the results compared to observation to see just how well the assumptions and simplifications stand up.

Molecular orbitals are, one would think, combinations of atomic orbitals, but the combination is not a simple matter of adding them up. As is taught in high school chemistry they combine to different degrees giving rise to ionic, covalent and polar covalent molecular orbitals. One way to interpret this mathematically is to say that the molecular wavefunction, ψ, is given by

$$\psi = c_A\psi_A + c_B\psi_B \tag{8.168}$$

where ψ_A and ψ_B are the atomic orbital wavefunctions for atoms A and B respectively. c_A and c_B are coefficients which tell us what contribution the individual wavefunctions make to the result. An ionic bond would have one coefficient equal to zero and the other equal to one whilst a polar bond would have non-equal non-zero values of c_A and c_B.

This equation is just the beginning, if we can combine two orbitals in this way why not more? Students are probably familiar with alternating series of double bonds (in benzene for example) and how these are supposedly more stable than a set of isolated double bonds. In extending this idea of combining atomic orbitals it is possible to arrive at a solution which shows that this lowering of energy is the case—even when making gross simplifications.

The derivations look difficult but even though they involve several integrals and wavefunctions most rearrangements are algebraic and along the way the integrals are substituted by variables. In the following equations we will find α_A and α_B which are the so-called 'Coulomb integrals' and β, a resonance integral, and S, an overlap integral. We eventually arrive at what are called *secular equations* and their related *secular determinants*, e.g.,

$$\begin{vmatrix} \alpha_A - E & \beta - ES \\ \beta - ES & \alpha_B - E \end{vmatrix} \tag{8.169}$$

where E is the energy of the system. Physical chemistry texts will tell you that this only has a meaningful solution when it is equal to zero. The solution is by no means trivial but it expands to produce a quadratic in E from which values of the energy may be found. There are two results, one of which gives a lower energy, known as the *bonding energy*, and represents a bonding orbital.

$$E_{\text{bonding}} = \frac{(\alpha + \beta)}{1 + S} \tag{8.170}$$

That this is lower in energy than the constituent orbitals is rationalised by the fact that the Coulomb integrals are considered to be negative and the resonance integral is negative at normal bond lengths.

The other result is the *anti-bonding energy* and this represents an anti-bonding orbital.

$$E_{\text{anti/bonding}} = \frac{(\alpha - \beta)}{1 - S} \tag{8.171}$$

We have not yet seen how any of this works in practice but the examples below show a little of what might be expected.

We can derive secular determinants for systems larger than two orbitals but the resulting determinant is more complicated—there is a resonance integral and an overlap integral for each pair of orbitals in the system. In ethene there is a system of sigma bonds and each carbon atom has a non-hybridised p orbital. To describe the pi bond we ignore the sigma framework and consider only these two orbitals—another simplification, the *Hückel approximation.* As expected there is a Coulomb integral for each orbital, and a resonance and an overlap integral. If we extend the system to butadiene with four carbon atoms then in addition to the Coulomb integrals there is a resonance integral and an overlap integral for each pair of orbitals. This results in a 4×4 determinant

$$\begin{vmatrix} \alpha - E & \beta_{AB} - ES_{AB} & \beta_{AC} - ES_{AC} & \beta_{AD} - ES_{AD} \\ \beta_{BA} - ES_{BA} & \alpha - E & \beta_{BC} - ES_{BC} & \beta_{BD} - ES_{BD} \\ \beta_{CA} - ES_{CA} & \beta_{CB} - ES_{CB} & \alpha - E & \beta_{CD} - ES_{CD} \\ \beta_{DA} - ES_{DA} & \beta_{DB} - ES_{DB} & \beta_{DC} - ES_{DC} & \alpha - E \end{vmatrix} \tag{8.172}$$

where A, B, C and D are labels for the four atoms. Again this can only be solved meaningfully if it is equal to zero. This may be difficult to solve but we can make further simplifications (also called *Hückel simplifications*). The first is that all overlap integrals are set to zero, making all the ES terms disappear. The second is that resonance integrals between non-near neighbours (AC, AD etc.) are zero and all remaining resonance integrals are equal to β. Thus (8.172) becomes

$$\begin{vmatrix} \alpha - E & \beta & 0 & 0 \\ \beta & \alpha - E & \beta & 0 \\ 0 & \beta & \alpha - E & \beta \\ 0 & 0 & \beta & \alpha - E \end{vmatrix} = 0 \tag{8.173}$$

Although now a gross simplification it presents something which is possible to solve and gives an idea of the relative energies of the molecular orbitals.

Example—Hückel approximation on linear combination of two orbitals

Applying the above simplifications to determinant (8.169) we obtain

$$\begin{vmatrix} \alpha - E & \beta \\ \beta & \alpha - E \end{vmatrix} = 0 \tag{8.174}$$

Although it is not too difficult to solve this we will show typical methodology before attempting a more difficult problem. First we divide each column by β and then replace $(\alpha - E)/\beta$ by x (a procedure we will reverse once we have solved the determinant) to obtain

$$\begin{vmatrix} x & 1 \\ 1 & x \end{vmatrix} = 0 \tag{8.175}$$

The expansion of this is

$$\begin{vmatrix} x & 1 \\ 1 & x \end{vmatrix} = x^2 - 1 = 0 \tag{8.176}$$

This has the solution $x = \pm 1$ or substituting for x

$$\frac{\alpha - E}{\beta} = \pm 1 \tag{8.177}$$

One solution is

$$\begin{aligned} \alpha - E &= \beta \\ \therefore E &= \alpha - \beta \end{aligned} \tag{8.178}$$

The other is

$$\begin{aligned} \alpha - E &= -\beta \\ \therefore E &= \alpha + \beta \end{aligned} \tag{8.179}$$

This is in agreement with our previous statement that the combination of orbitals has two solutions, one resulting from the addition and one from the difference of these integrals.

Example—Hückel MO for combination of four orbitals, model of butadiene

In this we combine four identical p orbitals from the four carbon atoms of butadiene. Taking determinant (8.173) we simplify its appearance by factoring β from each column again and then replacing $(\alpha - E)/\beta$ by x.

$$\begin{vmatrix} x & 1 & 0 & 0 \\ 1 & x & 1 & 0 \\ 0 & 1 & x & 1 \\ 0 & 0 & 1 & x \end{vmatrix} = 0 \tag{8.180}$$

Expansion of this is not as tedious as might first appear as the two columns with zero at the head yield zero result. The expansion goes as follows.

$$\begin{vmatrix} x & 1 & 0 & 0 \\ 1 & x & 1 & 0 \\ 0 & 1 & x & 1 \\ 0 & 0 & 1 & x \end{vmatrix} = x\begin{vmatrix} x & 1 & 0 \\ 1 & x & 1 \\ 0 & 1 & x \end{vmatrix} - 1\begin{vmatrix} 1 & 1 & 0 \\ 0 & x & 1 \\ 0 & 1 & x \end{vmatrix} \tag{8.181}$$

$$= x\left\{x\begin{vmatrix} x & 1 \\ 1 & x \end{vmatrix} - 1\begin{vmatrix} 1 & 1 \\ 0 & x \end{vmatrix}\right\} - 1\left\{1\begin{vmatrix} x & 1 \\ 1 & x \end{vmatrix} - 1\begin{vmatrix} 0 & 1 \\ 0 & x \end{vmatrix}\right\}$$

The final line expands to produce

$$\begin{aligned} \begin{vmatrix} x & 1 & 0 & 0 \\ 1 & x & 1 & 0 \\ 0 & 1 & x & 1 \\ 0 & 0 & 1 & x \end{vmatrix} &= x\{x(x^2 - 1) - 1(x - 0)\} - 1\{1(x^2 - 1) - 1(0 - 0)\} \\ &= x\{x^3 - x - x\} - 1\{x^2 - 1\} \\ &= x^4 - 2x^2 - x^2 + 1 \\ &= x^4 - 3x^2 + 1 \end{aligned} \tag{8.182}$$

Reminding ourselves that this is equal to zero

$$x^4 - 3x^2 + 1 = 0 \tag{8.183}$$

we have, in effect a quadratic equation in x^2. That is, if we let $z = x^2$ then equation (8.183) becomes

$$z^2 - 3z + 1 = 0 \tag{8.184}$$

We can solve this using equation (1.151) given in Chapter 1 where

$$\begin{aligned} a &= 1 \\ b &= -3 \\ c &= 1 \end{aligned} \tag{8.185}$$

and

$$
\begin{aligned}
z &= \frac{3 \pm \sqrt{3^2 - 4(1 \times 1)}}{2 \times 1} \\
&= \frac{3 \pm \sqrt{5}}{2}
\end{aligned} \tag{8.186}
$$

This gives $z = 2.618$ and $z = 0.382$. Taking the square roots to find x we have

$$
\begin{aligned}
x &= \pm\sqrt{2.618} = \pm 1.618 \\
x &= \pm\sqrt{0.382} = \pm 0.618
\end{aligned} \tag{8.187}
$$

Thus there are four solutions and we can now substitute back for $(\alpha - E)/\beta$ in each one to find a value for the energy. These are

$$
\begin{aligned}
E_1 &= \alpha + 1.618\beta \\
E_2 &= \alpha + 0.618\beta \\
E_3 &= \alpha - 1.618\beta \\
E_4 &= \alpha - 0.618\beta
\end{aligned} \tag{8.188}
$$

These are the energies of the four Hückel molecular orbitals for butadiene. Since α and β are negative E_1 is the lowest and E_3 is the highest. There are four electrons and two will fill each of the two lowest energy orbitals. The bonding energy of butadiene is estimated to be

$$
\begin{aligned}
E &= 2(\alpha + 1.618\beta) + 2(\alpha + 0.618\beta) \\
&= 4\alpha + 4.472\beta
\end{aligned} \tag{8.189}
$$

Such methods may be used to estimate the energy of bonds in any conjugated system, for example benzene which has three conjugated π bonds comprising six p orbitals. The significance is that we can compare bonding energies for the MO which combines four orbitals with the bond energy for isolated π bonds, each of which combines two orbitals. The case of having two isolated p–p molecular orbitals each of which contains two electrons results in an energy given by

$$
\begin{aligned}
E &= 2 \times 2(\alpha + \beta) \\
&= 4\alpha + 4\beta
\end{aligned} \tag{8.190}
$$

Once again reminding ourselves that α and β are negative this means that the four orbital MO is more stable by 0.472β. We find that as the length of conjugation increases so the stability of the MO increases in comparison with isolated double bonds.

8.5 Matrices

A matrix is a rectangular array of numbers similar in appearance to a determinant, the differences being that they are not constrained to be square and that they are bounded by brackets rather than straight lines. For example,

$$\begin{pmatrix} 1 & 2 \\ 3 & 4 \end{pmatrix}, \quad \begin{pmatrix} 1 \\ 2 \\ 3 \end{pmatrix}, \quad (a \quad b \quad c \quad d), \quad \begin{pmatrix} a & b \\ c & d \\ e & f \end{pmatrix} \tag{8.191}$$

are all matrices. Unlike determinants they do not have a value although for a square matrix it is possible to calculate the determinant value. They are used in several areas in chemistry, most of which are beyond the scope of an introductory text such as this. They are usually used where series of linear equations arise as we have seen for solution of secular and other simultaneous equations. Another use is the analysis of complex reaction schemes comprising series of equilibria (which also give rise to series of equations). They are also used in representing transformations from one coordinate system to another—a method used in representing symmetry operations on molecules. Many of these uses are found in more advanced chemistry courses but students are likely to come across them sooner or later. As with much of this chapter the chemistry is often more difficult to understand than the pure mathematics.

8.5.1 Matrix Terminology and Simple Algebra

A matrix is described by its size, the number of rows (m) *by* the number of columns (n). This is written $m \times n$. Thus

$$\boldsymbol{A} = \begin{pmatrix} a & b \\ c & d \\ e & f \end{pmatrix} \tag{8.192}$$

is a 3×2 matrix, having three rows and two columns. A matrix such as

$$\begin{pmatrix} 1 & 3 \\ 2 & 4 \end{pmatrix} \tag{8.193}$$

where $m = n$ (2×2 in this case) is called a *square* matrix. We can find the determinant of matrix $\boldsymbol{A}$, labelled $\det \boldsymbol{A}$ or $|\boldsymbol{A}|$, by treating it as a determinant, e.g.,

$$\det \begin{pmatrix} 2 & 3 \\ 1 & 3 \end{pmatrix} = \begin{vmatrix} 2 & 3 \\ 1 & 3 \end{vmatrix} = 6 - 3 = 3 \tag{8.194}$$

A matrix with only one column may be referred to as a *column matrix* or *column vector*. Likewise a matrix with only one row is called a *row matrix* or *row vector*. The *transpose* of a matrix is found by exchanging its rows and columns, i.e., from an $m \times n$ matrix to an $n \times m$ matrix. The transpose of the matrix $\boldsymbol{A}$ in (8.195) below,

$$\boldsymbol{A} = \begin{pmatrix} 1 & 2 & 3 & 4 \\ 4 & 3 & 2 & 1 \end{pmatrix} \tag{8.195}$$

labelled $\boldsymbol{A}^{\mathrm{T}}$ or $\tilde{\boldsymbol{A}}$, is found by making row 1 into column, 1, row 2 into column 2 etc.

$$\boldsymbol{A}^{\mathrm{T}} = \begin{pmatrix} 1 & 4 \\ 2 & 3 \\ 3 & 2 \\ 4 & 1 \end{pmatrix} \tag{8.196}$$

Therefore the transpose of a column vector is a row vector and the transpose of a row vector is a column vector.

$$\boldsymbol{A} = \begin{pmatrix} a \\ b \\ c \end{pmatrix} \qquad \therefore \boldsymbol{A}^{\mathrm{T}} = (a \quad b \quad c) \tag{8.197}$$

A square matrix which has all of its elements except those in diagonal positions (left to right) equal to zero is called a *diagonal* matrix, e.g.,

$$\begin{pmatrix} 1 & 0 & 0 & 0 \\ 0 & 4 & 0 & 0 \\ 0 & 0 & 6 & 0 \\ 0 & 0 & 0 & 2 \end{pmatrix} \tag{8.198}$$

The diagonal matrix whose non-zero elements are equal to one is called the *unit* or *identity* matrix,

$$\begin{pmatrix} 1 & 0 & 0 & 0 \\ 0 & 1 & 0 & 0 \\ 0 & 0 & 1 & 0 \\ 0 & 0 & 0 & 1 \end{pmatrix} \tag{8.199}$$

Matrix (8.199) is a unit matrix of *order* 4, since this is a 4×4 matrix.

Matrices may be added or subtracted from each other only if they are of the same order. Thus a 3×2 matrix may be added to a 3×2 matrix but not a 2×2 or any other matrix. To add two matrices together we add the elements in identical position (row and column). For example,

$$\begin{pmatrix} a_1 & b_1 & c_1 \\ d_1 & e_1 & f_1 \end{pmatrix} + \begin{pmatrix} a_2 & b_2 & c_2 \\ d_2 & e_2 & f_2 \end{pmatrix} = \begin{pmatrix} a_1 + a_2 & b_1 + b_2 & c_1 + c_2 \\ d_1 + d_2 & e_1 + e_2 & f_1 + f_2 \end{pmatrix} \tag{8.200}$$

It is possible to multiple a matrix by some scalar quantity, in which case all the elements are multiplied by that quantity.

$$2 \times \begin{pmatrix} a_1 & b_1 & c_1 \\ d_1 & e_1 & f_1 \end{pmatrix} = \begin{pmatrix} 2a_1 & 2b_1 & 2c_1 \\ 2d_1 & 2e_1 & 2f_1 \end{pmatrix} \tag{8.201}$$

The product of two matrices may only be found if an important criterion is fulfilled. This is that the number of columns of the first is equal to the number of rows of the second. We can multiply a 3×2 matrix and a 2×2 matrix, but not a 3×2 and a 4×4, or even a 3×2 and another 3×2. Neither can we multiply a 2×2 and a 3×2 matrix—the order is important! The product of an $m \times n$ and an $n \times p$ matrix is a matrix of dimensions $m \times p$. The multiplication is as follows.

We label the elements according to their position as rows, label i, and columns, label j. The element $a_{ij} = a_{12}$ is the first row, second column. For example, a 3 × 2 matrix would appear as

$$\begin{pmatrix} a_{11} & a_{12} \\ a_{21} & a_{22} \\ a_{31} & a_{32} \end{pmatrix} \tag{8.202}$$

To multiply two 2 × 2 matrices, $\boldsymbol{A}$ and $\boldsymbol{B}$,

$$\boldsymbol{A} = \begin{pmatrix} a_{11} & a_{12} \\ a_{21} & a_{22} \end{pmatrix}, \quad \boldsymbol{B} = \begin{pmatrix} b_{11} & b_{12} \\ b_{21} & b_{22} \end{pmatrix} \tag{8.203}$$

we first note that the product is another 2 × 2 matrix, $\boldsymbol{C}$.

$$\boldsymbol{C} = \begin{pmatrix} c_{11} & c_{12} \\ c_{21} & c_{22} \end{pmatrix} \tag{8.204}$$

The individual elements are calculated using the equations

$$c_{11} = a_{11}b_{11} + a_{12}b_{21} \tag{8.205a}$$

$$c_{12} = a_{11}b_{12} + a_{12}b_{22} \tag{8.205b}$$

$$c_{21} = a_{21}b_{11} + a_{22}b_{21} \tag{8.205c}$$

$$c_{22} = a_{21}b_{12} + a_{22}b_{22} \tag{8.205d}$$

Although this appears complicated it is merely repetitive. The result for element c_{11} is effectively the sum of products of the first row (first matrix) and first column (second matrix). The element c_{21} comes from the sum of the products of the second row (first matrix) and the first row (second matrix). The position of the resultant element is dictated by the rows and columns of the first and second matrices.

The matrix (8.202) may be multiplied by a 2 × 2 matrix showing that this pattern is repeated for bigger matrices:

$$\boldsymbol{AB} = \begin{pmatrix} a_{11} & a_{12} \\ a_{21} & a_{22} \\ a_{31} & a_{32} \end{pmatrix} \begin{pmatrix} b_{11} & b_{12} \\ b_{21} & b_{22} \end{pmatrix} = \begin{pmatrix} a_{11}b_{11} + a_{12}b_{21} & a_{11}b_{12} + a_{12}b_{22} \\ a_{21}b_{11} + a_{22}b_{21} & a_{21}b_{12} + a_{22}b_{22} \\ a_{31}b_{11} + a_{32}b_{21} & a_{31}b_{12} + a_{32}b_{22} \end{pmatrix} \tag{8.206}$$

or

$$\boldsymbol{AB} = \begin{matrix} \sum \text{1st row} \times \text{1st col} & \sum \text{1st row} \times \text{2nd col} \\ \sum \text{2nd row} \times \text{1st col} & \sum \text{2nd row} \times \text{2nd col} \\ \sum \text{3rd row} \times \text{1st col} & \sum \text{3rd row} \times \text{2nd col} \end{matrix} \tag{8.207}$$

The sums are the same as the subscripts for the resultant matrix.

Example—multiplication of two matrices

Two matrices are defined as follows,

$$\boldsymbol{A} = \begin{pmatrix} 1 & 2 \\ 1 & 2 \end{pmatrix}, \quad \boldsymbol{B} = \begin{pmatrix} 1 & 2 \\ 3 & 4 \end{pmatrix} \tag{8.208}$$

Find $\boldsymbol{AB}$ and $\boldsymbol{BA}$.

$\boldsymbol{AB}$ is given by

$$\begin{aligned} \boldsymbol{AB} &= \begin{pmatrix} 1 & 2 \\ 1 & 2 \end{pmatrix} \begin{pmatrix} 1 & 2 \\ 3 & 4 \end{pmatrix} = \begin{pmatrix} 1+6 & 2+8 \\ 1+6 & 2+8 \end{pmatrix} \\ &= \begin{pmatrix} 7 & 10 \\ 7 & 10 \end{pmatrix} \end{aligned} \tag{8.209}$$

$\boldsymbol{BA}$ is given by

$$\begin{aligned} \boldsymbol{BA} &= \begin{pmatrix} 1 & 2 \\ 3 & 4 \end{pmatrix} \quad \begin{pmatrix} 1 & 2 \\ 1 & 2 \end{pmatrix} = \begin{pmatrix} 1+2 & 2+4 \\ 3+4 & 6+8 \end{pmatrix} \\ &= \begin{pmatrix} 3 & 6 \\ 7 & 14 \end{pmatrix} \end{aligned} \tag{8.210}$$

Thus it can be seen that $\boldsymbol{AB}$ is not the same as $\boldsymbol{BA}$. A more startling demonstration is if we multiply a 1×3 row matrix by a 3×1 column matrix.

$$(1 \quad 2 \quad 3) \quad \begin{pmatrix} 1 \\ 2 \\ 3 \end{pmatrix} = (1 + 4 + 9) = (14) \tag{8.211}$$

The result is a 1×1 matrix. In contrast the product of a 3×1 column matrix and a 1×3 row matrix is a 3×3 matrix.

$$\begin{aligned} \begin{pmatrix} 1 \\ 2 \\ 1 \end{pmatrix} \quad (2 \quad 1 \quad 3 &= \begin{pmatrix} 1\times 2 & 1\times 1 & 1\times 3 \\ 2\times 2 & 2\times 1 & 2\times 3 \\ 1\times 2 & 1\times 1 & 1\times 3 \end{pmatrix} \\ &= \begin{pmatrix} 2 & 1 & 3 \\ 4 & 2 & 6 \\ 2 & 1 & 3 \end{pmatrix} \end{aligned} \tag{8.212}$$

Remember, in finding a product of $(m \times n)$ and $(n \times p)$ then the innermost numbers (n) must match and the product is defined by the outermost numbers $(m \times p)$.

A matrix is unchanged when it is multiplied by the unit matrix. For example,

$$\begin{aligned} \begin{pmatrix} 1 & 0 \\ 0 & 1 \end{pmatrix} \begin{pmatrix} 1 & 3 \\ 2 & 4 \end{pmatrix} &= \begin{pmatrix} 1\times 1+0\times 2 & 1\times 3+0\times 4 \\ 0\times 1+1\times 2 & 0\times 3+1\times 4 \end{pmatrix} \\ &= \begin{pmatrix} 1 & 3 \\ 2 & 4 \end{pmatrix} \end{aligned} \tag{8.213}$$

In this case the same result is obtained if the unit matrix is the second matrix.

$$\begin{pmatrix} 1 & 3 \\ 2 & 4 \end{pmatrix}\begin{pmatrix} 1 & 0 \\ 0 & 1 \end{pmatrix} = \begin{pmatrix} 1 \times 1 + 3 \times 0 & 1 \times 0 + 3 \times 1 \\ 2 \times 1 + 4 \times 0 & 2 \times 0 + 4 \times 1 \end{pmatrix} = \begin{pmatrix} 1 & 3 \\ 2 & 4 \end{pmatrix} \quad (8.214)$$

8.5.2 The Inverse Matrix

The inverse of a matrix is such that the product of a matrix and its inverse is the unit matrix. The inverse of a matrix $\boldsymbol{A}$ is labelled $\boldsymbol{A}^{-1}$ and if the unit matrix is simply '1', then

$$\boldsymbol{A}^{-1}\boldsymbol{A} = \boldsymbol{A}\boldsymbol{A}^{-1} = 1 \quad (8.215)$$

Finding the number which when multiplied by another number is equal to one is almost trivial (2 multiplied by 1/2, 3 multiplied by 1/3 etc.) but finding the inverse matrix is much less so. For a 2×2 matrix the formula is

$$\boldsymbol{A} = \begin{pmatrix} a_{11} & a_{12} \\ a_{21} & a_{22} \end{pmatrix}, \quad \text{then } \boldsymbol{A}^{-1} = \frac{\begin{pmatrix} a_{22} & -a_{12} \\ -a_{21} & a_{11} \end{pmatrix}}{a_{11}a_{22} - a_{12}a_{21}} \quad (8.216)$$

The denominator of the equation for $\boldsymbol{A}^{-1}$ may be recognised as the determinant of the matrix in question. The numerator is called the *adjoint* matrix (labelled $\hat{\boldsymbol{A}}$) and the derivation of this lies in a manipulation of the determinant not discussed in the previous section. Needless to say the derivation of this is lengthy rather than difficult and is to be found in more advanced texts.

Inverse matrices are used to find the solution of simultaneous equations. If we take the equations

$$a_1x + b_1y = c_1 \quad (8.217)$$

$$a_2x + b_2y = c_2 \quad (8.218)$$

these may be represented in terms of matrices.

$$\begin{pmatrix} a_1 & b_1 \\ a_2 & b_2 \end{pmatrix}\begin{pmatrix} x \\ y \end{pmatrix} = \begin{pmatrix} c_1 \\ c_2 \end{pmatrix} \quad (8.219)$$

As an exercise you should multiply out the matrices on the left to show that they do indeed form the left hand sides of the pair of simultaneous equations. We can represent the 2×2 matrix by $\boldsymbol{A}$, the 2×1 matrix by $\boldsymbol{X}$ and the matrix on the right by $\boldsymbol{C}$. Then

$$\boldsymbol{AX} = \boldsymbol{C} \quad (8.220)$$

If we multiply both sides by $\boldsymbol{A}^{-1}$ then we get

$$\boldsymbol{A}^{-1}\boldsymbol{AX} = \boldsymbol{A}^{-1}\boldsymbol{C} \quad (8.221)$$

Since $\boldsymbol{A}^{-1}\boldsymbol{A}$ is the unit matrix and the product of this and $\boldsymbol{X}$ is $\boldsymbol{X}$ the result is

$$\boldsymbol{X} = \boldsymbol{A}^{-1}\boldsymbol{C} \quad (8.222)$$

Thus to find the solution, $\boldsymbol{X}$, or in the case of general equations (8.217) and (8.218) we need to find the inverse of the matrix.

To do this we need the determinant of the matrix which is

$$\det \boldsymbol{A} = \begin{vmatrix} a_1 & b_1 \\ a_2 & b_2 \end{vmatrix} = a_1 b_2 - a_2 b_1 \tag{8.223}$$

and the adjoint, which is

$$\hat{\boldsymbol{A}} = \begin{pmatrix} b_2 & -b_1 \\ -a_1 & a_1 \end{pmatrix} \tag{8.224}$$

Thus $\boldsymbol{A}^{-1}$ is

$$\begin{aligned} \boldsymbol{A}^{-1} &= \frac{1}{a_1 b_2 - a_2 b_1} \begin{pmatrix} b_2 & -b_1 \\ -a_1 & a_1 \end{pmatrix} \\ &= \begin{pmatrix} \dfrac{b_2}{a_1 b_2 - a_2 b_1} & \dfrac{-b_1}{a_1 b_2 - a_2 b_1} \\ \dfrac{-a_2}{a_1 b_2 - a_2 b_1} & \dfrac{a_1}{a_1 b_2 - a_2 b_1} \end{pmatrix} \end{aligned} \tag{8.225}$$

When this is multiplied by $\boldsymbol{C}$ above then we end up with a matrix corresponding to equations (8.217) and (8.218).

$$\begin{pmatrix} x \\ y \end{pmatrix} = \begin{pmatrix} \dfrac{c_1 b_2 - c_2 b_1}{a_1 b_2 - a_2 b_1} \\ \dfrac{a_1 c_2 - a_2 c_1}{a_1 b_2 - a_2 b_1} \end{pmatrix} \tag{8.226}$$

You should follow this through yourself to prove that this is the case. The method is, however, most easily illustrated with an example.

Example—matrix solution of simultaneous (linear) equations

We will again use the equations

$$3x - y = -1 \tag{8.227}$$

$$2x - y = -2 \tag{8.228}$$

In matrix form these are

$$\begin{pmatrix} 3 & -1 \\ 2 & -1 \end{pmatrix} \begin{pmatrix} x \\ y \end{pmatrix} = \begin{pmatrix} -1 \\ -2 \end{pmatrix} \tag{8.229}$$

We need to find the inverse of the 2×2 matrix on the left ($\boldsymbol{A}^{-1}$) which is found as

$$A^{-1} = \frac{1}{-3+2}\begin{pmatrix} -1 & 1 \\ -2 & 3 \end{pmatrix}$$
$$= -1\begin{pmatrix} -1 & 1 \\ -2 & 2 \end{pmatrix} \tag{8.230}$$
$$= \begin{pmatrix} 1 & -1 \\ 2 & -3 \end{pmatrix}$$

We multiply this by the right hand side and obtain

$$\begin{pmatrix} x \\ y \end{pmatrix} = \begin{pmatrix} 1 & -1 \\ 2 & -3 \end{pmatrix}\begin{pmatrix} -1 \\ -2 \end{pmatrix} = \begin{pmatrix} -1+2 \\ -2+6 \end{pmatrix} = \begin{pmatrix} 1 \\ 4 \end{pmatrix} \tag{8.231}$$

As before we reach the result $x = 1, y = 4$.

The solution of several simultaneous linear equations is an important topic in quantum mechanics and in solving problems of complex reaction mechanisms. Therefore a systematic method such as this for providing these solutions is highly desirable.

8.5.3 Symmetry Transformations

The study of the symmetry of molecules is not confined to ideas about their shapes but covers ideas about many of their properties—their motion, wavefunctions, dipoles, reactivity and spectroscopic properties. This is another topic which can only be touched on briefly in this book but as symmetry operations may be described by matrix notation and matrix algebra then it deserves some mention.

We are familiar with the concept of symmetry in that an object is symmetrical if it has identical halves. If one part is reflected through an imaginary mirror down the centre of the object then it matches the opposite side exactly. In chemistry this idea is taken much further, in that any operation on an object (which can include a molecule, its wavefunc-

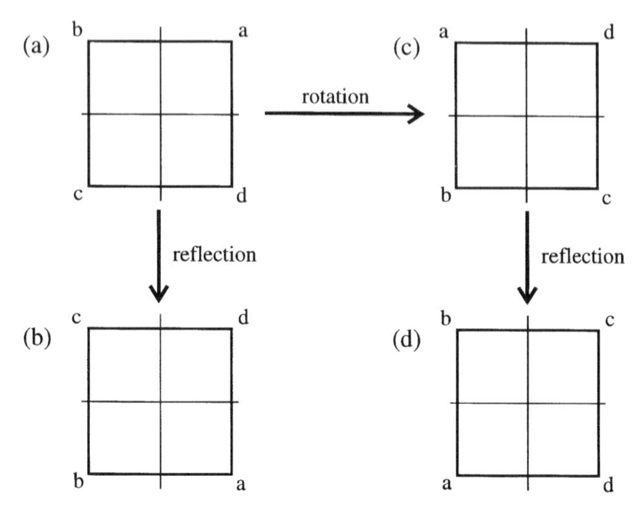

Figure 8.13 Symmetry operations on the square abcd. Each operation leaves the square ostensibly unchanged although the labelled corners move

tion, related vectors or molecular motions) which leaves the object essentially the same is a symmetry operation. For example, the square in Figure 8.13, centred on the origin, may undergo certain transformations and appear, at the end, just as it does at the beginning. It may be rotated about the origin through 90° and it may be reflected through mirror planes through the x and y axes (coming out of the paper). The vertices may also be inverted through the origin to become their counterparts in the opposite quadrant. This is therefore a highly symmetrical object. How such transformations may be represented by matrices is described in the following examples.

Example—rotation about an axis of symmetry

This example refers to the square in Figure 8.13. The corners are labelled a, b, c and d starting in the positive quadrant and moving anticlockwise. These points (which may be thought of as atoms in a square planar molecule) can be represented by a column vector.

$$\boldsymbol{A} = \begin{pmatrix} a \\ b \\ c \\ d \end{pmatrix} \tag{8.232}$$

Some texts use row vectors but this representation is perhaps more logical as the matrix which describes the transformation must go before the column and therefore appears more as the symmetry operator. If as shown in Figure 8.13 this undergoes a 90° rotation in the anticlockwise sense the shape is identical but the labels have moved. The column vector is now

$$\boldsymbol{A}' = \begin{pmatrix} d \\ a \\ b \\ c \end{pmatrix} \tag{8.233}$$

The problem is to find a matrix which will turn column vector $\boldsymbol{A}$ into $\boldsymbol{A}'$. In this case this is not as difficult as it first appears. Recalling the rules on matrix multiplication, the product of an $m \times n$ and an $n \times p$ matrix is an $m \times p$ matrix. We start with a 4×1 matrix and the transforming matrix must therefore have four columns. The result will be another 4×1 matrix and therefore it must have four rows ((4×4) by (4×1) gives a (4×1) matrix). How this is going to work can be seen just by looking at the first row multiplication to give the first element of the resultant. If we leave the rest of this matrix blank for the moment we have

$$\begin{pmatrix} d \\ a \\ b \\ c \end{pmatrix} = \begin{pmatrix} x_1 & x_2 & x_3 & x_3 \\ — & — & — & — \\ — & — & — & — \\ — & — & — & — \end{pmatrix} \begin{pmatrix} a \\ b \\ c \\ d \end{pmatrix} \tag{8.234}$$

where x_1, x_2 etc. are unknowns. Then

$$d = x_1 a + x_2 b + x_3 c + x_3 d \tag{8.235}$$

If we make the x values equal to zero or one then the result becomes more obvious. x_1, x_2 and x_3 must be zero and x_4 must be one.

$$d = 0 \times a + 0 \times b + 0 \times c + 1 \times d \tag{8.236}$$

Performing the same exercise for the second row

$$\begin{pmatrix} d \\ a \\ b \\ c \end{pmatrix} = \begin{pmatrix} - & - & - & - \\ x_1 & x_2 & x_3 & x_4 \\ - & - & - & - \\ - & - & - & - \end{pmatrix} \begin{pmatrix} a \\ b \\ c \\ d \end{pmatrix} \tag{8.237}$$

and

$$a = x_1 a + x_2 b + x_3 c + x_3 d \tag{8.238}$$

In this case x_1 is one and the rest are zero. Repeating this procedure—which amounts to deciding which of the elements is to be multiplied by one and the rest by zero—gives the result

$$\begin{pmatrix} d \\ a \\ b \\ c \end{pmatrix} = \begin{pmatrix} 0 & 0 & 0 & 1 \\ 1 & 0 & 0 & 0 \\ 0 & 1 & 0 & 0 \\ 0 & 0 & 1 & 0 \end{pmatrix} \begin{pmatrix} a \\ b \\ c \\ d \end{pmatrix} \tag{8.239}$$

Example—reflection in a plane

The rectangle in Figure 8.14 is now reflected in a plane along the x axis. Thus a exchanges places with d, and b with c. The transformation is

$$\begin{pmatrix} a \\ b \\ c \\ d \end{pmatrix} \rightarrow \begin{pmatrix} d \\ c \\ b \\ a \end{pmatrix} \tag{8.240}$$

We can follow through the previous methodology to get the result

$$\begin{pmatrix} d \\ c \\ b \\ a \end{pmatrix} = \begin{pmatrix} 0 & 0 & 0 & 1 \\ 0 & 0 & 1 & 0 \\ 0 & 1 & 0 & 0 \\ 1 & 0 & 0 & 0 \end{pmatrix} \begin{pmatrix} a \\ b \\ c \\ d \end{pmatrix} \tag{8.241}$$

Example—combinations of transformations

What happens if we follow the rotation above by the reflection and how is this represented? The rotation gave us one column vector and reflection of this in the x plane gives

$$\begin{pmatrix} a \\ b \\ c \\ d \end{pmatrix} \xrightarrow{\text{rotation}} \begin{pmatrix} d \\ a \\ b \\ c \end{pmatrix} \xrightarrow{\text{reflection}} \begin{pmatrix} c \\ b \\ a \\ b \end{pmatrix} \tag{8.242}$$

The transformation matrix may be found in two ways. The first, directly from this result, is

$$\begin{pmatrix} c \\ b \\ a \\ d \end{pmatrix} = \begin{pmatrix} 0 & 0 & 1 & 0 \\ 0 & 1 & 0 & 0 \\ 1 & 0 & 0 & 0 \\ 0 & 0 & 0 & 1 \end{pmatrix} \begin{pmatrix} a \\ b \\ c \\ d \end{pmatrix} \tag{8.243}$$

The second way is to multiply the two matrices corresponding to the two operations. Of course the order in which we do this is important but it helps to think of them as operators. Thus the first operation is rotation on the matrix $\boldsymbol{A}$

$$\text{rotation}\,(\boldsymbol{A}) \tag{8.244}$$

and the reflection is performed on this,

$$\text{reflection}\,(\text{rotation}\,(\boldsymbol{A})) \tag{2.245}$$

This is the way the matrices are written to multiply them. Therefore we write and produce the result

$$\begin{matrix} (\text{reflection}) & \times & (\text{rotation}) & = & (\text{resultant}) \\ \downarrow & & \downarrow & & \downarrow \end{matrix}$$

$$\begin{pmatrix} 0 & 0 & 0 & 1 \\ 0 & 0 & 1 & 0 \\ 0 & 1 & 0 & 0 \\ 1 & 0 & 0 & 0 \end{pmatrix} \begin{pmatrix} 0 & 0 & 0 & 1 \\ 1 & 0 & 0 & 0 \\ 0 & 1 & 0 & 0 \\ 0 & 0 & 1 & 0 \end{pmatrix} = \begin{pmatrix} 0 & 0 & 1 & 0 \\ 0 & 1 & 0 & 0 \\ 1 & 0 & 0 & 0 \\ 0 & 0 & 0 & 1 \end{pmatrix} \tag{8.246}$$

8.6 Concluding Remarks

In this chapter we have introduced some slightly more advanced ideas in mathematics. Although there are some new ideas and new forms of algebra the individual operations which constitute this algebra are no more advanced than the material covered in Chapters 1 and 2. An understanding of complex numbers, vectors, determinants and matrices is highly beneficial as these are all used in advanced chemistry courses and while the chemistry may be abstract in nature the mathematical basis is based on relatively simple principles.

Problems

8.1 For the following complex numbers find $z_1 + z_2$, $z_1 - z_2$, $z_1 \times z_2$ and $z_1 \div z_2$,

(a) $z_1 = 2x + 4i, z_2 = 3x + 4i$
(b) $z_1 = i - 2x, z_2 = 2x + i$
(c) $z_1 = 10 - 2i, z_2 = 10 + 2i$
(d) $z_1 = i, z_1 = 1 - i$

8.2 Two vectors $\boldsymbol{a}$ and $\boldsymbol{b}$ are defined as
$\boldsymbol{a}$ = magnitude 2 with direction 30°
$\boldsymbol{b}$ = magnitude 1 with direction 45°
What are the sum, difference, and scalar and vector products of these two vectors?

8.3 What are the scalar and vector products of the following vectors?

(a) $\boldsymbol{a} = \boldsymbol{i} + 2\boldsymbol{j} - 2\boldsymbol{k}, \boldsymbol{b} = \boldsymbol{a} = 2\boldsymbol{i} - \boldsymbol{j} + 3\boldsymbol{k}$
(b) $\boldsymbol{a} = 2\boldsymbol{i} + 2\boldsymbol{j} + 2\boldsymbol{k}, \boldsymbol{b} = \boldsymbol{a} = \boldsymbol{i} + \boldsymbol{j} + \boldsymbol{k}$

8.4 Calculate the values of the following determinants.

(a) $\begin{vmatrix} 1 & 3 \\ 2 & 4 \end{vmatrix}$ (b) $\begin{vmatrix} 4 & 1 \\ 2 & 3 \end{vmatrix}$ (c) $\begin{vmatrix} 1 & 0 \\ 0 & 1 \end{vmatrix}$ (d) $\begin{vmatrix} 1 & 0 \\ 0 & 1 \end{vmatrix}$

(e) $\begin{vmatrix} 1 & 1 & 0 \\ 2 & 3 & 2 \\ 3 & 2 & 3 \end{vmatrix}$ (f) $\begin{vmatrix} 1 & 0 & 0 \\ 0 & 0 & 1 \\ 1 & 0 & 1 \end{vmatrix}$ (g) $\begin{vmatrix} 1 & 1 & 1 \\ 1 & 1 & 0 \\ 1 & 0 & 1 \end{vmatrix}$

8.5 Solve the simultaneous equations using the method of determinants.

(a) $2x + y = 4, x + 2y = 6$
(b) $x + 3y = 1, 2x + 3y = 2, 3x + 4y = 3$

8.6 Evaluate the following products.

(a) $\begin{pmatrix} 1 & 1 \\ 2 & 3 \end{pmatrix}\begin{pmatrix} 2 & 3 \\ 1 & 2 \end{pmatrix}$ (b) $(1 \quad 0 \quad 2 \quad 1)\begin{pmatrix} 2 \\ 1 \\ 1 \\ 2 \end{pmatrix}$ (c) $\begin{pmatrix} 2 \\ 3 \end{pmatrix}(1 \quad 2)$

8.7 Find the transform of the matrices.

(a) $\begin{pmatrix} 1 & 2 \\ 2 & 1 \end{pmatrix}$ (b) $\begin{pmatrix} 1 & 0 & 1 & 0 \\ 2 & 0 & 2 & 0 \end{pmatrix}$

8.8 Find the inverse of the following matrices.

(a) $\begin{pmatrix} 1 & 2 \\ 0 & 1 \end{pmatrix}$ (b) $\begin{pmatrix} 2 & 2 \\ 1 & 3 \end{pmatrix}$ (c) $\begin{pmatrix} 1 & 0 \\ 0 & 1 \end{pmatrix}$

8.9 Solve the simultaneous equations using the matrix method.

$x + 2y = 3, x + y = 4$

8.10 An equilateral triangle has corners labelled A, B and C in an anticlockwise fashion. Write matrices for the following symmetry operations.

(a) A rotation through 120° so that A is transformed onto B, B onto C and C onto A
(b) A reflection through the mirror plane through A and the line joining B and C
(c) The operation in (a) followed by that in (b)
(d) The operation in (b) followed by that in (a)

Project

The layout of a spreadsheet is tailor-made for solving determinants and matrices. Develop spreadsheets for the following purposes.

(a) To calculate two by two determinants
(b) To calculate three by three determinants
(c) To solve sets of two and three simultaneous equations
(d) To calculate the products of matrices of different dimensions
(e) To calculate transform and inverse matrices
(f) To solve simultaneous equations using the matrix method

Appendix 1
The Greek Alphabet

Greek name	Lower case	Upper case
Alpha	α	A
Beta	β	B
Gamma	γ	Γ
Delta	δ	Δ
Epsilon	ε	E
Zeta	ζ	Z
Eta	η	H
Theta	θ	Θ
Iota	ι	I
Kappa	κ	K
Lambda	λ	Λ
Mu	μ	M
Nu	ν	N
Xi	ξ	Ξ
Omicron	o	O
Pi	π	Π
Rho	ρ	P
Sigma	σ	Σ
Tau	τ	T
Upsilon	υ	Y
Phi	r	Φ
Chi	χ	X
Psi	ψ	Ψ
Omega	ω	Ω

Appendix 2 Solutions and Hints for Selected Problems

Chapter 1

1.1 $1/x + 2$ may be written as $1/x + 2x/x$ and then as one fraction $(1 + 2x)/x$. The whole fraction may then be written as the right hand side.

1.2 The left hand side is split into two fractions, $x/y + y/y$. As y/y is one this leaves only one y and so the equation is rearranged to give $y = xb/(a + b)$.

1.3 (a) If rate is zero then $k_1A = k_2B$ and so $B/A = k_1/k_2$.
(b) As $B = A_0 - A$ then $k_1/k_2 = (A_0 - A)/A$. Then follow the same rearrangement as in 1.2 to give $A = A_0k_2/(k_1 + k_2)$.

1.4 **(a)** $B = K_1A$ and $C = K_2A$
(b) $A_0 = A + B + C$
(c) $A_0 = A + K_1A + K_2A$
(d) $A_0 = A(1 + K_1 + K_2)$, $\therefore A = A_0/(1 + K_1 + K_2)$
(e) Substitute for A in (a) above, $B = K_1A_0/(1 + K_1 + K_2)$ and $C = K_2A_0/(1 + K_1 + K_2)$.

1.5 This is the central equation used when doing calculations for titrations—when the number of moles of one species is the same as the number of moles of another. $N_1 = C_1V_1$ and $N_2 = C_2V_2$ then $C_1V_1 = C_2V_2$ and $C_1 = C_2V_2/V_1$.

1.6 Invert both sides and then split the fraction into two, $1/\theta - \theta/\theta$ or $1/\theta - 1$. The rearrangement is then straightforward to give $P = k_1(1/\theta + 1)/k_2$.

1.7 (a) Divide each by three and remove outside the bracket to give $3(x + y)$
(b) Divide each by x to give $x(x + 1)$
(c) This is given in Chapter 1 as $(x + 1)^2$

(d) Even cases which are not, at first glance, easy to factorise may be rearranged by removing one variable. In this case divide through by x to give $x(1 + 1/x)$.
(e) $(x - y)(x + y)$
(f) This is similar to (e) but with $y^2 = 4$, so the answer is $(x - 2)(x + 2)$

1.8 These can all be found using the rules on combinations of indices.

(a) $y = x$
(b) $y = x^{2.5}$
(c) $y = x^{(1/3-5)} = x^{-14/3}$
(d) $y = x^{(-1-0.5)} = x^{-3.2}$
(e) $y = x^2 + z^2$
(f) $y = x^{(6-2-3/2)} = x^{5.2}$

1.9 Use equation (1.150) to identify the coefficients a, b and c and equation (1.151) to determine the ?????.

Chapter 2

2.1 (a) 1 (b) 4 (c) 1 (d) 4 (e) 2 (f) − 34 (g) 0 (h) 0

2.2 (a) 0 (b) 0.301 (c)7 (d) 13

2.3 Use $\log(x^{1/2}) = (1/2)\log(x)$, then calculate the antilogarithm.

2.4 Calculate logs for both and plot log(initial rate) vs. log(initial concentration). Nearest integer to slope is 2, therefore second order.

2.6 (a) 1.105 (b) 1.101 (c) 0.99

2.7 Hint: when $x \gg \mu$ then μ effectively disappears from the exponential terms and the equation simplifies accordingly. If $x \ll \mu$ then x disappears from the exponential term. If they are equal then the exponential term becomes e^0.

2.8 Again we have an exponential term as e^0 and hence the populations are equal.

2.12 Use the Bragg equation, result 130 pm.

Chapter 3

3.1 Mean = 25.2, range = 16.3, median = 25.1, standard deviation = 3.94

3.2 Range = 9.7, mean = 13.08, std deviation = 2.74

3.3 Range = 12.6, mean = 110.51, std deviation = 2.74

3.5 (a) 1.23 (b) 12.3 (c) 124 (d) 0.00123 (e) 1.10
(f) 101 (g) 10.3 (h) 10.4 (i) $1.00 \times 10_3$ (j) − 0.608

3.6 (a) 7 ± 0.3 (b) 25 ± 1 (c) 109.5 ± 0.2 (d) 0.06 ± 0.04
(f) 1.32 ± 0.04 (g) 75 ± 6 (h) Hint: calculate the quotient and then the logarithm!

Chapter 4

4.1 (a) 0 (b) 6 (c) $12x + 2$ (d) $-1/x_2$ (e) $1/(2x^{1/2})$

4.2 (a) $1/x$ (b) $-c/x^2$ (c) $\exp(x)$ (d) $\cos(x)$
(e) $2/x$ since $\ln(x^2) = 2\ln(x)$. Can also solve using the chain rule.

4.3 (a) Solve as product of two functions, $\sin(x)$ and $\cos(x)$.
(b) Again a product, $\sin(x)$ and $\sin(x)$.
(c) Quotient: $\tan(x) = \sin(x)/\cos(x)$
(d) Follow the hint!
(e) Another product, x and $\ln(x)$.

4.4 These are all functions of functions and can be solved using the chain rule.

4.7 Find the second derivative and then factorise—the result should be equal to $-k^2x$, i.e., the original equation is a general solution to the second order differential equation

$$\frac{d^2x}{dt^2} = -kx^2$$

This is the general equation describing harmonic (wave) motion and is also one form of the time-dependent Schrödinger equation.

Chapter 5

5.1 (a) $4x + C$ (b) $x^2/2 + C$ (c) $2x^3 + C$ (d) $x^3 + 2x^2 + C$
(e) $-1/x$

5.2 (a) 12 (b) 1 (c) 1 (d) $1/a_0 - 1/a$

5.10 (a) Substitute $u = 5x$ then $dx = du/5$.
(b) Substitute $u = \ln(x)$ then $du/dx = 1/x$ and $du = dx/x$ so the integrand becomes du/u.
(c) Substitute $u = -kx$.

Chapter 7

7.1 (a) 1 (b) 3 (c) 3

7.2 The minimum score is 2 (1 and 1), and the maximum is 12 (6 + 6). As it is only the full score that matters and it is the same die then 1 + 2 is the same as 2 + 1 etc. The scores represent configurations and the ways of getting them, the microstates. These are:

Total	2	3	4	5	6	7	8	9	10	11	12
	1 + 1	1 + 2	1 + 3 2 + 2	1 + 4 2 + 3	1 + 5 2 + 4 3 + 3	1 + 6 2 + 5 3 + 4	2 + 6 3 + 5 4 + 4	3 + 6 4 + 5	4 + 6 5 + 5	5 + 6	6 + 6
No. of ways	1	1	2	2	3	3	3	2	2	1	1

Chapter 8

8.1 For each of z_1 and z_2 add the real parts and then the imaginary parts and write the result as the sum. The differences are found in much the same way, only subtraction is used, obviously. The products are found as if you were expanding the product of a pair of bracketed terms, as shown in Chapter 1. In the final result remember that i^2 is equal to -1. To find the quotients multiply the quotient by the complex conjugate of the denominator divided by itself (z^*/z^*). This has the effect of turning the denominator into a real number.

8.2 Sum: $\theta = 15.9°$ magnitude $= 1.064$
Difference: 1.93
Scalar product: magnitude 0.518, as the two vectors are in the xy plane then the
Vector product: product is in the z direction (up)

8.4 (a) -2
(b) 10
(c) 1
(d) -1 (note (c) and (d) have columns interchanged and the sign reverses)
(e) 5
(f) 0
(g) 1

8.6 (a) $\begin{pmatrix} 3 & 5 \\ 7 & 120 \end{pmatrix}$ (b) (6) (c) $\begin{pmatrix} 2 & 4 \\ 3 & 6 \end{pmatrix}$

8.10 (a) $\begin{pmatrix} 0 & 0 & 1 \\ 1 & 0 & 0 \\ 0 & 1 & 0 \end{pmatrix} \begin{pmatrix} A \\ B \\ C \end{pmatrix} = \begin{pmatrix} C \\ A \\ B \end{pmatrix}$ (in the rotation they all move around one)

(b) $\begin{pmatrix} 1 & 0 & 0 \\ 0 & 0 & 1 \\ 0 & 1 & 0 \end{pmatrix} \begin{pmatrix} A \\ B \\ C \end{pmatrix} = \begin{pmatrix} A \\ C \\ B \end{pmatrix}$ (in the reflection A stays static and C and B swap places)

(c) The product of both changes is the product of the matrices. The result is:

$$\begin{pmatrix} 0 & 1 & 0 \\ 1 & 0 & 0 \\ 0 & 0 & 1 \end{pmatrix}$$

Index